Problem Books in Mathematics

T0142205

Edited by K.A. Bencsáth
P.R. Halmos

Springer
New York
Berlin
Heidelberg
Hong Kong
London
Milan
Paris
Tokyo

Problem Books in Mathematics

Series Editors: K.A. Bencsáth and P.R. Halmos

Pell's Equation
by *Edward J. Barbeau*

Polynomials
by *Edward J. Barbeau*

Problems in Geometry
by *Marcel Berger, Pierre Pansu, Jean-Pic Berry, and Xavier Saint-Raymond*

Problem Book for First Year Calculus
by *George W. Bluman*

Exercises in Probability
by *T. Cacoullos*

Probability Through Problems
by *Marek Capiński and Tomasz Zastawniak*

An Introduction to Hilbert Space and Quantum Logic
by *David W. Cohen*

Unsolved Problems in Geometry
by *Hallard T. Croft, Kenneth J. Falconer, and Richard K. Guy*

Berkeley Problems in Mathematics (2nd ed.)
by *Paulo Ney de Souza and Jorge-Nuno Silva*

Problem-Solving Strategies
by *Arthur Engel*

Problems in Analysis
by *Bernard R. Gelbaum*

Problems in Real and Complex Analysis
by *Bernard R. Gelbaum*

Theorems and Counterexamples in Mathematics
by *Bernard R. Gelbaum and John M.H. Olmsted*

Exercises in Integration
by *Claude George*

(continued after index)

T.Y. Lam

Exercises in Classical Ring Theory

Second Edition

Springer

T.Y. Lam
Department of Mathematics
University of California, Berkeley
Berkeley, CA 94720-0001
USA
lam@math.berkeley.edu

Series Editors:
Katalin A. Bencsáth
Mathematics
School of Science
Manhattan College
Riverdale, NY 10471
USA
katalin.bencsath@manhattan.edu

Paul R. Halmos
Department of Mathematics
Santa Clara University
Santa Clara, CA 95053
USA
phalmos@scuacc.scu.edu

Mathematics Subject Classification (2000): 00A07, 13-01, 16-01

Library of Congress Cataloging-in-Publication Data

Lam, T.Y. (Tsit-Yuen), 1942–
    Exercises in classical ring theory / T.Y. Lam.—2nd ed.
       p. cm.—(Problem books in mathematics)
    Includes indexes.

    1. Rings (Algebra)   I. Title. II. Series.

    QA247.L26 2003
    512′.4—dc21

                                                    2003042429

ISBN 978-1-4419-1829-1   e-ISBN 978-0-387-21771-0

Printed in the United States of America.

9 8 7 6 5 4 3 2 1

www.springer-ny.com

Springer-Verlag   New York Berlin Heidelberg
A member of BertelsmannSpringer Science+Business Media GmbH

To Chee King

Juwen, Fumei, Juleen, and Dee-Dee

Preface to the Second Edition

The four hundred problems in the first edition of this book were largely based on the original collection of exercises in my Springer Graduate Text *A First Course in Noncommutative Rings*, ca. 1991. A second edition of this ring theory text has since come out in 2001. Among the special features of this edition was the inclusion of a large number of newly designed exercises, many of which have not appeared in other books before.

It has been my intention to make the solutions to these new exercises available. Since *Exercises in Classical Ring Theory* has also gone out of print recently, this seemed a propitious time to issue a new edition of our Problem Book. In this second edition, typographical errors were corrected, various improvements on problem solutions were made, and the *Comments* on many individual problems have been expanded and updated. All in all, we have added eighty-five exercises to the first edition, some of which are quite challenging. In particular, all exercises in the second edition of *First Course* are solved here, with essentially the same reference numbers. As before, we envisage this book to be useful in at least three ways: (1) as a companion to *First Course* (second edition), (2) as a source book for self-study in problem-solving, and (3) as a convenient reference for much of the folklore in classical ring theory that is not easily available elsewhere.

Hearty thanks are due to my U.C. colleagues K. Goodearl and H.W. Lenstra, Jr. who suggested several delightful exercises for this new edition.

While we have tried our best to ensure the accuracy of the text, occasional slips are perhaps inevitable. I'll welcome comments from my readers on the problems covered in this book, and invite them to send me their corrections and suggestions for further improvements at the address lam@math.berkeley.edu.

Berkeley, California   T.Y.L.
01/02/03

Preface to the First Edition

This is a book I wished I had found when, many years ago, I first learned the subject of ring theory. All those years have flown by, but I still did not find *that* book. So finally I decided to write it myself.

All the books I have written so far were developed from my lectures; this one is no exception. After writing *A First Course in Noncommutative Rings* (Springer-Verlag GTM **131**, hereafter referred to as "*FC*"), I taught ring theory in Berkeley again in the fall of 1993, using *FC* as text. Since the main theory is already fully developed in *FC*, I asked my students to read the book at home, so that we could use part of the class time for doing the exercises from *FC*. The combination of lectures and problem sessions turned out to be a great success. By the end of the course, we covered a significant portion of *FC* and solved a good number of problems. There were 329 exercises in *FC*; while teaching the course, I compiled 71 additional ones. The resulting four hundred exercises, with their full solutions, comprise this ring theory problem book.

There are many good reasons for a problem book to be written in ring theory, or for that matter in any subject of mathematics. First, the solutions to different exercises serve to illustrate the problem-solving process and show how general theorems in ring theory are applied in special situations. Second, the compilation of solutions to interesting and unusual exercises extends and completes the standard treatment of the subject in textbooks. Last, but not least, a problem book provides a natural place in which to record leisurely some of the folklore of the subject: the "tricks of the trade" in ring theory, which are well known to the experts in the field but may not be familiar to others, and for which there is usually no good reference. With all of the above objectives in mind, I offer this modest problem book for the use and enjoyment of students, teachers and researchers in ring theory and other closely allied fields.

This book is organized in the same way as *FC*, in eight chapters and twenty-five sections. It deals mainly with the "classical" parts of ring theory, starting with the Wedderburn-Artin theory of semisimple rings, Jacobson's theory of the radical, and the representation theory of groups and algebras, then continuing with prime and semiprime rings, primitive and semiprimitive rings, division rings, ordered rings, local and semilocal rings, and the theory of idempotents, and ending with perfect and semiperfect rings. For the reader's information, we should note that this book does not include problems in the vast areas of module theory (e.g., projectivity, injectivity, and flatness), category theory (e.g., equivalences and dualities), or rings of quotients (e.g., Ore rings and Goldie rings). A selection of exercises in these areas will appear later in the author's *Lectures on Modules and Rings*.

While many problems in this book are chosen from *FC*, an effort has been made to render them as independent as possible from the latter. In particular, the statements of all problems are complete and self-contained and should be accessible to readers familiar with the subject at hand, either through *FC* or another ring theory text at the same level. But of course, solving ring theory problems requires a considerable tool kit of theorems and results. For such, I find it convenient to rely on the basic exposition in *FC*. (Results therein will be referred to in the form *FC*-(x.y).) For readers who may be using this book independently of *FC*, an additional challenge will be, indeed, to try to follow the proposed solutions and to figure out along the way *exactly what* are the theorems in *FC* needed to justify the different steps in a problem solution! Very possibly, meeting this challenge will be as rewarding an experience as solving the problem itself.

For the reader's convenience, each section in this book begins with a short introduction giving the general background and the theoretical basis for the problems that follow. All problems are solved in full, in most cases with a lot more details than can be found in the original sources (if they exist). A majority of the solutions are accompanied by a *Comment* section giving relevant bibliographical, historical or anecdotal information, pointing out relations to other exercises, or offering ideas on further improvements and generalizations. These *Comment* sections rounding out the solutions of the exercises are intended to be a particularly useful feature of this problem book.

The exercises in this book are of varying degrees of difficulty. Some are fairly routine and can be solved in a few lines. Others might require a good deal of thought and take up to a page for their solution. A handful of problems are chosen from research papers in the literature; the solutions of some of these might take a couple of pages. Problems of this latter kind are usually identified by giving an attribution to their sources. A majority of the other problems are from the folklore of the subject; with these, no attempt is made to trace the results to their origin. Thus, the lack of a reference for any particular problem only reflects my opinion that the problem is "in the public domain," and should in no case be construed as a claim to originality. On the other hand, the responsibility for any errors or flaws in the presentation of the solutions to any problems remains squarely my own. In the future, I would indeed very much like to receive from my readers communications concerning misprints, corrections, alternative solutions, etc., so that these can be taken into account in case later editions are possible.

Writing solutions to 400 ring-theoretic exercises was a daunting task, even though I had the advantage of choosing them in the first place. The arduous process of working out and checking these solutions could not have been completed without the help of others. Notes on many of the problems were distributed to my Berkeley class of fall 1993; I thank all students in this class for reading and checking my notes and making contributions toward the solutions. Dan Shapiro and Jean-Pierre Tignol have both given their time generously to this project, not only by checking some of my solutions but also by making many valuable suggestions for improvements. Their mathematical insights have greatly enhanced the quality of this work. Other colleagues have helped by providing examples and counterexamples, suggesting alternative solutions, pointing out references and answering my mathematical queries: among them, I should especially thank George Bergman, Rosa Camps, Keith Conrad, Warren Dicks, Kenneth Goodearl, Martin Isaacs, Irving Kaplansky, Hendrik Lenstra, André Leroy, Alun Morris and Barbara Osofsky. From start to finish, Tom von Foerster at Springer-Verlag has guided this project with a gentle hand; I remain deeply appreciative of his editorial acumen and thank him heartily for his kind cooperation.

As usual, members of my family deserve major credit for the timely completion of my work. The writing of this book called for no small sacrifices on the part of my wife Chee-King and our four children, Juwen, Fumei, Juleen, and Dee-Dee; it is thus only fitting that I dedicate this modest volume to them in appreciation for their patience, understanding and unswerving support.

Berkeley, California   T.Y.L.
April 1994

Contents

Notes to the Reader

The four hundred and eighty-five (485) exercises in the eight chapters of this book are organized into twenty-five consecutively numbered sections. As we have explained in the Preface, many of these exercises are chosen from the author's *A First Course in Noncommutative Rings*, (2nd edition), hereafter referred to as *FC*. A cross-reference such as *FC*-(12.7) refers to the result (12.7) in *FC*. Exercise 12.7 will refer to the exercise so labeled in §12 in this book. In referring to an exercise appearing (or to appear) in the same section, we shall sometimes drop the section number from the reference. Thus, when we refer to "Exercise 7" *within* §12, we shall mean Exercise 12.7.

The ring theory conventions used in this book are the same as those introduced in *FC*. Thus, a ring R means a ring with identity (unless otherwise specified). A subring of R means a subring containing the identity of R (unless otherwise specified). The word "ideal" always means a two-sided ideal; an adjective such as "noetherian" likewise means both right and left noetherian. A ring homomorphism from R to S is supposed to take the identity of R to that of S. Left and right R-modules are always assumed to be unital; homomorphisms between modules are (usually) written on the opposite side of the scalars. "Semisimple rings" are in the sense of Wedderburn, Noether and Artin: these are rings R that are semisimple as (left or right) modules over themselves. Rings with Jacobson radical zero are called Jacobson semisimple (or semiprimitive) rings.

Throughout the text, we use the standard notations of modern mathematics. For the reader's convenience, a partial list of the notations commonly used in basic algebra and ring theory is given on the following pages.

Some Frequently Used Notations

$\mathbb{Z}$	ring of integers
$\mathbb{Z}_n$	integers modulo n (or $\mathbb{Z}/n\mathbb{Z}$)
$\mathbb{Q}$	field of rational numbers
$\mathbb{R}$	field of real numbers
$\mathbb{C}$	field of complex numbers
$\mathbb{F}_q$	finite field with q elements
$\mathbb{M}_n(S)$	set of $n \times n$ matrices with entries from S
S_n	symmetric group on $\{1, 2, \ldots, n\}$
A_n	alternating group on $\{1, 2, \ldots, n\}$
$\subset, \subseteq$	used interchangeably for inclusion
$\subsetneq, \subsetneqq$	strict inclusions
$\|A\|$, Card A	used interchangeably for the cardinality of the set A
$A \backslash B$	set-theoretic difference
$A \twoheadrightarrow B$	surjective mapping from A onto B
δ_{ij}	Kronecker deltas
E_{ij}	matrix units
tr	trace (of a matrix or a field element)
det	determinant of a matrix
$\langle x \rangle$	cyclic group generated by x
$Z(G)$	center of the group (or the ring) G
$C_G(A)$	centralizer of A in G
$H \lhd G$	H is a normal subgroup of G
$[G : H]$	index of subgroup H in a group G
$[K : F]$	field extension degree
$[K : D]_\ell$, $[K : D]_r$	left, right dimensions of $K \supseteq D$ as D-vector space
K^G	G-fixed points on K
$\mathrm{Gal}(K/F)$	Galois group of the field extension K/F
M_R, $_RN$	right R-module M, left R-module N
$M \oplus N$	direct sum of M and N
$M \otimes_R N$	tensor product of M_R and $_RN$
$\mathrm{Hom}_R(M, N)$	group of R-homomorphisms from M to N
$\mathrm{End}_R(M)$	ring of R-endomorphisms of M
$\mathrm{soc}(M)$	socle of M
$\mathrm{length}(M)$, $\ell(M)$	(composition) length of M
nM (or M^n)	$M \oplus \cdots \oplus M$ (n times)
$\prod_i R_i$	direct product of the rings $\{R_i\}$
char R	characteristic of the ring R
R^{op}	opposite ring of R
$\mathrm{U}(R)$, R^*	group of units of the ring R
$\mathrm{U}(D)$, D^*, $\dot{D}$	multiplicative group of the division ring D

$GL(V)$	group of linear automorphisms of a vector space V
$GL_n(R)$	group of invertible $n \times n$ matrices over R
$SL_n(R)$	group of $n \times n$ matrices of determinant 1 over a commutative ring R
rad R	Jacobson radical of R
$\text{Nil}^*(R)$	upper nilradical of R
$\text{Nil}_*(R)$	lower nilradical (or prime radical) of R
Nil (R)	ideal of nilpotent elements in a commutative ring R
$\text{soc}(_RR)$, $\text{soc}(R_R)$	socle of R as left, right R-module
$\text{ann}_\ell(S)$, $\text{ann}_r(S)$	left, right annihilators of the set S
kG, $k[G]$	(semi)group ring of the (semi)group G over the ring k
$k[x_i : i \in I]$	polynomial ring over k with (commuting) variables $\{x_i : i \in I\}$
$k \langle x_i : i \in I \rangle$	free ring over k generated by $\{x_i : i \in I\}$
$k[x; \sigma]$	skew polynomial ring with respect to an endomorphism σ on k
$k[x; \delta]$	differential polynomial ring with respect to a derivation δ on k
$[G, G]$	commutator subgroup of the group G
$[R, R]$	additive subgroup of the ring R generated by all $[a, b] = ab - ba$
f.g.	finitely generated
ACC	ascending chain condition
DCC	descending chain condition
LHS	left-hand side
RHS	right-hand side

Chapter 1
Wedderburn-Artin Theory

§1. Basic Terminology and Examples

The exercises in this beginning section cover the basic aspects of rings, ideals (both 1-sided and 2-sided), zero-divisors and units, isomorphisms of modules and rings, the chain conditions, and Dedekind-finiteness. A ring R is said to be *Dedekind-finite* if $ab = 1$ in R implies that $ba = 1$. The chain conditions are the usual noetherian (ACC) or artinian (DCC) conditions which can be imposed on submodules of a module, or on 1-sided or 2-sided ideals of a ring.

Some of the exercises in this section lie at the foundations of noncommutative ring theory, and will be used freely in all later exercises. These include, for instance, the computation of the center of a matrix ring (Exercise 9), the computation of the endomorphism ring for n (identical) copies of a module (Exercise 20), and the basic facts pertaining to direct decompositions of a ring into 1-sided or 2-sided ideals (Exercises 7 and 8).

Throughout these exercises, the word "ring" means an associative (but not necessarily commutative) ring with an identity element 1. (On a few isolated occasions, we shall deal with rings without an identity.[1] Whenever this happens, it will be clearly stated.) The word "subring" always means a subring containing the identity element of the larger ring. If $R = \{0\}$, R is called the *zero ring*; note that this is the case iff $1 = 0$ in R. If $R \neq \{0\}$ and $ab = 0 \Rightarrow a = 0$ or $b = 0$, R is said to be a *domain*.

Without exception, the word "ideal" refers to a 2-sided ideal. One-sided ideals are referred to as left ideals or right ideals. The *units* in a ring R are

---

[1] Rings without identities are dubbed "rngs" by Louis Rowen.

the elements of R with both left and right inverses (which must be equal). The set

$$U(R) = \{a \in R : a \text{ is a unit}\}$$

is a group under multiplication, and is called the *group of units* of R. If $R \neq \{0\}$ and $U(R) = R\backslash\{0\}$, R is said to be a *division ring*. To verify that a nonzero ring R is a division ring, it suffices to check that every $a \in R\backslash\{0\}$ is *right*-invertible: see Exercise 2 below.

All (say, left) R-modules $_RM$ are assumed to be *unital*, that is, $1 \cdot m = m$ for all $m \in M$.

Exercises for §1

Ex. 1.1. Let $(R, +, \times)$ be a system satisfying all axioms of a ring with identity, except possibly $a + b = b + a$. Show that $a + b = b + a$ for all $a, b \in R$, so R is indeed a ring.

Solution. By the two distributive laws, we have

$$(a + b)(1 + 1) = a(1 + 1) + b(1 + 1) = a + a + b + b,$$
$$(a + b)(1 + 1) = (a + b)1 + (a + b)1 = a + b + a + b.$$

Using the additive group laws, we deduce that $a + b = b + a$ for all $a, b \in R$.

Ex. 1.2. It was mentioned above that a nonzero ring R is a division ring iff every $a \in R\backslash\{0\}$ is right-invertible. Supply a proof for this statement.

Solution. For the "if" part, it suffices to show that $ab = 1 \Longrightarrow ba = 1$ in R. From $ab = 1$, we have $b \neq 0$, so $bc = 1$ for some $c \in R$. Now left multiplication by a shows $c = a$, so indeed $ba = 1$.

Ex. 1.3. Show that the characteristic of a domain is either 0 or a prime number.

Solution. Suppose the domain R has characteristic $n \neq 0$. If n is not a prime, then $n = n_1 n_2$ where $1 < n_i < n$. But then $n_i 1 \neq 0$, and $(n_1 1)(n_2 1) = n1 = 0$ contradicts the fact that R is a domain.

Ex. 1.4. True or False: "If ab is a unit, then a, b are units"? Show the following for any ring R:

(a) If a^n is a unit in R, then a is a unit in R.
(b) If a is left-invertible and not a right 0-divisor, then a is a unit in R.
(c) If R is a domain, then R is Dedekind-finite.

Solution. The statement in quotes is false in general. If R is a ring that is not Dedekind-finite, then, for suitable $a, b \in R$, we have $ab = 1 \neq ba$. Here, ab is a unit, but neither a nor b is a unit. (Conversely, it is easy to see that, if R is Dedekind-finite, then the statement becomes true.)

For (a), note that if $a^n c = c a^n = 1$, then a has a right inverse $a^{n-1}c$ and a left inverse ca^{n-1}, so $a \in U(R)$. For (b), say $ba = 1$. Then

$$(ab - 1)a = a - a = 0.$$

If a is not a right 0-divisor, then $ab = 1$ and so $a \in U(R)$. (c) follows immediately from (b).

Ex. 1.4*. Let $a \in R$, where R is any ring.

(1) Show that if a has a left inverse, then a is not a left 0-divisor.
(2) Show that the converse holds if $a \in aRa$.

Solution. (1) Say $ba = 1$. Then $ac = 0$ implies $c = (ba)c = b(ac) = 0$.
(2) Write $a = ara$, and assume a is not a left 0-divisor. Then $a(1 - ra) = 0$ yields $ra = 1$, so a has left inverse r.

Comment. In general, an element $a \in R$ is called (von Neumann) regular if $a \in aRa$. If every $a \in R$ is regular, R is said to be a von Neumann regular ring.

Ex. 1.5. Give an example of an element x in a ring R such that $Rx \subsetneq xR$.

Solution. Let R be the ring of 2×2 upper triangular matrices over a nonzero ring k, and let $x = \begin{pmatrix} 1 & 0 \\ 0 & 0 \end{pmatrix}$. A simple calculation shows that

$$Rx = \begin{pmatrix} k & 0 \\ 0 & 0 \end{pmatrix}, \quad \text{and} \quad xR = \begin{pmatrix} k & k \\ 0 & 0 \end{pmatrix}.$$

Therefore, $Rx \subsetneq xR$. Alternatively, we can take R to be any non-Dedekind-finite ring, say with $xy = 1 \neq yx$. Then xR contains 1 so $xR = R$, but $1 \notin Rx$ implies that $Rx \subsetneq R = xR$.

Ex. 1.6. Let a, b be elements in a ring R. If $1 - ba$ is left-invertible (resp. invertible), show that $1 - ab$ is left-invertible (resp. invertible), and construct a left inverse (resp. inverse) for it explicitly.

Solution. The left ideal $R(1 - ab)$ contains

$$Rb(1 - ab) = R(1 - ba)b = Rb,$$

so it also contains $(1 - ab) + ab = 1$. This shows that $1 - ab$ is left-invertible. This proof lends itself easily to an explicit construction: if $u(1 - ba) = 1$, then

$$b = u(1 - ba)b = ub(1 - ab), \quad \text{so}$$
$$1 = 1 - ab + ab = 1 - ab + aub(1 - ab) = (1 + aub)(1 - ab).$$

Hence,

$$(1 - ab)^{-1} = 1 + a(1 - ba)^{-1}b,$$

where x^{-1} denotes "a left inverse" of x. The case when $1 - ba$ is invertible follows by combining the "left-invertible" and "right-invertible" cases.

Comment. The formula for $(1-ab)^{-1}$ above occurs often in linear algebra books (for $n \times n$ matrices). Kaplansky taught me a way in which you can always rediscover this formula, "even if you are thrown up on a desert island with all your books and papers lost." Using the formal expression for inverting $1-x$ in a power series, one writes (on sand):

$$(1-ab)^{-1} = 1 + ab + abab + ababab + \cdots$$
$$= 1 + a(1 + ba + baba + \cdots)b$$
$$= 1 + a(1-ba)^{-1}b.$$

Once you hit on this correct formula, a direct verification (for 1-sided or 2-sided inverses) is a breeze.[2]

For an analogue of this exercise for rings possibly without an identity, see Exercise 4.2.

Ex. 1.7. Let $B_1, \ldots, B_n$ be left ideals (resp. ideals) in a ring R. Show that $R = B_1 \oplus \cdots \oplus B_n$ iff there exist idempotents (resp. central idempotents) $e_1, \ldots, e_n$ with sum 1 such that $e_i e_j = 0$ whenever $i \neq j$, and $B_i = Re_i$ for all i. In the case where the B_i's are ideals, if $R = B_1 \oplus \cdots \oplus B_n$, then each B_i is a ring with identity e_i, and we have an isomorphism between R and the direct product of rings $B_1 \times \cdots \times B_n$. Show that any isomorphism of R with a finite direct product of rings arises in this way.

Solution. Suppose e_i's are idempotents with the given properties, and $B_i = Re_i$. Then, whenever

$$a_1 e_1 + \cdots + a_n e_n = 0,$$

right multiplication by e_i shows that $a_i e_i = 0$. This shows that we have a decomposition

$$R = B_1 \oplus \cdots \oplus B_n.$$

Conversely, suppose we are given such a decomposition, where each B_i is a left ideal. Write $1 = e_1 + \cdots + e_n$, where $e_i \in B_i$. Left multiplying by e_i, we get

$$e_i = e_i e_1 + \cdots + e_i e_n.$$

This shows that $e_i = e_i^2$, and $e_i e_j = 0$ for $i \neq j$. For any $b \in B_i$, we also have

$$b = be_1 + \cdots + be_i + \cdots + be_n,$$

so $b = be_i \in Re_i$. This shows that $B_i = Re_i$. If each e_i is central, $B_i = Re_i = e_i R$ is clearly an ideal. Conversely, if each B_i is an ideal, the above work shows that $b = be_i = e_i b$ for any $b \in B_i$, so B_i is a ring with identity e_i. Since $B_i B_j = 0$ for $i \neq j$, it follows that each e_i is central. We finish easily

---

[2] This trick was also mentioned in an article of P.R. Halmos in Math. Intelligencer *3* (1981), 147–153. Halmos attributed the trick to N. Jacobson.

by showing that

$$B_1 \times \cdots \times B_n \to R$$

defined by

$$(b_1, \ldots, b_n) \mapsto b_1 + \cdots + b_n \in R$$

is an isomorphism of rings. The last part of the Exercise is now routine: we omit it.

Ex. 1.8. Let $R = B_1 \oplus \cdots \oplus B_n$, where the B_i's are ideals of R. Show that any left ideal (resp. ideal) I of R has the form $I = I_1 \oplus \cdots \oplus I_n$ where, for each i, I_i is a left ideal (resp. ideal) of the ring B_i.

Solution. Let $I_i = I \cap B_i$, and write $B_i = Re_i$ as in Exercise 1.7. We claim that $\bigoplus I_i \subseteq I$ is an equality. Indeed, for any $a \in I$, we have

$$a = 1a = e_1 a + \cdots + e_n a$$

with $e_i a = ae_i \in I \cap B_i = I_i$. Therefore, $I = \bigoplus I_i$, where I_i is a left ideal of R (and also of B_i). If I is an ideal of R, clearly I_i is an ideal of R (and also of B_i).

Ex. 1.9. Show that for any ring R, the center of the matrix ring $\mathbb{M}_n(R)$ consists of the diagonal matrices $r \cdot I_n$, where r belongs to the center of R.

Solution. Let E_{ij} be the matrix units. If $r \in Z(R)$, then

$$(r \cdot I_n)(aE_{ij}) = raE_{ij} = (aE_{ij})(rI_n),$$

so $r \cdot I_n \in Z(S)$, where $S = \mathbb{M}_n(R)$. Conversely, consider

$$M = \sum r_{ij} E_{ij} \in Z(S).$$

From $ME_{kk} = E_{kk}M$, we see easily that M is a diagonal matrix. This and $ME_{kl} = E_{kl}M$ together imply that $r_{kk} = r_{ll}$ for all k, l, so $M = r \cdot I_n$ for some $r \in R$. Since this commutes with all $a \cdot I_n (a \in R)$, we must have $r \in Z(R)$.

Ex. 1.10. Let p be a fixed prime.

(a) Show that any ring (with identity) of order p^2 is commutative.
(b) Show that there exists a noncommutative ring *without identity* of order p^2.
(c) Show that there exists a noncommutative ring (with identity) of order p^3.

Solution. (a) If the additive order of 1 is p^2, then $R = \mathbb{Z} \cdot 1$ is clearly commutative. If the additive order of 1 is p, then $R = \mathbb{Z} \cdot 1 \oplus \mathbb{Z} \cdot a$ for any $a \notin \mathbb{Z} \cdot 1$, and again R is commutative.

(c) The ring of 2×2 upper triangular matrices over $\mathbb{F}_p$ is clearly a non-commutative ring with identity of order p^3.

(b) For any nonzero ring k,

$$R = \left\{ \begin{pmatrix} a & b \\ 0 & 0 \end{pmatrix} : \quad a, b \in k \right\}$$

is a right ideal of $M_2(k)$. Therefore, $(R, +, \times)$ satisfies all the axioms of a ring, except perhaps the identity axiom. An easy verification shows that R has in fact no identity. Finally,

$$\begin{pmatrix} 1 & 0 \\ 0 & 0 \end{pmatrix} \begin{pmatrix} 0 & 1 \\ 0 & 0 \end{pmatrix} = \begin{pmatrix} 0 & 1 \\ 0 & 0 \end{pmatrix}, \quad \text{but} \quad \begin{pmatrix} 0 & 1 \\ 0 & 0 \end{pmatrix} \begin{pmatrix} 1 & 0 \\ 0 & 0 \end{pmatrix} = \begin{pmatrix} 0 & 0 \\ 0 & 0 \end{pmatrix},$$

so R is a *noncommutative* "rng". Note that, if we represent the elements of R in the simpler form (a, b) $(a, b \in k)$, then the multiplication in R can be expressed by

(1) $$(a, b)(c, d) = (ac, ad).$$

In particular, if $k = \mathbb{F}_p$, we get a noncommutative "rng" of cardinality p^2.

Comment. We can define another multiplication on $S = k^2$ by

(2) $$(a, b) * (c, d) = ((a + b)c, (a + b)d).$$

Then, $(S, +, *)$ is also a noncommutative "rng". This is just another "copy" of $(R, +, \times)$. In fact, $\varphi : S \to R$ given by

$$\varphi(a, b) = (a + b, b)$$

is a "rng" isomorphism from S to R. Note that S is also isomorphic to the "subrng" of $M_2(k)$ consisting of matrices of the form $\begin{pmatrix} a & b \\ a & b \end{pmatrix}$, with an isomorphism given by $(a, b) \mapsto \begin{pmatrix} a & b \\ a & b \end{pmatrix}$.

Ex. 1.11. Let R be a ring possibly without an identity. An element $e \in R$ is called a left (resp. right) identity for R if $ea = a$ (resp. $ae = a$) for every $a \in R$.

(a) Show that a left identity for R need not be a right identity.

(b) Show that if R has a unique left identity e, then e is also a right identity.

Solution. (a) For the "rng" R constructed in (b) of Exercise 10, any $(1, b) \in R$ is a left identity, but not a right identity.

(b) Suppose $e \in R$ is a *unique* left identity for R. Then for any $a, c \in R$,

$$(e + ae - a)c = c + ac - ac = c.$$

Therefore, $e + ae - a = e$, which implies $ae = a$ (for any $a \in R$).

Ex. 1.12. A left R-module M is said to be *hopfian* (after the topologist H. Hopf) if any surjective R-endomorphism of M is an automorphism.

(1) Show that any noetherian module M is hopfian.
(2) Show that the left regular module $_RR$ is hopfian iff R is Dedekind-finite.[3]
(3) Deduce from (1), (2) that any left noetherian ring R is Dedekind-finite.

Solution. (1) Let $\alpha : M \to M$ be surjective and M be noetherian. The ascending chain

$$\ker \alpha \subseteq \ker \alpha^2 \subseteq \cdots$$

must stop, so $\ker \alpha^i = \ker \alpha^{i+1}$ for some i. If $\alpha(m) = 0$, write $m = \alpha^i(m')$ for some $m' \in M$. Then

$$0 = \alpha(\alpha^i(m')) = \alpha^{i+1}(m')$$

implies that $0 = \alpha^i(m') = m$, so $\alpha \in \operatorname{Aut}_R(M)$.

(2) Suppose $_RR$ is hopfian, and suppose $ab = 1$. Then $x \mapsto xb$ defines a surjective endomorphism α of $_RR$. Therefore, α is an automorphism. Since

$$\alpha(ba) = bab = b = \alpha(1),$$

we must have $ba = 1$, so R is Dedekind-finite. The converse is proved by reversing this argument.

(3) Since $_RR$ is a noetherian module, it is hopfian by (1), so R is Dedekind-finite by (2).

Comment. (a) In the proof of (3), a weaker ascending chain condition would have sufficed. In fact, an endomorphism $\alpha : {}_RR \to {}_RR$ is given by right multiplication by $b = \alpha(1)$, and α^i is given by right multiplication by b^i. Thus, $\ker (\alpha^i) = \operatorname{ann}_\ell(b^i)$ and we need only assume that any left annihilator chain

$$\operatorname{ann}_\ell(b) \subseteq \operatorname{ann}_\ell(b^2) \subseteq \cdots$$

stabilizes in order to ensure that $_RR$ is hopfian by the argument used for (1) in the exercise.

(b) By (1) of the exercise, any left noetherian ring R has the property that any finitely generated left R-module is hopfian. There are other classes of rings with this property too: see the *Comments* on Exercises 4.16 and 20.9. The *Comment* on the latter contains also a characterization of rings whose finitely generated modules are all hopfian.

(c) The hopfian property can also be studied in the category of groups: a group G is said to be hopfian if any surjective endomorphism of G is an

---

[3] In particular, R being hopfian is a left-right symmetric notion.

automorphism. The same proof used in (1) above shows that, *if the normal subgroups of a group G satisfy the ACC, then G is hopfian.* For instance, all polycyclic groups are hopfian (see p. 394 of Sims' book "Computation with Finitely Presented Groups," Cambridge Univ. Press, 1994), and all free groups of finite rank are also hopfian (see p. 109 of the book "Combinatorial Group Theory" by Magnus, Karrass, and Solitar, J. Wiley Interscience, 1966).

(d) One can also formulate a dual version of "hopfian," and obtain the notion of "cohopfian": see Exercise 4.16 below.

Ex. 1.13. Let A be an algebra over a field k such that every element of A is algebraic over k.

(a) Show that A is Dedekind-finite.
(b) Show that a left 0-divisor of A is also a right 0-divisor.
(c) Show that a nonzero element of A is a unit iff it is not a 0-divisor.
(d) Let B be a subalgebra of A, and $b \in B$. Show that b is a unit in B iff it is a unit in A.

Solution. Consider any nonzero element $b \in A$, and let

$$a_n b^n + \cdots + a_m b^m = 0 \quad (a_i \in k, \ a_n \neq 0 \neq a_m, n \geq m)$$

be a polynomial of smallest degree satisfied by b. If $m > 0$, then

$$c = a_n b^{n-1} + \cdots + a_m b^{m-1} \neq 0,$$

and we have $cb = bc = 0$. In this case, b is both a left 0-divisor and a right 0-divisor. If $m = 0$, then, for

$$d = a_n b^{n-1} + \cdots + a_1,$$

we have $db = bd = -a_0 \in k^*$. In this case, b is a unit in A. This gives (a), (b) and (c). If $b \in B$ is as in (d) and b is a unit in A, the above also shows that

$$b^{-1} = -a_0^{-1} d \in k[b] \subseteq B.$$

Comment. An algebra satisfying the hypothesis of this exercise is called an *algebraic algebra* over k. For other interesting properties of such an algebra, see Exercises 12.6B, 13.11, and 23.6(2).

Ex. 1.14. (Kaplansky) Suppose an element a in a ring has a right inverse b but no left inverse. Show that a has infinitely many right inverses. (In particular, if a ring is finite, it must be Dedekind-finite.)

Solution. Suppose we have already constructed n distinct right inverses $b_1, \ldots, b_n$ for a. We shall show that there exist at least $n+1$ distinct right inverses for a. Indeed, consider the elements

$$c_i = 1 - b_i a \quad (1 \leq i \leq n),$$

which have the property that

$$ac_i = a(1 - b_i a) = 0.$$

If $c_i = c_j$, then $b_i a = b_j a$, and right multiplication by b_1 shows that $b_i = b_j$. This shows that $c_1, \ldots, c_n$ are distinct. Also, each $c_i \neq 0$, since a has no left inverse. Therefore,

$$\{b_1, b_1 + c_1, \ldots, b_1 + c_n\}$$

are $n + 1$ distinct right inverses for a.

Alternative Solution. Consider the elements

$$d_j = (1 - ba)a^j \quad (j \geq 1),$$

which have the property that $ad_j = 0$. We claim that the d_j's are distinct. Indeed, if $d_i = d_j$ for some $i > j$, then, right multiplying $(1 - ba)a^i = (1 - ba)a^j$ by b^j, we get

$$(1 - ba)a^{i-j} = 1 - ba \qquad (\text{since } a^j b^j = 1).$$

But then

$$[(1 - ba)a^{i-j-1} + b]a = 1,$$

a contradiction. Since the d_j's are distinct, $\{d_j + b : j \geq 1\}$ is an infinite set of right inverses for a.

Comment. The first solution above is essentially that given in C.W. Bitzer's paper, "Inverses in rings with unity," Amer. Math. Monthly *70* (1963), 315. As for the second solution, a more elaborate construction is possible. Given $ab = 1 \neq ba$ in a ring, Jacobson has shown that the elements $e_{ij} = b^i(1 - ba)a^j$ give a set of "matrix units" in the sense that $e_{ij}\, e_{kl} = \delta_{jk}\, e_{il}$ (where δ_{jk} are the Kronecker deltas). For more details, see FC-(21.26).

Ex. 1.15. Let $A = \mathbb{C}[x; \sigma]$, where σ denotes complex conjugation on $\mathbb{C}$.
(a) Show that $Z(A) = \mathbb{R}[x^2]$.
(b) Show that $A/A \cdot (x^2 + 1)$ is isomorphic to $\mathbb{H}$, the division ring of real quaternions.
(c) Show that $A/A \cdot (x^4 + 1)$ is isomorphic to $\mathbb{M}_2(\mathbb{C})$.

Solution. (a) Here A is the ring of all "skew polynomials" $\sum a_i x^i$ $(a_i \in \mathbb{C})$, multiplied according to the rule $xa = \sigma(a)x$ for $a \in \mathbb{C}$. Since $\sigma^2 = 1$,

$$x^2 a = \sigma^2(a)x^2 = ax^2 \quad \text{for all} \quad a \in \mathbb{C}.$$

This shows that $\mathbb{R}[x^2] \subseteq Z(A)$. Conversely, consider any $f = \sum a_r x^r \in Z(A)$. From $fa = af$, we have $a_r \sigma^r(a) = aa_r$ for all $a \in \mathbb{C}$. Setting $a = i$, we see that $a_r = 0$ for odd r. Therefore, $f = \sum a_{2s} x^{2s}$. From $fx = xf$, we see further that $\sigma(a_{2s}) = a_{2s}$, so $f \in \mathbb{R}[x^4]$.

(b) Since $x^2 + 1 \in Z(A)$, $A \cdot (x^2 + 1)$ is an ideal, so we can form the quotient ring

$$\bar{A} = A/A \cdot (x^2 + 1).$$

Expressing the ring of real quaternions in the form $\mathbb{H} = \mathbb{C} \oplus \mathbb{C}j$, we can define $\varphi : A \to \mathbb{H}$ by $\varphi(x) = j$, and $\varphi(a) = a$ for all $a \in \mathbb{C}$. Since $ja = \sigma(a)j$ in $\mathbb{H}$ for any $a \in \mathbb{C}$, φ gives a ring homomorphism from A to $\mathbb{H}$. This induces a ring homomorphism $\bar{\varphi} : \bar{A} \to \mathbb{H}$, since $\varphi(x^2 + 1) = j^2 + 1 = 0$. In view of

$$\bar{\varphi}(\overline{a + bx}) = a + bj \qquad (\text{for } a, b \in \mathbb{C}),$$

$\bar{\varphi}$ is clearly an isomorphism.

(c) Changing notations, we write here $\bar{A} = A/A \cdot (x^4 + 1)$. Define $\varphi : A \to M_2(\mathbb{C})$ by

$$\varphi(x) = \begin{pmatrix} 0 & i \\ 1 & 0 \end{pmatrix} \quad \text{and} \quad \varphi(a) = \begin{pmatrix} a & 0 \\ 0 & \sigma(a) \end{pmatrix} \quad \text{for} \quad a \in \mathbb{C}.$$

Since

$$\begin{pmatrix} 0 & i \\ 1 & 0 \end{pmatrix} \begin{pmatrix} a & 0 \\ 0 & \sigma(a) \end{pmatrix} = \begin{pmatrix} 0 & i\sigma(a) \\ a & 0 \end{pmatrix} = \begin{pmatrix} \sigma(a) & 0 \\ 0 & a \end{pmatrix} \begin{pmatrix} 0 & i \\ 1 & 0 \end{pmatrix},$$

φ gives a ring homomorphism from A to $M_2(\mathbb{C})$. Again, φ induces a ring homomorphism $\bar{\varphi} : \bar{A} \to M_2(\mathbb{C})$, since

$$\varphi(x^4 + 1) = \begin{pmatrix} 0 & i \\ 1 & 0 \end{pmatrix}^4 + I = \begin{pmatrix} i & 0 \\ 0 & i \end{pmatrix}^2 + I = 0.$$

By a straightforward computation, for $b_k,\ c_k \in \mathbb{R}$:

$$\bar{\varphi}\left(\sum_{k=0}^{3} \overline{(b_k + ic_k)x^k} \right)$$

$$= \begin{pmatrix} b_0 + c_0 i + i(b_2 + c_2 i) & -(b_3 + ic_3) + i(b_1 + c_1 i) \\ b_1 - c_1 i + i(b_3 - c_3 i) & b_0 - ic_0 + i(b_2 - c_2 i) \end{pmatrix}.$$

Clearly, this is the zero matrix only if all $b_k,\ c_k = 0$. Therefore, $\bar{\varphi}$ is one-one. Since $\bar{\varphi}$ is an $\mathbb{R}$-homomorphism and both $\bar{A}$ and $M_2(\mathbb{C})$ have dimension 8 over $\mathbb{R}$, it follows that $\bar{\varphi}$ is an isomorphism. (Note that $\bar{\varphi}$ here is *not* a homomorphism of left $\mathbb{C}$-vector spaces, since $\bar{\varphi}$ is not the identity map from $\mathbb{C} \subseteq \bar{A}$ to $\mathbb{C} \subseteq M_2(\mathbb{C})$!)

Ex. 1.16. Let K be a division ring with center k.

(1) Show that the center of the polynomial ring $R = K[x]$ is $k[x]$.

(2) For any $a \in K \backslash k$, show that the ideal generated by $x - a$ in $K[x]$ is the unit ideal.

(3) Show that any ideal $I \subseteq R$ has the form $R \cdot h$ where $h \in k[x]$.

Solution. (1) Clearly $k[x] \subseteq Z(R)$. Conversely, if

$$f = \sum a_i x^i \in Z(R),$$

then $fa = af$ for all $a \in K$ shows that each $a_i \in Z(K) = k$, and hence $f \in k[x]$.

(2) Fix $b \in K$ such that $ab \neq ba$. Then $(x - a)$ (the ideal generated by $x - a$) contains

$$b(x - a) - (x - a)b = ab - ba \in U(K),$$

so $(x - a) = R$.

(3) We may assume $I \neq 0$, and fix a monic polynomial of the least degree in I. By the usual Euclidean algorithm argument, we see that $I = R \cdot h$. For any $a \in K$, we have $ha \in I = R \cdot h$, so $ha = rh$ for some $r \in R$. By comparing the leading terms, we see that $r \in K$, and in fact $r = a$. Thus, $ha = ah$ for any $a \in K$, which means that $h \in k[x]$.

Ex. 1.17. Let x, y be elements in a ring R such that $Rx = Ry$. Show that there exists a right R-module isomorphism $f : xR \to yR$ such that $f(x) = y$.

Solution. Define $f(xr) = yr$ for any $r \in R$. Note that if $xr = xr'$, then $x(r - r') = 0$ implies that

$$y(r - r') \in Rx(r - r') = 0.$$

This shows that $yr = yr'$, so f is well-defined. Clearly, f is an R-module epimorphism from xR to yR, with $f(x) = y$. Finally, if $xr \in \ker(f)$, then $yr = 0$ and we have $xr \in (Ry)r = 0$. Therefore, f is the isomorphism we want.

Ex. 1.18. For any ring k, let

$$A = \left\{ \begin{pmatrix} a & b \\ c & d \end{pmatrix} : a + c = b + d \in k \right\}.$$

Show that A is a subring of $\mathbb{M}_2(k)$, and that it is isomorphic to the ring R of 2×2 lower triangular matrices over k.

Solution. Let $\alpha = \begin{pmatrix} 1 & 1 \\ 0 & 1 \end{pmatrix}$. Then $\alpha^{-1} R \alpha$ consists of the matrices

$$\begin{pmatrix} 1 & -1 \\ 0 & 1 \end{pmatrix} \begin{pmatrix} x & 0 \\ y & z \end{pmatrix} \begin{pmatrix} 1 & 1 \\ 0 & 1 \end{pmatrix} = \begin{pmatrix} x - y & x - y - z \\ y & y + z \end{pmatrix}.$$

We see easily that the set of these matrices is exactly A. Therefore, A is just a "conjugate" of the subring R in the ring $\mathbb{M}_2(k)$. In particular, $A \cong R$.

(We can define mutually inverse isomorphisms $f : R \to A$ and $g : A \to R$ explicitly by

$$f \begin{pmatrix} x & 0 \\ y & z \end{pmatrix} = \begin{pmatrix} x - y & x - y - z \\ y & y + z \end{pmatrix} \in A, \quad \text{and}$$

$$g \begin{pmatrix} a & b \\ c & d \end{pmatrix} = \begin{pmatrix} a + c & 0 \\ c & d - c \end{pmatrix} \in R,$$

where $a + c = b + d$.)

Comment. A similar construction shows that A is also isomorphic to the ring S of 2×2 *upper* triangular matrices over k. (The fact that $R \cong S$ is a special case of Exercise 1.22(1) below.)

Ex. 1.19. Let R be a domain. If R has a minimal left ideal, show that R is a division ring. (In particular, a left artinian domain must be a division ring.)

Solution. Let $I \subseteq R$ be a minimal left ideal, and fix an element $a \neq 0$ in I. Then $I = Ra = Ra^2$. In particular, $a = ra^2$ for some $r \in R$. Cancelling a, we have $1 = ra \in I$, so $I = R$. The minimality of I shows that R has no left ideals other than (0) and R, so R is a division ring (cf. Exercise 1.2).

Ex. 1.20. Let $E = \mathrm{End}_R(M)$ be the ring of endomorphisms of an R-module M, and let nM denote the direct sum of n copies of M. Show that $\mathrm{End}_R(nM)$ is isomorphic to $\mathbb{M}_n(E)$ (the ring of $n \times n$ matrices over E).

Solution. Say M is a right R-module, and we write endomorphisms on the left. Let $\varepsilon_j : M \to nM$ be the jth inclusion, and $\pi_i : nM \to M$ be the ith projection. For any endomorphism $F : nM \to nM$, let f_{ij} be the composition $\pi_i F \varepsilon_j \in E$. Define a map

$$\alpha : \mathrm{End}_R(nM) \to \mathbb{M}_n(E)$$

by $\alpha(F) = (f_{ij})$. Routine calculations show that α is an isomorphism of rings.

Ex. 1.21. Let R be a finite ring. Show that there exists an infinite sequence $n_1 < n_2 < n_3 < \cdots$ of natural numbers such that, for any $x \in R$, we have $x^{n_1} = x^{n_2} = x^{n_3} = \cdots$.

Solution. Label the elements of R as $x_1, x_2, \ldots, x_k$. Since there are at most k distinct elements in the set $\{x_1, x_1^2, x_1^3, \ldots\}$, there must exist $r_1 < r_2 < \cdots$ such that

$$x_1^{r_1} = x_1^{r_2} = \cdots.$$

By considering $\{x_2^{r_1}, x_2^{r_2}, \ldots\}$, we see similarly that there exist a subsequence $s_1 < s_2 < \cdots$ of $\{r_i\}$ such that

$$x_2^{s_1} = x_2^{s_2} = \cdots.$$

Repeating this construction a finite number of times, we arrive at a sequence $n_1 < n_2 < \ldots$ such that

$$x_i^{n_1} = x_i^{n_2} = \cdots \qquad \text{for } 1 \le i \le k.$$

For the next two problems, note that an *anti-isomorphism* $\varepsilon : k \to k'$ (from a ring k to another ring k') is an additive isomorphism with $\varepsilon(ab) = \varepsilon(b)\varepsilon(a)$ for all $a, b \in k$ (and hence $\varepsilon(1) = 1$). An *involution* ε on a ring k is an anti-automorphism $\varepsilon : k \to k$ with $\varepsilon^2 = \mathrm{Id}_k$.

Ex. 1.22. For any ring k, let $A = \mathbb{M}_n(k)$ and let R (resp. S) denote the ring of $n \times n$ upper (resp. lower) triangular matrices over k.

(1) Show that $R \cong S$.
(2) Suppose k has an anti-automorphism (resp. involution). Show that the same is true for A, R and S.
(3) Under the assumption of (2), show that $R, S, R^{\mathrm{op}}, S^{\mathrm{op}}$ are all isomorphic.

Solution. To simplify the notations, we shall work in the (sufficiently typical) case $n = 3$.

(1) Let $E = \begin{pmatrix} 0 & 0 & 1 \\ 0 & 1 & 0 \\ 1 & 0 & 0 \end{pmatrix}$ and let α be the inner automorphism of A defined by E (with $\alpha^2 = \mathrm{Id}_A$). An easy calculation shows that

$$\alpha \begin{pmatrix} a & b & c \\ d & e & f \\ g & h & i \end{pmatrix} = \begin{pmatrix} i & h & g \\ f & e & d \\ c & b & a \end{pmatrix}.$$

In particular, α restricts to a ring isomorphism from R to S.

(2) Suppose $\varepsilon : k \to k$ is an anti-automorphism (resp. involution). Composing the transpose map with ε on matrix entries, we can define $\delta_0 : A \to A$ with

$$\delta_0 \begin{pmatrix} a & b & c \\ d & e & f \\ g & h & i \end{pmatrix} = \begin{pmatrix} \varepsilon(a) & \varepsilon(d) & \varepsilon(g) \\ \varepsilon(b) & \varepsilon(e) & \varepsilon(h) \\ \varepsilon(c) & \varepsilon(f) & \varepsilon(i) \end{pmatrix}.$$

It is easy to check that this δ_0 is an anti-automorphism (resp. involution) of A, and therefore so is $\delta := \alpha \circ \delta_0$ given by

$$\delta \begin{pmatrix} a & b & c \\ d & e & f \\ g & h & i \end{pmatrix} = \begin{pmatrix} \varepsilon(i) & \varepsilon(f) & \varepsilon(c) \\ \varepsilon(h) & \varepsilon(e) & \varepsilon(b) \\ \varepsilon(g) & \varepsilon(d) & \varepsilon(a) \end{pmatrix}.$$

By inspection, we see that this δ restricts to anti-automorphisms (resp. involutions) on the subrings R and S of A.

(3) follows immediately from (1) and (2). (Conversely, one can also show that $R \cong R^{\mathrm{op}} \Longrightarrow k \cong k^{\mathrm{op}}$.)

Comment. In (2) above, α and δ_0 *commute* as operators on A. Thus, we have a commutative diagram

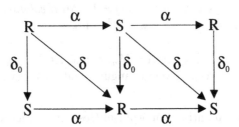

In particular, (R, δ) and (S, δ) are isomorphic "as rings with involutions," with isomorphism defined by α.

Ex. 1.22A. Let R be the upper triangular ring $\begin{pmatrix} \mathbb{Z} & \mathbb{Z}_2 \\ 0 & \mathbb{Z}_2 \end{pmatrix}$.

(1) Show that every right 0-divisor in R is a left 0-divisor.
(2) Using (1), show that R is not isomorphic to its opposite ring R^{op}.

Solution. (1) Suppose $\alpha = \begin{pmatrix} m & \bar{a} \\ 0 & \bar{n} \end{pmatrix}$ is not a left 0-divisor. Then m must be odd, for otherwise $\begin{pmatrix} m & \bar{a} \\ 0 & \bar{n} \end{pmatrix}$ is right annihilated by $\begin{pmatrix} 0 & \bar{1} \\ 0 & 0 \end{pmatrix}$. In addition, we must have $\bar{n} = \bar{1}$, for otherwise $\bar{n} = 0$, and α would be right annihilated by $\begin{pmatrix} 0 & \bar{a} \\ 0 & \bar{1} \end{pmatrix}$ since $m\bar{a} + \bar{a} \in 2\mathbb{Z} \cdot \bar{a} = 0$. But then α is also not a right 0-divisor, for

$$0 = \begin{pmatrix} x & \bar{z} \\ 0 & \bar{y} \end{pmatrix} \begin{pmatrix} m & \bar{a} \\ 0 & \bar{1} \end{pmatrix} = \begin{pmatrix} xm & x\bar{a} + \bar{z} \\ 0 & \bar{y} \end{pmatrix} \Longrightarrow x = 0, \ \bar{y} = \bar{z} = 0.$$

(2) If $R \cong R^{\mathrm{op}}$, (1) would imply that every left 0-divisor in R is a right 0-divisor. Now $\beta = \begin{pmatrix} 2 & 0 \\ 0 & \bar{1} \end{pmatrix}$ is a left 0-divisor since $\begin{pmatrix} 2 & 0 \\ 0 & \bar{1} \end{pmatrix} \begin{pmatrix} 0 & \bar{1} \\ 0 & 0 \end{pmatrix} = 0$, but

$$0 = \begin{pmatrix} x & \bar{z} \\ 0 & \bar{y} \end{pmatrix} \begin{pmatrix} 2 & 0 \\ 0 & \bar{1} \end{pmatrix} = \begin{pmatrix} 2x & \bar{z} \\ 0 & \bar{y} \end{pmatrix} \Longrightarrow x = 0, \ \bar{y} = \bar{z} = 0$$

shows that β is *not* a right 0-divisor—a contradiction.

Ex. 1.22B. Let R be the upper triangular ring $\begin{pmatrix} \mathbb{Z}_{2^k} & \mathbb{Z}_2 \\ 0 & \mathbb{Z}_2 \end{pmatrix}$, where $k \geq 2$.

(1) Compute the set N of nilpotent elements of R.

(2) By comparing the left and right annihilators of N, show that $R \not\cong R^{\mathrm{op}}$.

Solution. (1) Let $A = 2\mathbb{Z}_{2^k} \subseteq \mathbb{Z}_{2^k}$. Then $\begin{pmatrix} A & \mathbb{Z}_2 \\ 0 & 0 \end{pmatrix} \subseteq N$, since any element in the former has kth power equal to zero. Conversely, if $\begin{pmatrix} \overline{m} & \overline{a} \\ 0 & \overline{n} \end{pmatrix} \in N$, then $\overline{m} \in \mathbb{Z}_{2^k}$ and $\overline{n} \in \mathbb{Z}_2$ must both be nilpotent, so $\overline{m} \in A$ and $\overline{n} = 0$. This shows that $N = \begin{pmatrix} A & \mathbb{Z}_2 \\ 0 & 0 \end{pmatrix}$.

(2) Let $B = 2^{k-1}\mathbb{Z}_{2^k} \subseteq \mathbb{Z}_{2^k}$. From

$$\begin{pmatrix} \overline{2b} & \overline{c} \\ 0 & 0 \end{pmatrix} \begin{pmatrix} \overline{x} & \overline{z} \\ 0 & \overline{y} \end{pmatrix} = \begin{pmatrix} \overline{2bx} & \overline{cy} \\ 0 & 0 \end{pmatrix},$$

it follows that $\mathrm{ann}_r(N) = \begin{pmatrix} B & \mathbb{Z}_2 \\ 0 & 0 \end{pmatrix}$. On the other hand, from

$$\begin{pmatrix} \overline{x} & \overline{z} \\ 0 & \overline{y} \end{pmatrix} \begin{pmatrix} \overline{2b} & \overline{c} \\ 0 & 0 \end{pmatrix} = \begin{pmatrix} \overline{2xb} & \overline{xc} \\ 0 & 0 \end{pmatrix},$$

and the fact that $k \geq 2$, we see easily that $\mathrm{ann}_\ell(N) = \begin{pmatrix} B & \mathbb{Z}_2 \\ 0 & \mathbb{Z}_2 \end{pmatrix}$. Thus we have $|\mathrm{ann}_r(N)| = 4$, and $|\mathrm{ann}_\ell(N)| = 8$. Since the nilpotent elements of R correspond to those of R^{op} (under the one-one correspondence $a \leftrightarrow a^{\mathrm{op}}$), this surely implies that $R \not\cong R^{\mathrm{op}}$.

Comment. I thank H.W. Lenstra, Jr. for suggesting this finite modification of Ex. 1.22A. In later parlance (from Ch. 3), N here is exactly the Jacobson radical of R, and $\mathrm{ann}_\ell(N)$ and $\mathrm{ann}_r(N)$ are respectively the right and left socles of the ring R: see Ex. 4.20. We note, however, that the assumption $k \geq 2$ is essential since, for $k = 1$, the ring R admits an involution $\begin{pmatrix} \overline{x} & \overline{z} \\ 0 & \overline{y} \end{pmatrix} \mapsto \begin{pmatrix} \overline{y} & \overline{z} \\ 0 & \overline{x} \end{pmatrix}$ by (the solution to) Ex. 1.22, so in particular, $R \cong R^{\mathrm{op}}$, and the example no longer works.

It seems likely that, by taking $k = 2$ above, the ring of 16 elements obtained in this exercise is the *smallest* ring R such that $R \not\cong R^{\mathrm{op}}$. We leave this matter to the curious reader.

Ex. 1.23. For a fixed $n \geq 1$, let $R = \begin{pmatrix} \mathbb{Z} & n\mathbb{Z} \\ \mathbb{Z} & \mathbb{Z} \end{pmatrix}$ and $S = \begin{pmatrix} \mathbb{Z} & \mathbb{Z} \\ n\mathbb{Z} & \mathbb{Z} \end{pmatrix}$. Show that $R \cong S$, and that these are rings with involutions.

Solution. As in the solution to Exercise 22, the inner automorphism of $M_2(\mathbb{Z})$ defined by $\begin{pmatrix} 0 & 1 \\ 1 & 0 \end{pmatrix}$ sends $\begin{pmatrix} a & b \\ c & d \end{pmatrix}$ to $\begin{pmatrix} d & c \\ b & a \end{pmatrix}$, so it restricts to a ring isomorphism α from R to S. Next, applying Exercise 22(2) with $\varepsilon = \mathrm{Id}_{\mathbb{Z}}$, we get an involution δ on $M_2(\mathbb{Z})$ sending $\begin{pmatrix} a & b \\ c & d \end{pmatrix}$ to $\begin{pmatrix} d & b \\ c & a \end{pmatrix}$. By restriction, δ defines involutions on the subrings R and S of $M_2(\mathbb{Z})$.

Alternative Solution. Define $\beta : R \to R$ by

$$\beta \begin{pmatrix} a & b \\ c & d \end{pmatrix} = \begin{pmatrix} a & nc \\ b/n & d \end{pmatrix}.$$

A direct calculation shows that β is an involution on R. An involution γ on S can be defined similarly. Now consider the transpose map $t : R \to S$, which is an anti-isomorphism. By composing $R \overset{\beta}{\to} R \overset{t}{\to} S$, we obtain an isomorphism $\alpha' : R \to S$ given by

$$\alpha' \begin{pmatrix} a & b \\ c & d \end{pmatrix} = \begin{pmatrix} a & b/n \\ nc & d \end{pmatrix}.$$

Comment. Here, $(R, \delta), (S, \delta)$ are isomorphic as rings with involutions (with isomorphism defined by α), and $(R, \beta), (S, \gamma)$ are isomorphic as rings with involutions (with isomorphism defined by α'). How about (R, δ) and (R, β)?

Ex. 1.24. Let R be the ring defined in Exercise 23, where $n \geq 1$ is fixed.

(1) Show that $m \in \mathbb{Z}$ is a square in R iff m is a square in $\mathbb{Z}/n\mathbb{Z}$.

(2) Let $R = \begin{pmatrix} \mathbb{Z} & 2p\mathbb{Z} \\ \mathbb{Z} & \mathbb{Z} \end{pmatrix}$ where p is an odd prime. Show that $2p \in R^2$, $p \in R^2$, but $2 \in R^2$ iff 2 is a square in $\mathbb{Z}/p\mathbb{Z}$.

Solution. (1) First suppose m is a square in R. Then

(*)
$$mI = \begin{pmatrix} a & b \\ c & d \end{pmatrix}\begin{pmatrix} a & b \\ c & d \end{pmatrix} = \begin{pmatrix} a^2 + bc & b(a+d) \\ c(a+d) & d^2 + bc \end{pmatrix}$$

for some a, b, c, $d \in \mathbb{Z}$ with $n \mid b$. Therefore,

$$m = a^2 + bc \equiv a^2 \pmod{n}.$$

Conversely, if $m \equiv a^2 \pmod{n}$ for some $a \in \mathbb{Z}$, then $m = a^2 + nc$ for some c, and (*) holds with $b = n$ and $d = -a$.

(2) Applying (1) with $n = m = 2p$, we see that $2p \in R^2$. Next, we apply (1) with $n = 2p$ and $m = p$. Since

$$p^2 - p = p(p-1) \in 2p\mathbb{Z},$$

p is a square in $\mathbb{Z}/2p\mathbb{Z}$, so $p \in R^2$. Finally, we apply (1) with $n = 2p$ and $m = 2$. By (1), $2 \in R^2$ iff 2 is a square in $\mathbb{Z}/2p\mathbb{Z}$. Since

$$\mathbb{Z}/2p\mathbb{Z} \cong \mathbb{Z}/2\mathbb{Z} \times \mathbb{Z}/p\mathbb{Z},$$

this holds iff 2 is a square in $\mathbb{Z}/p\mathbb{Z}$.

Ex. 1.25. (Vaserstein) Let a, b, c be such that $ab + c = 1$ in a ring R. If there exists $x \in R$ such that $a + cx \in U(R)$, show that there exists $y \in R$ such that $b + yc \in U(R)$.

Solution. Write $u = a + cx \in U(R)$. We claim that the element $y := (1 - bx)u^{-1}$ works, i.e.

$$v := b + (1 - bx)u^{-1}c \in U(R).$$

To see this, note that

$$vx = bx + (1 - bx)u^{-1}(u - a) = 1 - (1 - bx)u^{-1}a,$$
$$vx(1 - ba) = 1 - ba - (1 - bx)u^{-1}a(1 - ba)$$
$$= 1 - ba - (1 - bx)u^{-1}(1 - ab)a$$
$$= 1 - [b + (1 - bx)u^{-1}c]a = 1 - va.$$

Therefore, for
$$w := a + x(1 - ba),$$

we have $vw = 1$. We finish by showing that $wv = 1$ (for then $v \in U(R)$). Note that

$$wb = ab + xb(1 - ab) = ab + xbc,$$
$$w(1 - bx) = a + x(1 - ba) - abx - xbcx$$
$$= a + (1 - ab)x - xb(a + cx)$$
$$= a + cx - xbu,$$
$$w(1 - bx)u^{-1}c = c - xbc.$$

Adding the first and the last equation yields $wv = ab + c = 1$, as desired.

Comment. This interesting exercise is a special case of a result of L.N. Vaserstein in algebraic K-theory (see his paper "Stable rank of rings and dimensionality of topological spaces", Funct. Anal. Appl. *5* (1971), 102–110). The above solution is an adaption of Vaserstein's proof, taken from K.R. Goodearl's paper "Cancellation of low-rank vector bundles," Pac. J. Math. *113* (1984), 289–302. To put this exercise in perspective, we need the following definition from §20: A ring R is said to have "right stable range 1" if, whenever $aR + cR = R$, the coset $a + cR$ contains a unit. "Left stable range 1" is defined similarly, using principal left ideals. The exercise above implies that these two conditions are in fact equivalent!

Ex. 1.26. For any right ideal A in a ring R, the *idealizer* of A is defined to be

$$\mathbb{I}_R(A) = \{r \in R : rA \subseteq A\}.$$

(1) Show that $\mathbb{I}_R(A)$ is the largest subring of R that contains A as an ideal.
(2) The ring $\mathbb{E}_R(A) := \mathbb{I}_R(A)/A$ is known as the *eigenring* of the right ideal A. Show that $\mathbb{E}_R(A) \cong \operatorname{End}_R(R/A)$ as rings. (Note that, in a way, this "computes" the endomorphism ring of an arbitrary cyclic module over any ring.)

Solution. (1) It is straightforward to check that $\mathbb{I}_R(A)$ is a subring of R. Since $A \cdot A \subseteq A$, $A \subseteq \mathbb{I}_R(A)$. Clearly A is an ideal in $\mathbb{I}_R(A)$. Conversely, if A is an ideal in some subring $S \subseteq R$, then $r \in S$ implies $rA \subseteq A$, so $r \in \mathbb{I}_R(A)$. This shows that $S \subseteq \mathbb{I}_R(A)$.

(2) Define $\lambda : \mathbb{I}_R(A) \to \operatorname{End}_R(R/A)$ by taking $\lambda(r)$ $(r \in \mathbb{I}_R(A))$ to be left multiplication by r. (Since $rA \subseteq A$, $\lambda(r)$ is a well-defined endomorphism of the module $(R/A)_R$.) Now $\lambda(r)$ is the zero endomorphism iff $rR \subseteq A$, that is, $r \in A$. Since λ is a ring homomorphism, it induces a ring embedding $\mathbb{E}_R(A) \to \operatorname{End}_R(R/A)$. We finish by showing that this map is *onto*. Given $\varphi \in \operatorname{End}_R(R/A)$, write $\varphi(\bar{1}) = \bar{r}$, where $r \in R$. Then

$$\varphi(\bar{x}) = \varphi(\bar{1} \cdot x) = \bar{r} \cdot x = \overline{rx}$$

for any $x \in R$. In particular, for $x \in A$, we see that $rx \in A$, so $r \in \mathbb{I}_R(A)$, and we have $\varphi = \lambda(r)$.

(It is of interest to note that, in case A is a *maximal* right ideal of R, then (2) coupled with Schur's Lemma *FC*-(3.6) implies that the eigenring $\mathbb{E}_R(A)$ is a division ring.)

Ex. 1.27. Let $R = \mathbb{M}_n(k)$ where k is a ring, and let A be the right ideal of R consisting of matrices whose first r rows are zero. Compute the idealizer $\mathbb{I}_R(A)$ and the eigenring $\mathbb{E}_R(A)$.

Solution. Write a matrix $\beta \in R$ in the block form $\begin{pmatrix} x & y \\ z & w \end{pmatrix}$, where $x \in \mathbb{M}_r(k)$, and similarly, write $\alpha \in A$ in the block form $\alpha = \begin{pmatrix} 0 & 0 \\ u & v \end{pmatrix}$. Since

$$\beta\alpha = \begin{pmatrix} x & y \\ z & w \end{pmatrix}\begin{pmatrix} 0 & 0 \\ u & v \end{pmatrix} = \begin{pmatrix} yu & yv \\ wu & wv \end{pmatrix},$$

the condition for $\beta A \subseteq A$ amounts to $y = 0$. Therefore, $\mathbb{I}_R(A)$ is given by the ring of "block lower-triangular" matrices $\left\{ \begin{pmatrix} x & 0 \\ z & w \end{pmatrix} \right\}$. Quotienting out the ideal $\left\{ \begin{pmatrix} 0 & 0 \\ z & w \end{pmatrix} \right\}$, we get the eigenring $\mathbb{E}_R(A) \cong \mathbb{M}_r(k)$.

Ex. 1.28. Do the same for the right ideal $A = xR$ in the free k-ring $R = k\langle x, y \rangle$, and for the right ideal $A = xR$ where $x = i + j + k$ in the ring

$$R = \mathbb{Z} \oplus \mathbb{Z}i \oplus \mathbb{Z}j \oplus \mathbb{Z}k$$

of quaternions with integer coefficients.

Solution. (1) Let $R = k\langle x, y \rangle$ and $A = xR$. For $r \in R$, we have $r \in \mathbb{I}_R(A)$ iff $r \cdot x \in xR$. Writing $r = r_0 + r_1$ where r_0 is the constant term of r, we see that

$$r \cdot x = r_0 x + r_1 x \in xR \quad \text{iff} \quad r_1 \in xR.$$

This shows that $\mathbb{I}_R(A) = k + xR$, from which we get

$$\mathbb{E}_R(A) = (k + xR)/xR \cong k.$$

(2) Let $R = \mathbb{Z} \oplus \mathbb{Z}i \oplus \mathbb{Z}j \oplus \mathbb{Z}k$ and $A = xR$, where $x = i + j + k$. Since $x^2 = -3$, we have $3R \subseteq xR$. Writing "bar" for the projection map $R \to \overline{R} = R/3R$, we check easily that the right annihilator of $\overline{x}$ in $\overline{R}$ has 9 elements. Since $|\overline{R}| = 3^4$, it follows that

$$(*) \qquad\qquad [R : xR] = \left[\overline{R} : \overline{x}\overline{R} \right] = 81/9 = 9.$$

Now xR is not an ideal in R, so we have

$$R \supsetneq \mathbb{I}_R(xR) \supseteq xR + \mathbb{Z} \supsetneq xR.$$

From (*), we see that $\mathbb{I}_R(xR) = xR + \mathbb{Z}$, and

$$\mathbb{E}_R(xR) = \mathbb{I}_R(xR)/xR \cong \mathbb{Z}/3\mathbb{Z}.$$

The next two exercises on idealizers are based on observations of H. Lenstra and the author.

Ex. 1.29A. For a right ideal A in a ring R, let $S = \mathbb{I}_R(A)$ be the idealizer of A. Among the following four conditions, show that (1), (2) are equivalent, and (3), (4) are equivalent:

(1) $\text{End}(_S(R/A))$ is a commutative ring.
(2) A is an ideal of R, and R/A is a commutative ring.
(3) A is a maximal right ideal and R/A is a cyclic left S-module.
(4) A is an ideal of R, and R/A is a division ring.

Solution. $(2) \Longrightarrow (1)$. Here, $S = R$ since A is an ideal. Thus,

$$\operatorname{End}(_S(R/A)) = \operatorname{End}(_R(R/A)) = \operatorname{End}(_{R/A}(R/A)) \cong R/A,$$

which is (by assumption) a commutative ring.

$(1) \Longrightarrow (2)$. For $r \in R$, let ρ_r denote right multiplication by r on R/A. This is meaningful since R/A is a right R-module, and we have $\rho_r \in \operatorname{End}(_S(R/A))$. Thus, for any $r, r' \in R$, (1) yields an equation $\rho_r \rho_{r'} = \rho_{r'} \rho_r$. Applying the two sides of this equation to the coset $1 + A \in R/A$, we get $rr' + A = r'r + A$, which clearly implies (2).

$(4) \Longrightarrow (3)$ is clear, since $S = R$ under the assumption (4).

$(3) \Longrightarrow (4)$. Under (3), $(R/A)_R$ is a simple module, so Schur's Lemma (FC-(3.6)) and Ex. 1.26 imply that the eigenring $E := S/A$ is a division ring. If $_S(R/A)$ is cyclic, R/A is a 1-dimensional left E-vector space. Since S/A is a nonzero E-subspace in R/A, we must have $S = R$, from which (4) follows immediately.

Ex. 1.29B. Let R be a ring with center C. Show that a right ideal A of R is an ideal if any of the following holds:

(5) $\operatorname{End}(_C(R/A))$ is a commutative ring;
(6) R/A is a cyclic left C-module;
(7) The factor group R/A is cyclic, or isomorphic to a subgroup of $\mathbb{Q}$.

Solution. Keep the notations in Ex. 1.29A. Since $S \supseteq C$, $\operatorname{End}(_S(R/A))$ is a subring of $\operatorname{End}(_C(R/A))$. Therefore, we have $(5) \Longrightarrow (1) \Longrightarrow (2)$.

Next, suppose (6) holds. Then $_C(R/A)$ can be identified with C/I for some ideal I of C. Then

$$\operatorname{End}(_C(R/A)) \cong \operatorname{End}(_C(C/I)) \cong \operatorname{End}(_{C/I}(C/I)) \cong C/I$$

is a commutative ring, so we have $(6) \Longrightarrow (5) \Longrightarrow (2)$.

Finally, under (7), any $\mathbb{Z}$-endomorphism of R/A is induced by multiplication by an integer or a rational number. Since $\operatorname{End}(_S(R/A))$ is a subring of $\operatorname{End}(_\mathbb{Z}(R/A))$, we have $(7) \Longrightarrow (1) \Longrightarrow (2)$.

Comment. Note that the condition (7) can be further expanded, as there are many other abelian groups whose endomorphism rings are commutative. These include, for instance, the Prüfer p-group (the p-primary component of $\mathbb{Q}/\mathbb{Z}$), the group of p-adic integers, etc. Therefore, *if R/A is any one of these groups, A must be an ideal of R.* On the other hand, if $R/A \cong (\mathbb{Z}_m)^n$ or $\mathbb{Z}^n$ (as groups) where $n \geq 2$, A need not be an ideal any more. Easy examples are provided by Ex. 1.27 and Ex. 1.28.

Ex. 1.30. If $\mathfrak{m}$ and $\mathfrak{m}'$ are maximal right ideals in a ring R, show that the (simple) right R-modules $R/\mathfrak{m}$ and $R/\mathfrak{m}'$ are isomorphic iff there exists an element $r \in R \backslash \mathfrak{m}$ such that $r\mathfrak{m}' \subseteq \mathfrak{m}$. In this case, show that $\mathfrak{m}$ and $\mathfrak{m}'$ contain exactly the same ideals of R.

Solution. We first note the following generalization of the idea of an *idealizer* in Ex. 1.26. If A, A' are right ideals of R, we can form the additive group

$$I = I(A', A) := \{r \in R : rA' \subseteq A\},$$

which contains A as a subgroup. As before, we can check easily that sending $r \in I$ to the left multiplication by r defines a group isomorphism

$$\lambda : I/A \to \operatorname{Hom}_R(R/A', R/A).$$

It follows, therefore, that $R/A \cong R/A'$ as R-modules iff there exists an element $r \in I$ such that

$$rR + A = R, \quad \text{and} \quad rx \in A \Longrightarrow x \in A'.$$

The situation becomes a lot simpler if A, A' are replaced by the maximal right ideals $\mathfrak{m}$ and $\mathfrak{m}'$. Since $R/\mathfrak{m}$ and $R/\mathfrak{m}'$ are simple R-modules, they are isomorphic iff there is a *nonzero* homomorphism from $R/\mathfrak{m}'$ to $R/\mathfrak{m}$. Thus, the isomorphism criterion boils down to $I(\mathfrak{m}', \mathfrak{m}) \neq \mathfrak{m}$; that is, there exists an element $r \in R \backslash \mathfrak{m}$ such that $r\mathfrak{m}' \subseteq \mathfrak{m}$.

In this case, we have $rR + \mathfrak{m} = R$ (by the maximality of $\mathfrak{m}$). If J is any ideal contained in $\mathfrak{m}'$, then

$$J = (rR + \mathfrak{m})J = rJ + \mathfrak{m}J \subseteq r\mathfrak{m}' + \mathfrak{m} = \mathfrak{m}.$$

By symmetry, it follows that $\mathfrak{m}$ and $\mathfrak{m}'$ contain exactly the same ideals of R. Of course, this conclusion could also have been obtained from the observation that all ideals in $\mathfrak{m}$ are contained in a largest one, namely, the annihilator of the right module $R/\mathfrak{m}$. The extra argument given above, however, serves to illustrate the utility of the isomorphism criterion in this exercise.

Comment. A good example to keep in mind is $R = \mathbb{M}_n(\mathbb{Q})$, with $\mathfrak{m}$ (resp. $\mathfrak{m}'$) the maximal right ideal of matrices with first (resp. second) row zero. Here, $R/\mathfrak{m}$ and $R/\mathfrak{m}'$ are both isomorphic to the (simple) right R-module $\mathbb{Q}^n$. For the element r in the solution above, we can take *any* matrix with a first row $(0, 1, 0, \ldots, 0)$.

In general, if two maximal right ideals $\mathfrak{m}$, $\mathfrak{m}'$ in a ring R contain exactly the same ideals of R, it does not follow that $R/\mathfrak{m} \cong R/\mathfrak{m}'$. For instance, we can take R to be any simple ring with two nonisomorphic simple right R-modules $R/\mathfrak{m}$ and $R/\mathfrak{m}'$.

Ex. 1.31. Let I be a right ideal in a polynomial ring $A = R[x_1, \ldots, x_n]$, where R is any ring.

(1) If $I \cdot g = 0$ for some nonzero $g \in A$, show that $I \cdot r = 0$ for some nonzero $r \in R$.

(2) Assume R is a commutative ring or a reduced ring, and let $f \in A$. If $f \cdot g = 0$ for some nonzero $g \in A$, show that $f \cdot r = 0$ for some nonzero $r \in R$.

(3) Show that the conclusion of (2) need not hold if R is an arbitrary ring.

Solution. (1) We induct on n. First assume $n = 1$ and write $A = R[x]$. Let $g \in A\backslash\{0\}$ be of minimal degree such that $I \cdot g = 0$, say $g = bx^d + \cdots$, $b \neq 0$. If $d > 0$, then $f \cdot b \neq 0$ for some $f = \sum_i a_i x^i \in I$. We must have $a_i g \neq 0$ for some i (for otherwise $a_i b = 0$ for all i and hence $f \cdot b = 0$). Now pick j to be the largest integer such that $a_j g \neq 0$. From

$$0 = f \cdot g = (a_j x^j + \cdots + a_0) \cdot g,$$

we have $a_j b = 0$, so $\deg(a_j g) < d$. But $I \cdot (a_j g) \subseteq I \cdot g = 0$, which contradicts the choice of g. Thus, we must have $d = 0$ and $g \in R$.

For the inductive step, write $A = B[x_n]$, where $B = R[x_1, \ldots, x_{n-1}]$. Each polynomial $f \in A$ can be written in the form $\sum_i h_i x_n^i$ ($h_i \in B$). For convenience, let us call the h_i's the "coefficients" of f. Assume $I \cdot g = 0$, where $g \neq 0$ in A. By the paragraph above, we may assume that $g \in B$. Let I' be the right ideal in B generated by all "coefficients" of the polynomials in I. Since g does not involve x_n (and I is a right ideal), we see easily that $I \cdot g = 0 \Longrightarrow I' \cdot g = 0$. By the inductive hypothesis, there exists $r \in R\backslash\{0\}$ such that $I' \cdot r = 0$. From this, we have, of course, $I \cdot r = 0$, as desired.

(2) If R is commutative, we have $(fA)g = fgA = 0$, so (1) applies. If R is reduced, a leading term argument (plus induction) shows that A is also reduced. Thus $(gf)^2 = g(fg)f = 0$ implies that $gf = 0$, and hence $(fAg)^2 = 0$. It follows that $(fA)g = 0$, so (1) applies again.

(3) We look for a counterexample R, which, by (2), must be neither commutative nor reduced. An obvious choice is a matrix ring $R = \mathbb{M}_2(S)$, where S is a nonzero ring. Fortuitously, this does work! To see this, let $f = a + bx$ and $g = c + dx$, where

$$a = \begin{pmatrix} -1 & 0 \\ 0 & 0 \end{pmatrix}, \quad b = \begin{pmatrix} 0 & 1 \\ 0 & 0 \end{pmatrix}, \quad c = \begin{pmatrix} 0 & 0 \\ 1 & 1 \end{pmatrix}, \quad \text{and} \quad d = \begin{pmatrix} 1 & 1 \\ 0 & 0 \end{pmatrix}$$

in R. We have $ac = bd = 0 \in R$ and also

$$ad + bc = \begin{pmatrix} -1 & -1 \\ 0 & 0 \end{pmatrix} + \begin{pmatrix} 1 & 1 \\ 0 & 0 \end{pmatrix} = 0 \in R,$$

so $fg = 0 \in A$. However, *for any* $r \in R$, $f \cdot r = 0$ *implies* $r = 0$. To see this, let $r = \begin{pmatrix} p & q \\ s & t \end{pmatrix}$. From $f \cdot r = 0$, we have

$$0 = ar = \begin{pmatrix} -1 & 0 \\ 0 & 0 \end{pmatrix} \begin{pmatrix} p & q \\ s & t \end{pmatrix}, \quad \text{and} \quad 0 = br = \begin{pmatrix} 0 & 1 \\ 0 & 0 \end{pmatrix} \begin{pmatrix} p & q \\ s & t \end{pmatrix},$$

which forces $p = q = 0 = s = t$, so $r = 0 \in R$.

Comment. (1) of this exercise is due to N. McCoy; see his paper in the
MAA Monthly *64* (1957), 28–29. In the commutative case, (2) was obtained
earlier by McCoy in the MAA Monthly *49* (1942), 286–295. Part (3) was a
problem proposed by L.G. Jones and A. Rosenberg in the MAA Monthly *57*
(1950), 692. The solution above, given by L. Weiner, appeared in the MAA
Monthly *59* (1952), 336–337. Weiner also observed that the conclusion of
(2) need not hold for nonassociative (e.g. Jordan) algebras. More references
to related work in the literature were given in McCoy's 1957 paper.

It is natural to study also the 0-divisor problem for a *power series ring*
$B = R[[x]]$. D.E. Fields proved that, if R is commutative *and noetherian*,
the analogue of McCoy's result holds: if $f \in B$ is killed by some nonzero
$g \in B$, then f is killed by some nonzero $r \in R$; see his paper in Proc. A.M.S.
27 (1971), 427–433. An example in this paper, due to R. Gilmer, showed
that the noetherian assumption in this result cannot be removed.

Ex. 1.32. Show that, if $A = R[x]$ where R is a commutative ring, then for
$f = \sum_i a_i x^i \in A$:

$$(*) \qquad f \in U(A) \iff a_0 \in U(R) \text{ and } a_i \text{ is nilpotent for } i \geq 1.$$

Does either implication hold for an arbitrary ring R?

Solution. The "if" part is clear since, in any commutative ring, the sum
of a unit and a nilpotent element is always a unit. For the "only if" part,
a rather slick proof using basic commutative algebra (mainly the fact that
the intersection of prime ideals in R is equal to $\text{Nil}(R)$) is given in *FC*-pp.
67–68. For the sake of completeness, let us record below a proof that uses
a "bare-hands" approach, accessible to every beginning student in algebra.

Assume that $f \in U(A)$. It is easy to see that $a_0 \in U(R)$. We are done if
we can show that a_n *is nilpotent in case* $n \geq 1$ (for then $f - a_n x^n$ is also
a unit, and we can repeat the argument). Let

$$I = \{b \in R : a_n^t b = 0 \text{ for some } t \geq 1\}.$$

Say $fg = 1$, where $g = b_0 + \cdots + b_m x^m \in A$. *We claim that* $b_i \in I$ *for every*
i. If this is the case, there exists $t \geq 1$ such that $a_n^t b_i = 0$ for all i, and so
$a_n^t = f a_n^t g = 0$, as desired. The claim is certainly true if $i = m$. Inductively,
if $b_{i+1}, \ldots, b_m \in I$, then, comparing the coefficients of x^{n+i} in the equation
$fg = 1$, we have

$$a_n b_i + a_{n-1} b_{i+1} + a_{n-2} b_{i+2} + \cdots = 0.$$

For s sufficiently large, we then have $a_n^{s+1} b_i = -a_{n-1} a_n^s b_{i+1} - \cdots = 0$, so
$b_i \in I$ as claimed.

Note that the proofs above showed that $(*)$ holds as long as the a_i's
commute with each other: we don't really need the commutativity of R.

If the a_i's *do not* all commute, neither implication in $(*)$ is true (in general). To see this, let $R = \mathbb{M}_2(S)$ as in the solution to Ex. 1.31(3), where S is a nonzero ring. For a counterexample to "$\Longrightarrow$", take

$$a = \begin{pmatrix} 1 & 1 \\ 0 & 1 \end{pmatrix} \in U(R), \quad \text{and} \quad c = \begin{pmatrix} 0 & 0 \\ 1 & 0 \end{pmatrix} \in R.$$

Then, $b := ca = \begin{pmatrix} 0 & 0 \\ 1 & 1 \end{pmatrix}$ is not nilpotent (it is a nonzero idempotent), but since c is nilpotent,

$$a + bx = (1 + cx)a \in U(A) \cdot U(R) = U(A).$$

For a counterexample to "$\Longleftarrow$", let $f = a + cx$, where a and c are as above. We claim that $f \notin U(A)$. To see this, assume there exists an equation

$$(a + cx)(d_0 + d_1 x + \cdots + d_n x^n) = 1, \quad \text{where} \ d_i \in R.$$

Comparing coefficients of the two sides, we have

$$(\dagger) \quad ad_0 = 1, \quad ad_1 + cd_0 = 0, \quad \ldots, \quad ad_n + cd_{n-1} = 0, \quad \text{and} \quad cd_n = 0.$$

Solving successively for $d_0, d_1, \ldots, d_n$, we get

$$d_0 = a^{-1}, \quad d_1 = -a^{-1}ca^{-1}, \quad d_2 = a^{-1}ca^{-1}ca^{-1}, \quad \ldots ;$$

that is, $d_i = (-1)^i(a^{-1}c)^i a^{-1}$ for $i \le n$. Thus, the last equation in $(\dagger)$ gives $c(a^{-1}c)^n = 0$. But $a^{-1}c = \begin{pmatrix} -1 & 0 \\ 1 & 0 \end{pmatrix}$, so

$$c(a^{-1}c)^n = \begin{pmatrix} 0 & 0 \\ 1 & 0 \end{pmatrix} (-1)^n \begin{pmatrix} 1 & 0 \\ -1 & 0 \end{pmatrix} = (-1)^n \begin{pmatrix} 0 & 0 \\ 1 & 0 \end{pmatrix},$$

a contradiction!

Comment. While "$\Longleftarrow$" is not true in general, there is a closely related question on the units of $A = R[x]$ that has remained a difficult open problem in ring theory for years: *if $I \subseteq R$ is an ideal each of whose elements is nilpotent, is $1 + I[x] \subseteq U(A)$?* An affirmative answer to this question would amount to a solution to the famous Köthe Conjecture in classical ring theory: for more details on this, see Ex. (10.25) (especially statement (4)).

Ex. 1.33. Let k be a ring. Show that (1) the ring A generated over k by x, y with the relations $x^3 = 0$ and $xy + yx^2 = 1$ is the zero ring, and (2) the ring B generated over k by x, y with the relations $x^2 = 0$ and $xy + yx = 1$ is isomorphic to the matrix ring $\mathbb{M}_2(k[t])$.

Solution. (1) Right multiplying $xy + yx^2 = 1$ by x in A, we get $xyx = x$. Left multiplying the same equation by x^2, we get $x^2yx^2 = x^2$. Therefore, $x^2 = x(xyx)x = x^3 = 0$, and so $1 = xy + yx^2 = xy$. Left multiplying this by x yields $x = x^2y = 0$ and hence $1 = xy + yx^2 = 0 \in A$, proving that $A = (0)$.

(2) Define a ring homomorphism $\varphi : B \to \mathbb{M}_2(k[t])$ by $\varphi \mid k = \mathrm{Id}_k$, and

$$\varphi(x) = \begin{pmatrix} 0 & 0 \\ 1 & 0 \end{pmatrix}, \quad \varphi(y) = \begin{pmatrix} 0 & 1 \\ t & 0 \end{pmatrix}.$$

(It is easy to check that φ respects the relations $x^2 = 0$ and $xy + yx = 1$ on B.) We have $\varphi(y^2) = t \cdot I_2$, so

(*) $\varphi(y^{2n}) = t^n \cdot I_2, \quad \text{and} \quad \varphi(y^{2n+1}) = \begin{pmatrix} 0 & t^n \\ t^{n+1} & 0 \end{pmatrix}.$

Expressing B in the form $k[y] + k[y]x$, we can write an arbitrary element $\gamma \in B$ in the form $\alpha + \beta x$, where $\alpha = \sum a_i y^i$, and $\beta = \sum b_i y^i$ (with a_i, $b_i \in k$). In view of (*),

$$\varphi(\alpha) = \begin{pmatrix} a_0 + a_2 t + a_4 t^2 + \cdots & a_1 + a_3 t + a_5 t^2 + \cdots \\ a_1 t + a_3 t^2 + a_5 t^3 + \cdots & a_0 + a_2 t + a_4 t^2 + \cdots \end{pmatrix},$$

$$\varphi(\beta x) = \begin{pmatrix} b_1 t + b_3 t^2 + b_5 t^3 + \cdots & 0 \\ b_0 + b_2 t + b_4 t^2 + \cdots & 0 \end{pmatrix}.$$

If $\varphi(\gamma) = 0$, we must have all $a_i = 0$, and therefore all $b_i = 0$. This shows φ is one-one. The form of the matrices above also shows that φ is onto, so $B \cong \mathbb{M}_2(k[t])$.

Ex. 1.34. Let a be an element in a ring such that $ma = 0 = a^{2^r}$, where $m \geq 1$ and $r \geq 0$ are given integers. Show that $(1 + a)^{m^r} = 1$.

Solution. The proof is by induction on r. The case $r = 0$ being clear, we assume $r > 0$. Since $ma = 0$, the binomial theorem gives $(1 + a)^m = 1 + a^2 b$ where b is a polynomial in a with integer coefficients. Since

$$m(a^2 b) = 0 \quad \text{and} \quad (a^2 b)^{2^{r-1}} = a^{2^r} b^{2^{r-1}} = 0,$$

the inductive hypothesis (applied to the element $a^2 b$) implies that

$$1 = (1 + a^2 b)^{m^{r-1}} = [(1 + a)^m]^{m^{r-1}} = (1 + a)^{m^r},$$

as desired.

Comment. For a nice application of this exercise, see Ex. 5.11.

Ex. 1.35. Let A, B be left ideals in a ring R. Show that for any idempotent $e \in R$, we have the following two distributive laws:

(1) $eR \cap (A + B) = (eR \cap A) + (eR \cap B)$, and
(2) $eR + (A \cap B) = (eR + A) \cap (eR + B)$.
(3) Show that (1) and (2) no longer hold if eR is replaced by Re.

Solution. (1) We need only prove the inclusion "⊆". Let $x = a + b \in eR$, where $a \in A$ and $b \in B$. Then

$$x = ex = ea + eb \in (eR \cap A) + (eR \cap B),$$

since $ea \in eR \cap A$ and $eb \in eR \cap B$.

(2) Here, we need only prove the inclusion "⊇". For any element $x \in (eR + A) \cap (eR + B)$, we can write

$$x = er + a = es + b, \text{ where } r, s \in R, a \in A, \text{ and } b \in B.$$

Then $ex = er + ea = x - a + ea$, so $x - ex \in A$. Similarly, $x - ex \in B$. Thus,

$$x = ex + (x - ex) \in eR + (A \cap B),$$

as desired.

(3) To find the desired counterexamples, take $e = \begin{pmatrix} 1 & 0 \\ 0 & 0 \end{pmatrix}$ in the matrix ring $R = \mathbb{M}_2(\mathbb{Q})$, and

$$A = R \cdot \begin{pmatrix} 1 & 1 \\ 0 & 0 \end{pmatrix} = \left\{ \begin{pmatrix} a & a \\ b & b \end{pmatrix} \right\}, \qquad B = R \cdot \begin{pmatrix} 0 & 0 \\ 0 & 1 \end{pmatrix} = \left\{ \begin{pmatrix} 0 & c \\ 0 & d \end{pmatrix} \right\}.$$

Since $Re = \left\{ \begin{pmatrix} a & 0 \\ b & 0 \end{pmatrix} \right\}$, we have $Re \cap A = Re \cap B = (0)$. On the other hand, $A + B = R$ so $Re \cap (A + B) = Re$. This shows that (1) does not hold if eR is replaced by Re. Also, note that $Re + A = R$ and $Re + B = R$, so their intersection is R. On the other hand,

$$Re + (A \cap B) = Re + (0) = Re,$$

so (2) also fails if eR is replaced by Re.

§2. Semisimplicity

A left module M over a ring R is said to be *semisimple* if it has one of the following equivalent properties:

(1) Every submodule $N \subseteq M$ is a direct summand of M,
(2) M is the sum of a family of simple submodules, or
(3) M is the direct sum of a family of simple submodules.

Here, a *simple module* means a nonzero R-module M such that the only submodules of M are (0) and M. Each simple (left) R-module is isomorphic to $R/\mathfrak{m}$ for some maximal left ideal $\mathfrak{m}$ in R.

A general module is "seldom" semisimple. For instance, over $R = \mathbb{Z}$, $M = \mathbb{Z}/4\mathbb{Z}$ is not semisimple, since $N = 2\mathbb{Z}/4\mathbb{Z}$ is not a direct summand. Over the polynomial ring $R = \mathbb{Q}[x]$, the module $M = \mathbb{Q}^2$ on which x acts by the nilpotent matrix $\begin{pmatrix} 0 & 1 \\ 0 & 0 \end{pmatrix}$ ($xe_1 = 0$, $xe_2 = e_1$) is not semisimple, since the R-submodule $N = \mathbb{Q}e_1 \subseteq M$ is not a direct summand.

One of the most wonderful properties for a ring R to have is that all its (say, left) modules are semisimple. In this case, R is said to be a (*left*) *semisimple ring*. For this, it suffices to check that the left regular module $_RR$ be semisimple (FC-(2.5)), and there are other characterizations in terms of projective and injective modules (FC-(2.8), (2.9)). As it turns out, the definition of a semisimple ring is left-right symmetric (FC-(3.7)), so the "left", "right" adjectives may be ignored.

The exercises in this section deal with examples (and nonexamples) of semisimple rings, and some of their ideal-theoretic properties. Of fundamental importance is Exercise 7 which says all "reasonable" finiteness conditions on a semisimple module are equivalent (finite generation, finite length, noetherian, and artinian).

Exercises for §2

Ex. 2.1. Is any subring of a left semisimple ring left semisimple? Can any ring be embedded as a subring of a left semisimple ring?

Solution. The answer to the first question is "no": for instance, $\mathbb{Z}$ is a subring of the semisimple ring $\mathbb{Q}$, but $\mathbb{Z}$ is not a semisimple ring. The answer to the second question is also "no": for instance, $A = R_1 \times R_2 \times \cdots$ (where R_i are nonzero rings) cannot be embedded as a subring of a (left) semisimple ring. Indeed, if A is a subring of a ring R, then R will have nonzero idempotents $e_1, e_2, \ldots$ with $e_i e_j = 0$ for $i \neq j$. But then

$$R \supseteq Re_1 \oplus Re_2 \oplus \cdots,$$

and this implies that R is not left noetherian (let alone left semisimple). Similarly, if k is any nonzero ring, $A = k[x_1, x_2, \ldots]$ with the relations

$$x_i x_j = 0 \quad \text{(for all } i, j\text{)}$$

cannot be embedded in a left semisimple ring. Indeed, if A is a subring of a ring R, then

$$Rx_1 \subsetneq Rx_1 + Rx_2 \subsetneq \cdots$$

(by an easy proof), and again R is not left noetherian (let alone left semisimple).

Ex. 2.2. Let $\{F_i : i \in I\}$ be a family of fields. Show that the direct product $R = \prod_i F_i$ is a semisimple ring iff the indexing set I is finite.

Solution. First suppose R is semisimple. The solution of the previous exercise shows that I must be finite. Conversely, assume I is finite, and consider any ideal $A \subseteq R$. By Exercise 1.8, $A = \bigoplus_{i \in J} F_i$ for a subset $J \subseteq I$. Clearly, A is a direct summand of $_R R$, so $_R R$ is a semisimple module, as desired.

Ex. 2.3A. Determine which of the following are semisimple $\mathbb{Z}$-modules (where $\mathbb{Z}_n$ denotes $\mathbb{Z}/n\mathbb{Z}$):

$$\mathbb{Z}, \quad \mathbb{Q}, \quad \mathbb{Q}/\mathbb{Z}, \quad \mathbb{Z}_4, \quad \mathbb{Z}_6, \quad \mathbb{Z}_{12}, \quad \mathbb{Z}_2 \times \mathbb{Z}_2 \times \mathbb{Z}_2 \times \cdots, \quad \text{and}$$

$$\mathbb{Z}_2 \oplus \mathbb{Z}_3 \oplus \mathbb{Z}_5 \oplus \cdots, \quad \mathbb{Z}_2 \times \mathbb{Z}_3 \times \mathbb{Z}_5 \times \cdots.$$

Solution. $\mathbb{Z}$, $\mathbb{Q}$, and $\mathbb{Z}_2 \times \mathbb{Z}_3 \times \mathbb{Z}_5 \times \cdots$ are not torsion abelian groups, so they are not semisimple $\mathbb{Z}$-modules. $\mathbb{Z}_4$ is also not a semisimple $\mathbb{Z}$-module, since $2\mathbb{Z}_4$ is not a direct summand. Since $\mathbb{Z}_{12}$ and $\mathbb{Q}/\mathbb{Z}$ both contain a copy of $\mathbb{Z}_4$, they are also not semisimple. On the other hand, $\mathbb{Z}_6 \cong \mathbb{Z}_2 \oplus \mathbb{Z}_3$ and $\mathbb{Z}_2 \oplus \mathbb{Z}_3 \oplus \mathbb{Z}_5 \oplus \cdots$ are obviously semisimple. Finally, $\mathbb{Z}_2 \times \mathbb{Z}_2 \times \mathbb{Z}_2 \times \cdots$ is a $\mathbb{Z}_2$-vector space; this makes it semisimple over $\mathbb{Z}_2$, and hence over $\mathbb{Z}$.

Ex. 2.3B. What are the semisimple $\mathbb{Z}$-modules? (Characterize them in terms of their structure as abelian groups.)

Solution. Note that $\mathbb{Z}$-modules are just abelian groups, so the simple $\mathbb{Z}$-modules are the abelian simple groups, namely, the groups $\mathbb{Z}_p$, where p ranges over all prime numbers. Therefore, semisimple $\mathbb{Z}$-modules are just direct sums of copies of such $\mathbb{Z}_p$'s, and are, in particular torsion abelian groups. The ideas used in working with the examples in the previous exercise lead easily to a proof of the following: *an abelian group A is semisimple as a $\mathbb{Z}$-module iff A is torsion and its p-primary part is killed by p for each prime number p.*

Comment. The above description of semisimple modules also holds over any commutative principal ideal domain. (Simply replace the p's above by a "representative set" of the irreducible elements of the PID.)

Ex. 2.4. Let R be the (commutative) ring of all real-valued continuous functions on $[0, 1]$. Is R a semisimple ring?

Solution. The answer is "no". Indeed, assume R is semisimple. Consider the ideal

$$I = \{f \in R : f(0) = 0\}.$$

Then I is a direct summand of $_R R$, and Exercise 1.7 implies that $I = Re$ for some idempotent $e \in I$. But

$$e(x)^2 = e(x) \Longrightarrow e(x) \in \{0,1\}$$

for any $x \in [0,1]$. Since $e(0) = 0$, continuity of e forces e to be the 0-function, so $I = 0$; a contradiction.

Alternative Solution. Consider the ideals

$$I_n = \{f \in R : f([1/n, 1]) = 0\},$$

where $n \geq 1$. It is easy to see that $I_1 \supsetneq I_2 \supsetneq \cdots$, so R is not an artinian ring. In particular, R is not semisimple.

Ex. 2.5. Let R be a (left) semisimple ring. Show that, for any right ideal I and any left ideal J in R, $IJ = I \cap J$. If I, J, K are ideals in R, prove the following two distributive laws:

$$I \cap (J + K) = (I \cap J) + (I \cap K),$$
$$I + (J \cap K) = (I + J) \cap (I + K).$$

Solution. First, it is clear that $IJ \subseteq I$ and $IJ \subseteq J$, so $IJ \subseteq I \cap J$. To prove the reverse inclusion, write $J = Rf$ where $f = f^2 \in J$ (cf. Exercise 1.7). For any $a \in I \cap J$, write $a = rf \in I$, where $r \in R$. Then $a = rf \cdot f \in IJ$, as desired. Assuming that I, J, K are ideals, we have

$$I \cap (J + K) = I(J + K) = IJ + IK = (I \cap J) + (I \cap K),$$
$$(I + J) \cap (I + K) = (I + J)(I + K) = I + JK = I + (J \cap K).$$

Comment. The property $IJ = I \cap J$ in the first part of the Exercise turns out to characterize von Neumann regular rings: see Exercise 4.14 below. The fact that semisimple rings are von Neumann regular is explicitly noted in FC-(4.24).

In this exercise, if we assume the (equivalent) fact that R is right semisimple, then any right ideal $I \subseteq R$ has the form eR for some idempotent $e \in R$. Thus, according to Ex. 1.35, the two distributive laws above already hold for any right ideal I and left ideals J, K in R.

Ex. 2.6. Let R be a right semisimple ring. For $x, y \in R$, show that $Rx = Ry$ iff $x = uy$ for some unit $u \in U(R)$.

Solution. If $x = uy$ where $u \in U(R)$, then $Rx = Ruy = Ry$. Conversely, assume $Rx = Ry$. By Exercise 1.17, there exists a right R-isomorphism $f : yR \to xR$ such that $f(y) = x$. Write

$$R_R = yR \oplus A = xR \oplus B,$$

where A, B are right ideals. By considering the composition factors of R_R, yR and xR, we see that $A \cong B$ as right R-modules. Therefore, f can be extended to an automorphism g of R_R. Letting $u = g(1) \in U(R)$, we have

$$x = f(y) = g(y) = g(1y) = uy.$$

Comment. The conclusion of the Exercise is false in general if R is not a (right) semisimple ring. For instance, if R is a non-Dedekind-finite ring with elements x, z such that $zx = 1 \neq xz$, then $Rx = R1$ since Rx contains $zx = 1$, but $x \neq u1$ for any $u \in U(R)$. However, the conclusion of the Exercise is true if R is a domain: if we write $x = uy$ and $y = vx$ for u, $v \in R$, then (assuming $x \neq 0$), $x = uvx$ implies $uv = 1$, and similarly $y = vuy$ implies $vu = 1$. Thus, $u \in U(R)$.

Ex. 2.7. Show that for a semisimple module M over any ring R, the following conditions are equivalent:

(1) M is finitely generated;
(2) M is noetherian;
(3) M is artinian;
(4) M is a finite direct sum of simple modules.

Solution. $(1) \Rightarrow (4)$. Let $M = \bigoplus_{i \in I} M_i$ where the M_i's are simple modules. If M is generated by $m_1, \ldots, m_n$, we have $\{m_1, \ldots, m_n\} \subseteq \bigoplus_{i \in J} M_i$ for a finite subset $J \subseteq I$. Therefore, $M = \bigoplus_{i \in J} M_i$ (which of course implies that $J = I$).

$(4) \Rightarrow (2) \Rightarrow (1)$ and $(4) \Rightarrow (3)$ are trivial, so we are done if we can show $(3) \Rightarrow (4)$. Let $M = \bigoplus_{i \in I} M_i$ as above. If I is infinite, this decomposition of M would lead to a strictly descending chain of submodules of M. Therefore, I must be finite, and we have (4).

Ex. 2.8. Let M be a semisimple (say, left) module over a ring. Let $\{V_i : i \in I\}$ be a complete set of nonisomorphic simple modules which occur as submodules of M. Let M_i be the sum of all submodules of M which are isomorphic to V_i. It is easy to show that $M = \bigoplus_i M_i$: the M_i's are called the *isotypic components* of M. In the following, we assume that each M_i is finitely generated. By Exercise 7, this means that each $M_i \cong m_i V_i$ for suitable integers m_i. Let N be any submodule of M. Show that $N \cong \bigoplus_i n_i V_i$ for suitable $n_i \leq m_i$, and that $M/N \cong \bigoplus_i (m_i - n_i) V_i$.

Solution. Write $N = \bigoplus_i N_i$, where the N_i's are the isotypic components of N. Then $N_i \subseteq M_i$, so we have $N_i \cong n_i V_i$ for some $n_i \leq m_i$. By the Jordan-Hölder Theorem, $M_i/N_i \cong (m_i - n_i) V_i$. Therefore, we have $N \cong \bigoplus_i n_i V_i$ and

$$M/N \cong \bigoplus_i (M_i/N_i) \cong \bigoplus_i (m_i - n_i) V_i.$$

Comment. The exercises above can be equivalently stated in the form of a cancellation theorem: *If $N \oplus K \cong N' \oplus K'$ is a semisimple module whose isotypic components are all finitely generated, then $N \cong N'$ implies that $K \cong K'$.* Note that the finiteness assumption is essential to such a cancellation. For instance, if

$$M = V_1 \oplus V_1 \oplus \cdots,$$

the isomorphism type of a submodule $N \subseteq M$ is certainly *not* sufficient to determine the isomorphism type of M/N. (For $N = M$, $M/N = 0$; but for

$$N' = 0 \oplus V_1 \oplus V_1 \oplus \cdots,$$

$M/N' \cong V_1$. And yet $N \cong N'$.) For another cancellation result not requiring any semisimplicity assumption, see Exercise 20.2 below.

Ex. 2.9. Let U, V be modules over a commutative ring R. If U or V is semisimple, show that $U \otimes_R V$ (viewed as an R-module in the usual way) is also semisimple. How about the converse?

Solution. For the first statement, it suffices to handle the case where V is semisimple. Since tensor product distributes over direct sums, we may further assume that V is simple, say $V \cong R/\mathfrak{m}$ where $\mathfrak{m}$ is a maximal ideal of R. But then $\mathfrak{m} \cdot (U \otimes_R V) = 0$, so $U \otimes_R V$ is a vector space over the field $R/\mathfrak{m}$. Thus, $U \otimes_R V$ is semisimple over $R/\mathfrak{m}$, and also over R.

The converse statement does not hold in general. For instance, $\mathbb{Z}_4$ and $\mathbb{Z}_9$ are not semisimple over $\mathbb{Z}$, but $\mathbb{Z}_4 \otimes \mathbb{Z}_9 = 0$ is.

Comment. Information on another kind of tensor products (over group rings) is available from Ex. 8.29 and Ex. 8.30.

Ex. 2.10. Let U, V be semisimple modules over a commutative ring. Is $\mathrm{Hom}_R(U, V)$ (viewed as an R-module in the usual way) also semisimple?

Solution. If one of U, V is finitely generated, the answer is "yes", by the same argument given in the solution to Ex. 2.9. In general, however, the answer is "no", as we can show by the following example over $R = \mathbb{Z}$. Take $U = V = \bigoplus_p \mathbb{Z}_p$, where p ranges over all primes. Let $\varepsilon \in E := \mathrm{Hom}_{\mathbb{Z}}(V, V)$ be the identity map from V to V. We claim that ε has *infinite* additive order in E (which certainly implies that E is not a semisimple $\mathbb{Z}$-module). For any natural number n, take a prime $p > n$. Then

$$(n \cdot \varepsilon)(0, \ldots, \overline{1}, 0, \ldots) = (0, \ldots, \overline{n}, 0, \ldots) \neq 0$$

if the $\overline{1}$ appears in the coordinate corresponding to $\mathbb{Z}_p$. Therefore, $n \cdot \varepsilon \neq 0 \in E$, as claimed.

Ex. 2.11. If $_R V$ is a simple module over a commutative ring R, show that any direct product of copies of V is semisimple. Does this hold over an arbitrary ring R?

Solution. Representing V in the form $R/\mathfrak{m}$ where $\mathfrak{m}$ is a maximal ideal of R, we can solve the first part as in the solution to Ex. 2.9. However, the same conclusion does not hold (in general) without the commutativity assumption. To produce a counterexample, let V be a right vector space over a division ring D with a basis $\{e_1, e_2, \dots\}$, and let $R = \mathrm{End}(V_D)$. Then V is a simple left R-module. The map

$$\varphi: {}_RR \longrightarrow P := V \times V \times \cdots \quad \text{defined by} \quad \varphi(r) = (re_1, re_2, \dots)$$

is easily checked to be an R-module isomorphism. In particular, ${}_RP$ is cyclic. On the other hand, ${}_RP$ is obviously not noetherian, so it is not semisimple (by Ex. 7).

As a matter of fact, we can also easily name a nonsplit submodule of P. The direct sum $V \oplus V \oplus \cdots$ is nonsplit in P since it is not finitely generated, whereas a direct summand of P must be cyclic.

Comment. If we think of R as the ring of column-finite $\mathbb{N} \times \mathbb{N}$ matrices over D, the isomorphism φ corresponds to the decomposition of this matrix ring into a direct product of its "column spaces". Such a decomposition is quite useful. For instance, it explains why ${}_RR$ is not Dedekind-finite $({}_RR \oplus V \cong {}_RR)$, and it is also applicable toward Ex. 3.14. For more precise information on direct products of simple (and semisimple) modules, see the two later exercises Ex. 20.13 and Ex. 20.14.

Ex. 2.12. Let V be a left R-module with elements $e_1, e_2, \dots$ such that, for any n, there exists $r \in R$ such $re_n, re_{n+1}, \dots$ are almost all 0, but not all 0. Show that $S := V \oplus V \oplus \cdots$ is *not* a direct summand of $P := V \times V \times \cdots$ (so in particular, P is not a semisimple R-module).

Solution. This exercise is directly inspired by the last part of the solution to the previous exercise. Assume that $P = S \oplus T$ for some R-submodule $T \subseteq P$. Write

$$(e_1, e_2, \dots) = (s_1, s_2, \dots) + (t_1, t_2, \dots)$$

where $(s_1, s_2, \dots) \in S$, and $(t_1, t_2, \dots) \in T$. Then the exists an index n such that $t_i = e_i$ for all $i \geq n$. Let $r \in R$ be such that $re_n, re_{n+1}, \dots$ are almost all 0 but not all 0. Then

$$r(t_1, \dots, t_n, t_{n+1}, \dots) = (rt_1, \dots, rt_{n-1}, re_n, re_{n+1}, \dots) \neq 0$$

lies in S as well as in T, a contradiction.

§3. Structure of Semisimple Rings

The exercises in this section are mostly based on the Wedderburn-Artin Theorem, which classifies the semisimple rings. According to this theorem, a ring R is semisimple iff

$$R \cong \mathbb{M}_{n_1}(D_1) \times \cdots \times \mathbb{M}_{n_r}(D_r),$$

where the D_i's are division rings. Here, r is the number of distinct isomorphism types of the simple left R-modules, say, $M_1, \ldots, M_r$; n_i is the number of times M_i occurs as a composition factor in $_RR$; and D_i is the endomorphism ring of M_i, defined as a ring of *right* operators on M_i.

Each semisimple ring is uniquely decomposed into a direct sum of its simple components. The simple components which can occur are precisely the artinian simple rings, or equivalently, simple rings which have minimal left ideals. (Recall that a ring S is simple if $S \neq 0$ and the only ideals in S are (0) and S.) In general, there are many nonartinian simple rings. Various examples are given in FC-§3, but a general theory for such rings is decidedly beyond our reach.

Many exercises in this section are folklore in the subject, e.g. the Double Annihilator Property (Exercise 12) and the Dedekind-finiteness (Exercise 10) of a semisimple ring. Especially recommended is Exercise 6A, which says that if $_RM$ is a semisimple module, then so is M_E where $E = \text{End}(_RM)$. This is easy when $_RM$ is finitely generated (for then the ring E is semisimple), but quite a bit more tricky in general. Exercises 14, 15 and 16 offer useful basic information on the ring of linear transformations on a (possibly infinite-dimensional) vector space over a division ring.

In two additional exercises ((21) and (22)), we sketch a computation for the smallest number of elements $\mu(M)$ needed to generate a finitely generated module M over a semisimple ring. The fact that the expression for $\mu(M)$ (in the artinian simple ring case) involves Knuth's "ceiling function" of the composition length is a bit of a surprise to beginners in module theory. It certainly serves warning that everything is not the same as for vector spaces over fields!

Exercises for §3

Ex. 3.0. Show that, if M is a simple module over a ring R, then as an abelian group, M is isomorphic to a direct sum of copies of $\mathbb{Q}$, or a direct sum of copies of $\mathbb{Z}_p$ for some prime p.

Solution. By Schur's Lemma (FC-(3.6)), the R-endomorphism ring of M is a division ring D. Let F be the prime field of D. We may view M as a D-vector space, so M is also an F-vector space. As such, M is isomorphic to a direct sum of copies of F. This gives the desired conclusion since we have either $F \cong \mathbb{Q}$, or $F \cong \mathbb{Z}_p$ for some prime p.

Comment. The solution above offers perhaps the best "conceptual view" of this exercise. However, a direct proof *not* using Schur's Lemma is possible too. One observes that, if M has p-torsion for any prime p, then $pM = 0$; otherwise, argue that $pM = M$ for any prime p so M is a torsionfree divisible group. These remarks quickly lead us to the desired conclusions.

Ex. 3.1. Show that if R is semisimple, so is $\mathbb{M}_n(R)$.

Solution. Let $R \cong \prod_{i=1}^{r} \mathbb{M}_{n_i}(D_i)$, where the D_i's are suitable division rings. Then

$$\mathbb{M}_n(R) \cong \prod_i \mathbb{M}_n(\mathbb{M}_{n_i}(D_i)) \cong \prod_i \mathbb{M}_{nn_i}(D_i),$$

which is a semisimple ring.

Ex. 3.2. Let R be a domain. Show that if $\mathbb{M}_n(R)$ is semisimple, then R is a division ring.

Solution. Consider any chain $I_1 \supseteq I_2 \supseteq \cdots$ of left ideals in R. Then

$$\mathbb{M}_n(I_1) \supseteq \mathbb{M}_n(I_2) \supseteq \cdots$$

is a chain of left ideals in $\mathbb{M}_n(R)$, so it must become stationary. This implies that $I_1 \supseteq I_2 \supseteq \cdots$ also becomes stationary, so R is left artinian. By Exercise 1.19, R must be a division ring.

Ex. 3.3. Let R be a semisimple ring.

(a) Show that any ideal of R is a sum of simple components of R.
(b) Using (a), show that any quotient ring of R is semisimple.
(c) Show that a simple artinian ring S is isomorphic to a simple component of R iff there is a surjective ring homomorphism from R onto S.

Solution. (a) follows directly from Exercise 1.8.

(b) Write $R = R_1 \times \cdots \times R_r$ where the R_i's are the simple components of R. Consider any quotient ring R/I of R. After a reindexing, we may assume that $I = R_1 \oplus \cdots \oplus R_s$ for some $s \leq r$. Therefore,

$$R/I \cong R_{s+1} \times \cdots \times R_r$$

is a semisimple ring.

(c) If $S \cong R_i$, we can find a surjective ring homomorphism from R to S by utilizing the ith projection of $R = R_1 \times \cdots \times R_r$. Conversely, suppose $\varphi : R \to S$ is a surjective ring homomorphism. After a reindexing, we may assume that
$$\ker(\varphi) = R_1 \times \cdots \times R_{r-1}.$$
Therefore, $S \cong R/\ker(\varphi) \cong R_r$.

Comment. The solution to (c) shows that it remains true if we replace the words "simple artinian" by "indecomposable". (A nonzero ring is said to be *indecomposable* if it is not the direct product of two nonzero rings.)

Ex. 3.4. Show that the center of a simple ring is a field, and the center of a semisimple ring is a finite direct product of fields.

Solution. Suppose R is a simple ring, and let $0 \neq a \in Z(R)$. Then Ra is an ideal, so $Ra = R$. This implies that $a \in U(R)$. But clearly $a^{-1} \in Z(R)$, so $Z(R)$ is a field. Next, assume R is a semisimple ring, and let

$$R = \prod_{i=1}^{r} \mathbb{M}_{n_i}(D_i)$$

where the D_i's are division rings. By Exercise 1.9,

$$Z(D) = \prod_i Z\left(\mathbb{M}_{n_i}(D_i)\right) \cong \prod_i Z(D_i),$$

where the $Z(D_i)$'s are fields.

Comment. It follows in particular from this exercise that the center of a (semi)simple ring is (semi)simple. For results on the centers of other classes of rings, see Exercises 5.0 (J-semisimple rings), 10.0 (prime rings), 10.1 (semiprime rings), 11.11 (primitive rings), 19.6 (local rings), 20.12 (semilocal rings), 21.7 (von Neumann regular rings), and 21.25 (artinian and noetherian rings).

Ex. 3.5. Let M be a finitely generated left R-module and $E = \text{End}(_R M)$. Show that if R is semisimple (resp. simple artinian), then so is E.

Solution. First assume R is semisimple, and let $S_1, \ldots, S_r$ be a complete set of simple left R-modules. Then

$$M = M_1 \oplus \cdots \oplus M_r,$$

where M_i is the sum of all submodules of M which are isomorphic to S_i. Since M is finitely generated, $M_i \cong n_i S_i$ for suitable integers n_i. Therefore,

$$\text{End}_R M \cong \prod_i \text{End}_R M_i \cong \prod_i \mathbb{M}_{n_i}(\text{End}_R S_i)$$

by Exercise 1.20. By Schur's Lemma, all $\text{End}_R S_i$ are division rings, so $\text{End}_R M$ is semisimple. If, in fact, R is simple artinian, we have $r = 1$ in the calculation above. Therefore, $\text{End}_R M \cong \mathbb{M}_{n_1}(\text{End}_R S_1)$ is also a simple artinian ring.

Comment. In the case where $_R M$ is not finitely generated, E is no longer semisimple. However, it turns out that M is always finitely generated over E: see Ex. 11.6*.

Ex. 3.6A. Let M be a left R-module and $E = \text{End}(_R M)$. If $_R M$ is a semisimple R-module, show that M_E is a semisimple E-module.

Solution. Every nonzero element m of M can be written as $m_1 + \cdots + m_n$ where each Rm_i is simple. *We claim that each m_iE is a simple E-module.* Once this is proved, then m is contained in the semisimple E-module $\sum m_iE$, and we are done. To show that m_iE is a simple E-module, it suffices to check that, for any $e \in E$ such that $m_ie \neq 0$, m_ieE contains m_i. Consider the R-epimorphism $\varphi : Rm_i \to Rm_ie$ given by right multiplication by e. Since Rm_i is simple, φ is an isomorphism. Let $\psi : Rm_ie \to Rm_i$ be the inverse of φ, and extend ψ to an $f \in E$. (We can take f to be zero, for instance, on an R-module complement of Rm_ie.) Now

$$m_ief = (m_ie)\psi = (m_ie)\varphi^{-1} = m_i,$$

as desired.

Comment. This exercise came from Chevalley's book "Fundamental Concepts of Algebra" (Exercise 27 on p. 133). The hint to that exercise suggested a reduction to the case when the semisimple R-module M is "isotypic" (i.e. a sum of copies of a simple module). In this case, Chevalley claimed that M_E is a simple module. This is, unfortunately, incorrect. For instance, if k is any division ring, $R = \mathbb{M}_n(k)$, and M is the unique simple left R-module, then $_RM$ is certainly isotypic, but $E \cong k$ (see *FC*-p. 34), and M_E is not simple if $n > 1$. Note that the solution we offered for the exercise did not depend on the reduction to the isotypic case.

Ex. 3.6B. In the above Exercise, if M_E is a semisimple E-module, is $_RM$ necessarily a semisimple R-module?

Solution. The answer is "no" in general. To construct an example, let R be the ring of 2×2 upper triangular matrices over a field k, and let M be the left R-module k^2 with the R-action given by matrix multiplication from the left. An easy computation shows that $E = \text{End}(_RM) = k$ (cf. *FC*-p.110); in particular, M_E is a semisimple E-module. We claim that $_RM$ is not a semisimple R-module. In fact, consider the R-submodule

$$N = \left\{ \begin{pmatrix} a \\ 0 \end{pmatrix} : a \in k \right\} \subseteq M.$$

If $\begin{pmatrix} b \\ c \end{pmatrix} \notin N$, then $c \neq 0$, so $R \cdot \begin{pmatrix} b \\ c \end{pmatrix}$ contains

$$\begin{pmatrix} 0 & ac^{-1} \\ 0 & 0 \end{pmatrix} \begin{pmatrix} b \\ c \end{pmatrix} = \begin{pmatrix} a \\ 0 \end{pmatrix}$$

for all $a \in k$. This shows that N is not an R-direct summand of M, so $_RM$ is not semisimple.

Comment. The module M over the triangular matrix ring R can be used as a counterexample on quite a number of different occasions. Aside from the above exercise, we also use M and R in Exercises 1.5, 11.21, 21.5 and 22.3B.

Ex. 3.7. Let R be a simple ring that is finite-dimensional over its center k. (k is a field by Exercise 4 above.) Let M be a finitely generated left R-module and let $E = \text{End}(_RM)$. Show that

$$(\dim_k M)^2 = (\dim_k R)(\dim_k E).$$

Solution. Say $R = \mathbb{M}_n(D)$, where D is a division ring, with $Z(D) = Z(R) = k$. Let V be the unique simple left R-module, so $M \cong mV$ for some m. For $d = \dim_k D$, we have $\dim_k M = m \cdot \dim_k V = mnd$. Since

$$E \cong \mathbb{M}_m(\text{End}_R V) \cong \mathbb{M}_m(D),$$

it follows that

$$(\dim_k R)(\dim_k E) = (n^2 d)(m^2 d) = (mnd)^2 = (\dim_k M)^2.$$

Ex. 3.8. For R as in Exercise 7, show that R is isomorphic to a matrix algebra over its center k iff R has a nonzero left ideal $\mathfrak{A}$ with $(\dim_k \mathfrak{A})^2 \leq \dim_k R$.

Solution. If $R \cong \mathbb{M}_n(k)$, we can take $\mathfrak{A}$ to be $R \cdot E_{11}$ for which

$$(\dim_k \mathfrak{A})^2 = n^2 = \dim_k R.$$

Conversely, suppose R has a nonzero left ideal $\mathfrak{A}$ with $(\dim_k \mathfrak{A})^2 \leq \dim_k R$. Let us use the notations introduced in the solution to Exercise 7. Then

$$(nd)^2 = (\dim_k V)^2 \leq (\dim_k \mathfrak{A})^2 \leq \dim_k R = n^2 d.$$

This implies that $d = 1$, so $D = k$ and $R \cong \mathbb{M}_n(k)$.

Ex. 3.9. (a) Let R, S be rings such that $\mathbb{M}_m(R) \cong \mathbb{M}_n(S)$. Does this imply that $m = n$ and $R \cong S$? (b) Let us call a ring A a matrix ring if $A \cong \mathbb{M}_m(R)$ for some integer $m \geq 2$ and some ring R. True or False: "A homomorphic image of a matrix ring is also a matrix ring"?

Solution. (a) The answer is "no". For instance, for $R = \mathbb{M}_2(\mathbb{Q})$ and $S = \mathbb{M}_3(\mathbb{Q})$, we have $R \not\cong S$ (by dimension considerations), but $\mathbb{M}_3(R) \cong \mathbb{M}_2(S)$ since both rings are isomorphic to $\mathbb{M}_6(\mathbb{Q})$.

(b) The statement is indeed true. A homomorphic image S of $A = \mathbb{M}_m(R)$ has the form A/I where I is an ideal of A. But by FC-(3.1), $I = \mathbb{M}_m(\mathfrak{A})$ for some ideal $\mathfrak{A} \subseteq R$. Therefore,

$$S \cong \mathbb{M}_m(R)/\mathbb{M}_m(\mathfrak{A}) \cong \mathbb{M}_m(R/\mathfrak{A}).$$

If $m \geq 2$, then S is indeed a matrix ring.

Comment. In view of the counterexample we gave, it would be more reasonable to ask in (a): *Does* $\mathbb{M}_n(R) \cong \mathbb{M}_n(S)$ *imply* $R \cong S$? In a large number of cases, the answer to this new question is "yes." For instance, if R, S are *commutative* rings, $\mathbb{M}_n(R) \cong \mathbb{M}_n(S)$ implies

$$R \cong Z\left(\mathbb{M}_n(R)\right) \cong Z\left(\mathbb{M}_n(S)\right) \cong S,$$

by Exercise 1.9. If R, S are both left artinian rings, it can be shown that the answer is also "yes." However, if R, S are left noetherian rings, the answer is, in general, "no." In fact, there exist pairs of (left and right) noetherian rings $R \not\cong S$ such that $\mathbb{M}_n(R) \cong \mathbb{M}_n(S)$ for *every* $n \geq 2$. For a detailed discussion of this, see the author's booklet "Modules with Isomorphic Multiples, and Rings with Isomorphic Matrix Rings," Monog. No. 35, L'Enseig. Math., Geneva, Switzerland, 1999.

Ex. 3.10. Let R be any semisimple ring.

(1) Show that R is Dedekind-finite, i.e. $ab = 1$ implies $ba = 1$ in R.
(2) If $a \in R$ is such that $I = aR$ is an ideal in R, then $I = Ra$.
(3) Every element $a \in R$ can be written as a unit times an idempotent.

Solution. By the Wedderburn-Artin Theorem, we are reduced to the case when $R = \mathbb{M}_n(D)$ where D is a division ring.

(1) Think of R as $\text{End}(V_D)$ where V is the space of column n-tuples over D. If $ab = 1$ in R, then clearly $\ker(b) = 0$, and this implies that $b \in \text{Aut}(V_D)$. In particular, $ba = 1 \in R$.

(2) Since $R = \mathbb{M}_n(D)$ is now a simple ring, we have either $I = 0$ or $I = R$. If $I = 0$, then $a = 0$ and $Ra = I$ holds. If $I = R$, then $ab = 1$ for some $b \in R$. By (1), $ba = 1$, so $Ra = I$ also holds.

(3) By the theorem on Reduction to Echelon Forms, we can find invertible $n \times n$ matrices $b, c \in R$ such that

$$d := bac = \text{diag}(1, \ldots, 1, 0, \ldots, 0) \quad \text{(an idempotent).}$$

We have now

$$a = (b^{-1}c^{-1})(cdc^{-1}) = ue$$

where $u := b^{-1}c^{-1}$ is a unit and $e := cdc^{-1}$ is an idempotent.

Comment. (A) The conclusion in (1) also follows from Exercise 1.12, since a semisimple ring is always left noetherian. (B) The property (3) for semisimple rings goes back a long way, and was explicitly pointed out (and proved) by K. Asano, "Über Hauptidealringe mit Kettensatz", Osaka Math. J. *1* (1949), 52–61. In current terminology, (3) expresses the fact that semisimple rings are "unit-regular": for a more general view of this, see Exercises 4.14B and 4.14C.

Ex. 3.11. Let R be an n^2-dimensional algebra over a field k. Show that $R \cong \mathbb{M}_n(k)$ (as k-algebras) iff R is simple and has an element r whose minimal polynomial over k has the form $(x - a_1) \cdots (x - a_n)$ where $a_1, \ldots, a_n \in k$.

Solution. First, suppose $R = \mathbb{M}_n(k)$. Then R is simple, and the matrix with 1's on the line above the diagonal and 0's elsewhere has minimal polynomial x^n. Conversely, suppose R is simple and has an element r whose minimal polynomial over k is $(x - a_1) \cdots (x - a_n)$, where $a_1, \ldots, a_n \in k$. Then we have a *strictly descending* chain of left ideals

$$(*) \quad R \supsetneq R(r - a_1) \supsetneq R(r - a_1)(r - a_2) \supsetneq \cdots \supsetneq R(r - a_1) \cdots (r - a_n) = 0.$$

For, if $R(r - a_1) \cdots (r - a_i) = R(r - a_1) \cdots (r - a_{i+1})$ for some i, right multiplication by $(r - a_{i+2}) \cdots (r - a_n)$ would give

$$(r - a_1) \cdots (r - a_i)(r - a_{i+2}) \cdots (r - a_n) = 0 \in R,$$

which is impossible. Now, express R in the form $\mathbb{M}_m(D)$ where D is a division k-algebra of, say, dimension d. Then $n^2 = m^2 d$, and $_R R$ has composition length m. Appealing to $(*)$, we have $m^2 \geq n^2 = m^2 d$, so $d = 1$, $n = m$, and $R \cong \mathbb{M}_n(k)$.

Ex. 3.12. For a subset S in a ring R, let $\text{ann}_\ell(S) = \{a \in R : aS = 0\}$ and $\text{ann}_r(S) = \{a \in R : Sa = 0\}$. Let R be a semisimple ring, I be a left ideal and J be a right ideal in R. Show that $\text{ann}_\ell(\text{ann}_r(I)) = I$ and $\text{ann}_r(\text{ann}_\ell(J)) = J$.

Solution. By symmetry, it is sufficient to prove the above "Double Annihilator Property" for I. Let $I = Re$, where $e = e^2$ (see Exercise 1.7), and let $f = 1 - e$. We claim that $\text{ann}_r(I) = fR$. Indeed, since $I \cdot fR = RefR = 0$, we have $fR \subseteq \text{ann}_r(I)$. Conversely, if $a \in \text{ann}_r(I)$, then $ea = 0$ so $a = a - ea \in fR$. This proves $\text{ann}_r(I) = fR$, and hence similarly

$$\text{ann}_\ell(\text{ann}_r(I)) = \text{ann}_\ell(fR) = Re = I.$$

Comment. An artinian ring satisfying the Double Annihilator Properties in this Exercise is known as a *quasi-Frobenius ring*. The conclusion of the Exercise is therefore that *any semisimple ring is a quasi-Frobenius ring*.

The following is an interesting explicit illustration for the equation

$$\text{ann}_r(Re) = (1 - e)R$$

in a matrix ring $R = \mathbb{M}_n(k)$, where k is *any* ring. Take

$$e = E_{11} + E_{22} + \cdots + E_{rr}$$

where the E_{ij}'s are the matrix units. By explicit calculations, we see easily that

$$Re = \begin{pmatrix} k & \cdots & k & 0 & \cdots & 0 \\ \vdots & & \vdots & \vdots & & \vdots \\ k & \cdots & k & 0 & \cdots & 0 \end{pmatrix}, \quad (1-e)R = \left.\begin{pmatrix} 0 & \cdots\cdots & 0 \\ \vdots & & \vdots \\ 0 & \cdots\cdots & 0 \\ k & \cdots\cdots & k \\ \vdots & & \vdots \\ k & \cdots\cdots & k \end{pmatrix}\right\} n-r$$
$$\underbrace{\qquad\qquad}_{r}$$

and that the right annihilator of the former is precisely the latter.

Ex. 3.13. Let R be a simple, infinite-dimensional algebra over a field k. Show that any nonzero left R-module V is also infinite-dimensional over k.

Solution. The left action of R on V leads to a k-algebra homomorphism $\varphi : R \to \operatorname{End}(V_k)$. Since $V \neq 0$ and R is a simple ring, φ must be one-one. If $\dim_k V < \infty$, this would imply that $\dim_k R < \infty$. Therefore, V must be infinite-dimensional over k.

Ex. 3.14. (Over certain rings, the "rank" of a free module may not be defined.) Let D be a division ring, $V = \bigoplus_{i=1}^{\infty} e_i D$, and $E = \operatorname{End}(V_D)$. Define f_1, $f_2 \in E$ by $f_1(e_n) = e_{2n}$, $f_2(e_n) = e_{2n-1}$ for $n \geq 1$. Show that $\{f_1, f_2\}$ form a free E-basis for E_E. Therefore, as right E-modules, $E \cong E^2$; using this, show that $E^m \cong E^n$ for any finite $m, n > 0$!

Solution. Define $g_1, g_2 \in E$ by

$$g_1(e_{2n-1}) = 0, \quad g_1(e_{2n}) = e_n, \quad \text{and}$$
$$g_2(e_{2n-1}) = e_n, \quad g_2(e_{2n}) = 0$$

for $n \geq 1$. An easy calculation shows that

$$f_1 g_1 + f_2 g_2 = 1 \in E, \quad \text{and} \quad g_2 f_1 = g_1 f_2 = 0.$$

The former shows that $\{f_1, f_2\}$ span E_E. To show that f_1, f_2 are linearly independent in E_E, suppose $f_1 h_1 + f_2 h_2 = 0$, where $h_i \in E$. Then, for $h :$ $= f_1 h_1 = -f_2 h_2$, we have

$$h = (f_1 g_1 + f_2 g_2)h = (f_1 g_1)(-f_2 h_2) + (f_2 g_2)(f_1 h_1) = 0.$$

Since f_1, f_2 are *injective* maps, it follows that $h_1 = h_2 = 0$. Therefore, $E \cong E^2$ as right E-modules, and by induction, $E \cong E^n$ for all $n > 0$.

Alternatively, if we fix a D-isomorphism $\alpha : V \to V \oplus V$ and apply the functor $\operatorname{Hom}_D(V, -)$, we get a group isomorphism $\beta : E \to E \oplus E$. After checking that β respects the right E-action, we see that $E_E \cong E_E^2$. (Actually, this second solution "implies" the first solution. By choosing α suitably, we can get f_1, f_2 above as $\beta^{-1}(1, 0)$ and $\beta^{-1}(0, 1)$.)

For yet another view, recall from the solution to Ex. 2.11 that, viewing V naturally as a left E-module, we have an E-isomorphism $_EE \cong V \times V \times \cdots$. This clearly gives $_EE^2 \cong {}_EE$, and taking E-duals yields $E_E^2 \cong E_E$.

Comment. Taking endomorphism rings of the E-modules $E_E \cong E_E^n$ and applying Exercise 1.20, we deduce that

$$E \cong \operatorname{End}(E_E) \cong \operatorname{End}(E_E^n) \cong \mathbb{M}_n(E).$$

Thus, all matrix rings over E turn out to be isomorphic!

Ex. 3.15. Show that the ring E above has exactly three ideals: 0, E, and the ideal consisting of endomorphisms of finite rank.

We omit the solution since this exercise is only a special case of the next one. Nevertheless, the reader should solve this exercise first before attempting the more ambitious Ex. 16.

Ex. 3.16. Generalize the exercise above to the case of $E = \operatorname{End}(V_D)$, where $\dim_D V = \alpha$ is an arbitrary infinite cardinal.

Solution. For any infinite cardinal $\beta \leq \alpha$, let

$$E_\beta = \{f \in E : \operatorname{rank}(f) < \beta\},$$

where $\operatorname{rank}(f)$ denotes the cardinal number $\dim_D f(V)$. Since $\operatorname{rank}(g'fg) \leq \operatorname{rank}(f)$, E_β is an ideal of E. We claim that the ideals of E are (0), E and the E_β's. For this, we need the following crucial fact.

$(*)$ If $f, h \in E$ are such that $\operatorname{rank}(h) \leq \operatorname{rank}(f)$, then $h \in EfE$.

Indeed, write
$$V = \ker(h) \oplus V_1 = \ker(f) \oplus V_2.$$

Fix a basis $\{u_i : i \in I\}$ for V_1, and a basis $\{v_j : j \in J\}$ for V_2. We have

$$|I| = \operatorname{rank}(h) \leq \operatorname{rank}(f) = |J|,$$

so let us assume, for convenience, that $I \subseteq J$. Define $g \in E$ such that $g(\ker(h)) = 0$, and $g(u_i) = v_i$ for all $i \in I$. Noting that $\{f(v_j) : j \in J\}$ are linearly independent, we can also find $g' \in E$ such that $g'(f(v_i)) = h(u_i)$ for all $i \in I$. Then $h = g'fg$. In fact, both sides are zero on $\ker(h)$, and on $u_i \ (i \in I)$ we have
$$g'fg(u_i) = g'f(v_i) = h(u_i).$$

This proves $(*)$.

Now consider any ideal $\mathfrak{A} \neq 0, E$. Then, for any $f \in \mathfrak{A}$, $\operatorname{rank}(f) < \alpha$. For, if $\operatorname{rank}(f) = \alpha = \operatorname{rank}(\operatorname{Id})$, then $(*)$ implies $\operatorname{Id} \in EfE \subseteq \mathfrak{A}$, a contradiction. Since the class of cardinal numbers is well-ordered, there exists a cardinal

$\beta \leq \alpha$ which is least among cardinals larger than rank(f) for every $f \in \mathfrak{A}$. We leave to the reader the easy task of verifying that β is an infinite cardinal. Clearly, $\mathfrak{A} \subseteq E_\beta$ by the definition of E_β. We finish by showing that $E_\beta \subseteq \mathfrak{A}$. Let $h \in E_\beta$, so rank$(h) < \beta$. By the choice of β, we must have rank$(h) \leq$ rank(f) for some $f \in \mathfrak{A}$. But then, by $(*)$, $h \in EfE \subseteq \mathfrak{A}$, as desired.

Comment. This exercise implies that E_α is the unique maximal ideal of E; in particular, E/E_α is a simple ring. Another consequence of the exercise is that the ideals of E are linearly ordered with respect to inclusion.

Ex. 3.17. (K. A. Hirsch) Let k be a field of characteristic zero, and (a_{ij}) be an $m \times m$ skew symmetric matrix over k. Let R be the k-algebra generated by $x_1, \ldots, x_m$ with the relations $x_i x_j - x_j x_i = a_{ij}$ for all i, j. Show that R is a simple ring iff $\det(a_{ij}) \neq 0$. In particular, R is always nonsimple if m is odd.

Solution. Consider a linear change of variables given by $x_i' = \sum_j c_{ij} x_j$ where $C = (c_{ij}) \in \mathrm{GL}_m(k)$. We have

$$x_i' x_j' - x_j' x_i' = \sum_r c_{ir} x_r \sum_s c_{js} x_s - \sum_s c_{js} x_s \sum_r c_{ir} x_r$$

$$= \sum_{r,s} c_{ir} c_{js} (x_r x_s - x_s x_r)$$

$$= \sum_{r,s} c_{ir} a_{rs} c_{js}.$$

If we write

$$a_{ij}' = \sum_{r,s} c_{ir} a_{rs} c_{js},$$

then $x_i' x_j' - x_j' x_i' = a_{ij}'$, and we have $A' = CAC^T$ where $A = (a_{ij})$ and $A' = (a_{ij}')$, and "T" denotes the transpose. Therefore, we are free to perform any congruence transformation on A. After a suitable congruence transformation, we may therefore assume that A consists of a number of diagonal blocks $\begin{pmatrix} 0 & 1 \\ -1 & 0 \end{pmatrix}$, together with a zero block of size $t \geq 0$. If $t > 0$, then $\det(A) = 0$, and x_m generates a proper ideal in R. If $t = 0$, then $\det(A) \neq 0$ and $m = 2n$ for some n. Here, R is the nth Weyl algebra $A_n(k)$. Since k has characteristic zero, R is a simple ring by FC-(3.17).

Comment. Hirsch's paper "A note on non-commutative polynomials" appeared in J. London Math. Soc. *12* (1937), 264–266.

Ex. 3.18. (Quebbemann) Let k be a field of characteristic zero, and let R be the Weyl algebra $A_1(k)$ with generators x, y and relation $xy - yx = 1$. Let $p(y) \in k[y]$ be a fixed polynomial.

(a) Show that $R \cdot (x - p(y))$ is a maximal left ideal in R, and that the simple R-module $V = R/R \cdot (x - p(y))$ has R-endomorphism ring equal to k.

(b) Show that $R \to \operatorname{End}(V_k)$ is injective but not an isomorphism.

Solution. We can identify V as an abelian group with $k[y]$. In order to use this identification effectively, we must describe the action of y and x on "$V = k[y]$". Of course, the y action is just left multiplication by y. To describe the x action, consider any $v(y) \in k[y]$. Since

$$ x \cdot v(y) = v(y)x + \frac{dv}{dy} = v(y)\,(x - p(y)) + p(y)v(y) + \frac{dv}{dy}, $$

we see that $x - p(y)$ acts on $V = k[y]$ as differentiation with respect to y. To show that $_R V$ is simple, consider any $v(y) \neq 0$, say of degree m. Then $(x - p(y))^m \cdot v(y)$ is a nonzero constant in k. This shows that $R \cdot v(y) = V$, so V is simple. Now consider any $f \in \operatorname{End}_R V$ and let $f(1) = g(y)$. Then, in $V = k[y]$:

$$ 0 = f\left((x - p(y)) \cdot 1\right) = (x - p(y)) \cdot g(y) = \frac{dg}{dy}, $$

so $g \in k$. But then

$$ f\left(v(y)\right) = f\left(v(y) \cdot 1\right) = v(y)g, $$

so f is just multiplication by the constant $g \in k$. This completes the proof of (a). The natural map $\alpha : R \to \operatorname{End}(V_k)$ must be injective since R is a simple ring. However, since V_k is infinite-dimensional, $\operatorname{End}(V_k)$ is not a simple ring, by Exercise 15. Therefore, φ is not an isomorphism.

Comment. For any two rings R, S, an (R, S)-bimodule M is said to be *faithfully balanced* if the natural ring homomorphisms $R \to \operatorname{End}(M_S)$ and $S \to \operatorname{End}(_R M)$ giving the left and right actions are both isomorphisms. For instance, $_R R_R$ is a faithfully balanced (R, R)-bimodule. For any simple ring R, and any nonzero left ideal $\mathfrak{A} \subseteq R$ with endomorphism ring $D = \operatorname{End}(_R \mathfrak{A})$, Rieffel's Theorem FC-(3.11) says that $_R \mathfrak{A}_D$ is a faithfully balanced bimodule. The exercise above shows that Rieffel's result fails if one tries to replace the left ideal $\mathfrak{A}$ by a simple left R-module.

Quebbemann's paper on Weyl algebras appeared in J. Alg. *59* (1979), 311–312. The present exercise is only a special case of Quebbemann's main result, which says that *any* finite-dimensional division k-algebra D with $Z(D) = k$ is the ring of endomorphisms of a suitable simple left module U over $R = A_1(k)$ (if char $k = 0$). Thus, for any such D, there always exists an (R, D)-bimodule $_R U_D$ with $D \to \operatorname{End}(_R U)$ an isomorphism but $R \to \operatorname{End}(U_D)$ only a monomorphism, not an isomorphism.

Ex. 3.19. True or False: "If I is a minimal left ideal in a ring R, then $\mathbb{M}_n(I)$ is a minimal left ideal in $\mathbb{M}_n(R)$"?

Solution. In general, $\mathbb{M}_n(I)$ is a left ideal in $\mathbb{M}_n(R)$. However, it need not be a *minimal* left ideal. To construct an example, let $R = \mathbb{M}_2(k)$, where k is a field. Take I to be the minimal left ideal $\begin{pmatrix} k & 0 \\ k & 0 \end{pmatrix}$ in R. Then

$$\dim_k \mathbb{M}_n(I) = n^2 \dim_k I = 2n^2.$$

However, for $S = \mathbb{M}_n(R) \cong \mathbb{M}_{2n}(k)$, the unique simple left S-module M has k-dimension $2n$, so $\mathbb{M}_n(I) \cong n \cdot M$ as left S-modules. In particular, if $n > 1$, $\mathbb{M}_n(I)$ is *not* a minimal left ideal in $S = \mathbb{M}_n(R)$. Even more simply, k is a minimal left ideal of k, but $\mathbb{M}_n(k)$ is not a minimal left ideal of $\mathbb{M}_n(k)$ if $n > 1$.

Ex. 3.20. Let $\mathfrak{A}_i (1 \leq i \leq n)$ be ideals in a ring R, and let $\mathfrak{A} = \bigcap_i \mathfrak{A}_i$. If each $R/\mathfrak{A}_i$ is semisimple, show that $R/\mathfrak{A}$ is semisimple.

Solution. Each $\mathfrak{A}_i$ is a finite intersection of maximal ideals $\mathfrak{m} \subset R$ for which $R/\mathfrak{m}$ is simple artinian. Therefore, we may as well assume that each $R/\mathfrak{A}_i$ is simple artinian. In particular, each $\mathfrak{A}_i$ is a maximal ideal. We may further assume that $\mathfrak{A}_i \neq \mathfrak{A}_j$ for $i \neq j$. Then $\mathfrak{A}_i + \mathfrak{A}_j = R$ whenever $i \neq j$, and the Chinese Remainder Theorem implies that

$$R/\mathfrak{A} \cong \prod_{i=1}^{n} R/\mathfrak{A}_i,$$

which is a semisimple ring.

Comment. If $\{\mathfrak{A}_i\}$ is an *infinite* family of ideals, the exercise no longer holds. For instance, in the ring $R = \mathbb{Q} \times \mathbb{Q} \times \cdots$, let

$$\mathfrak{A}_i = \{(a_1, a_2, \ldots) : a_i = 0\} \quad (i \geq 1).$$

Then $R/\mathfrak{A}_i \cong \mathbb{Q}$ is semisimple, but $\mathfrak{A} = \bigcap \mathfrak{A}_i = 0$, and $R/\mathfrak{A} = R$ is *not* semisimple.

In the terminology of §12, this exercise says that a subdirect product of finitely many semisimple rings remains semisimple. For analogues of this for J-semisimple rings and for von Neumann regular rings, see Exercises 4.12A and 4.12D.

Ex. 3.21. For any finitely generated left module M over a ring R, let $\mu(M)$ denote the smallest number of elements that can be used to generate M. If R is an artinian simple ring, find a formula for $\mu(M)$ in terms of $\ell(M)$, the composition length of M.

Solution. Say $_R R \cong nV$ where V is the unique simple left R-module, and $n = \ell(_R R)$. We claim that

$$\mu(M) = \lceil \ell(M)/n \rceil,$$

where $\lceil \cdot \rceil$ is Knuth's notation for the "ceiling function." ($\lceil \alpha \rceil$ is defined to be the smallest integer $\geq \alpha$.) To prove this formula, let $\ell = \ell(M)$, and $k = \lceil \ell/n \rceil$. Since $\ell \leq kn$, there exists an epimorphism ${}_R R^k \to M$. Since ${}_R R^k$ can be generated by k elements, $\mu(M) \leq k$. If M can be generated by $k - 1$ elements, then there exists an epimorphism ${}_R R^{k-1} \to M$, and we get

$$\ell(M) \leq \ell(R^{k-1}) = (k-1)n.$$

This contradicts the definition of k, so we must have $\mu(M) = k$, as claimed.

Ex. 3.22. (1) Generalize the computation of $\mu(M)$ in the above exercise to the case of a finitely generated left module M over a semisimple ring R.
(2) Show that μ is subadditive, in the sense that $\mu(M \oplus N) \leq \mu(M) + \mu(N)$ for finitely generated R-modules M, N.
(3) Show that $N \subseteq M \Rightarrow \mu(N) \leq \mu(M)$.

Solution. We shall solve Part (1) by reducing the calculation of $\mu(M)$ to the case of modules over artinian simple rings. Let

$$M = M_1 \oplus \cdots \oplus M_r$$

be the decomposition of M into its isotypic components (cf. Exercise 2.8). Let

$$R = R_1 \oplus \cdots \oplus R_r,$$

where R_i is the simple component of R corresponding to M_i. Since $R_j M_i = 0$ for $j \neq i$, we have $\mu_R(M_i) = \mu_{R_i}(M_i)$. We now accomplish the desired reduction by proving that

$$(*) \qquad \mu(M) = \max\{\mu_{R_1}(M_1), \ldots, \mu_{R_r}(M_r)\}.$$

The inequality "$\geq$" is easy, since each M_i may be viewed as an epimorphic image of M. To prove the inequality "$\leq$", let us assume (in order to simplify the notation) that $r = 2$. Say $n = \mu_R(M_1)$, $m = \mu_R(M_2)$, with $n \geq m$. Let $\{x_1, \ldots, x_n\}$ be generators for M_1, and $\{y_1, \ldots, y_n\}$ be generators for M_2. We finish by showing that $x_1 + y_1, \ldots, x_n + y_n$ generate $M = M_1 \oplus M_2$. Indeed, for $(x, y) \in M_1 \oplus M_2$, write

$$x = \sum \alpha_i x_i \quad (\alpha_i \in R_1), \quad y = \sum \beta_i y_i \quad (\beta_i \in R_2).$$

Then, in the module M:

$$\sum (\alpha_i + \beta_i)(x_i + y_i) = \sum \alpha_i x_i + \sum \beta_i y_i = x + y,$$

so $\mu(M) \leq n$, as desired.

Part (2) of the exercise is trivial (and holds over any ring R). For Part (3), we can use Part (1) to reduce to the case of modules over artinian simple rings. In this case, the inequality $\mu(N) \leq \mu(M)$ in (3) follows immediately from the formula derived in Exercise 21, since $N \subseteq M$ implies $\ell(N) \leq \ell(M)$. Alternatively, since R is semisimple, $N \subseteq M$ implies that there exists an R-epimorphism $M \to N$, which gives $\mu(N) \leq \mu(M)$ right away.

Ex. 3.23. Show that a nonzero ring R is simple iff each simple left R-module is faithful.

Solution. First assume R is simple and let M be a simple left R-module. Then $\operatorname{ann}(M) \neq R$. Since $\operatorname{ann}(M)$ is an ideal, we must have $\operatorname{ann}(M) = 0$, so M is faithful. Conversely, assume that R is not simple. Then R has an ideal $I \neq 0, R$. Let $\mathfrak{m}$ be a maximal left ideal containing I and let $M = R/\mathfrak{m}$. Then $_R M$ is simple, but it is not faithful since $IM = 0$.

Ex. 3.24. (Jacobson) A subset S in a ring R is said to be *nil* (resp. *nilpotent*) if every $s \in S$ is nilpotent (resp. if $S^m = 0$ for some m, where S^m denotes the set of all products $s_1 \cdots s_m$ with $s_i \in S$).

(1) Let $R = \mathbb{M}_n(D)$ where D is a division ring. Let $S \subseteq R$ be a nonempty nil set which is closed under multiplication. Show that $S^n = 0$.
(2) Let R be any semisimple ring. Show that any nonempty nil set $S \subseteq R$ closed under multiplication is nilpotent.

Solution. (2) follows from (1) by applying the Wedderburn-Artin Theorem and by projecting S to the simple components of R. In the following, we assume $R = \mathbb{M}_n(D)$ as in (1), and identify R with $\operatorname{End}(V_D)$ where $\dim_D V = n$. Note that, in this case, any nilpotent set $S_o \subseteq R$ will automatically satisfy $S_o^n = 0$. This follows readily by considering the chain of D-subspaces

$$V \supseteq S_o V \supseteq S_o^2 V \supseteq \cdots .$$

(For any subset $T \subseteq R$, TV denotes the D-subspace $\{\sum_i t_i v_i : t_i \in T, v_i \in V\}$.)

Clearly, the set $S \subseteq R$ in (1) contains 0. Consider all nilpotent subsets $S_i \subseteq S$ (e.g. $\{0, s\}$ for any $s \in S$). Since $S_i^n = 0$ for all i, Zorn's Lemma can be applied to show the existence of a *maximal* nilpotent subset $S_o \subseteq S$. We see easily that $\{0\} \subsetneqq S_o$. Let $U = S_o V$. Then $0 \neq U \neq V$, so $\dim_D U$, $\dim_D V/U$ are both $< n$. Consider

$$S_1 := \{s \in S : \ sU \subseteq U\}.$$

Clearly $S_1 \supseteq S_o$, and $S_1^2 \subseteq S_1$. Invoking an inductive hypothesis at this point, we may assume S_1 is nilpotent on U and on V/U. Then S_1 itself is nilpotent, and so $S_1 = S_o$. In particular, for any $s \in S \backslash S_o$, we have $sS_o \not\subseteq S_o$ (for otherwise

$$sU = sS_o V \subseteq S_o V = U$$

implies $s \in S_1 = S_o$).

Assume, for the moment, that $S_o \neq S$. Take $s \in S \backslash S_o$. Then $ss_1 \notin S_o$ for some $s_1 \in S_o$, and, since $ss_1 \in S$, $(ss_1)s_2 \notin S_o$ for some $s_2 \in S_o$, etc. But then we get

$$s(s_1 s_2 \cdots s_n) \notin S_c$$

where all $s_i \in S_o$, contradicting the fact that

$$s(s_1 s_2 \cdots s_n) = 0 \in S_o.$$

Therefore, we must have $S = S_o$, and so $S^n = 0$.

Comment. In the ring theory literature, there are *many* generalizations of (2) in the above exercise to other rings, due to Hopkins, Levitzki, Goldie, Herstein, Small, Lanski, J.W. Fisher, and others. Loosely speaking, if R satisfies certain finiteness conditions, then one hopes to prove that a nil set $S \subseteq R$ closed under multiplication is nilpotent. For instance, the result of Lanski says that this is true in any right Goldie ring, and in particular in any right noetherian ring; see his paper in Canadian J. Math. *21* (1969), 904–907.

Ex. 3.25. Suppose a ring R has exactly seven left ideals $\neq 0$, R. Show that one of these must be an ideal (and give some examples of such a ring R).

Solution. Assume otherwise. Then R has no nontrivial ideals, so it is a simple ring. Having only a finite number of left ideals, R must also be left artinian. Therefore, by Wedderburn's theorem, $R = \mathbb{M}_n(D)$ for some division ring D, where, of course, $n \geq 2$. If $n \geq 3$, it is easy to produce (by column constructions) more than seven nontrivial left ideals in R. Thus, we must have $n = 2$. In this case, consider the map

$$\alpha : \{\text{lines through 0 in } (D^2)_D\} \to \{\text{nontrivial left ideals in } R\},$$

defined by taking annihilators. Using linear algebra in the plane, we check readily that α is a 1–1 correspondence. Therefore, the number of nontrivial left ideals in R is $|D| + 1$. Thus, we have $|D| = 6$, which is impossible!

Two obvious examples of the rings R under consideration are the commutative rings $\mathbb{Z}_4 \times \mathbb{Z}_4$ and $\mathbb{Z}_{2^8}$. For a noncommutative example, consider the ring $R = \begin{pmatrix} k & k \\ 0 & k \end{pmatrix}$, where k is a finite field. Using FC-(1.17)(1), one computes that R has exactly $|k| + 3$ nontrivial left ideals, so taking $k = \mathbb{F}_4$ gives what we want.

Comment. The fact that α is a bijection, which is particularly easy to see for $(D^2)_D$, generalizes nicely to the finite-dimensional vector space $(D^n)_D$. Here, taking annihilators defines a bijection (in fact a lattice anti-isomorphism) from the subspaces of D^n to the left ideals in $\mathbb{M}_n(D)$. (This can be shown, e.g. using the semisimplicity of $\mathbb{M}_n(D)$; for more information on this, see Ex. 11.15.) Applying this also for $n = 3$ (along with $n = 2$), we see that the number 7 is only the "tip of the iceberg": the exercise remains equally valid if we replace 7 by 11, 13, 15, 16, 19, 23, 25, 27, $\ldots$, and many other numbers!

I learned this exercise from K. Goodearl, who informed me that (an equivalent form of) it once appeared as a question in a written qualifying examination at the University of Washington. The original form of the question was: "Show that a simple ring cannot have exactly nine left ideals." I recast this question into its present form (Ex. 3.25) so as to "hide away" the fact that the exercise is really all about simple artinian rings and the Artin-Wedderburn Theorem! Hopefully, a reader will have more fun and joy figuring this out on his or her own, and solving the exercise accordingly.

Chapter 2
Jacobson Radical Theory

§4. The Jacobson Radical

The Jacobson radical of a ring R, denoted by rad R, is the intersection of the maximal left ideals of R. This notion is left-right symmetric; in particular, rad R is an ideal of R. A good way to understand rad R is to think of it as the ideal of elements annihilating all left (resp. right) simple R-modules. The Jacobson radical is also closely tied in with $U(R)$, the group of units of R. In fact, rad R is the largest ideal I such that $1 + I \subseteq U(R)$.

The definition of rad R led N. Jacobson to a new notion of semisimplicity: a ring R is called *Jacobson semisimple* (or *J-semisimple* for short) if rad $R = 0$. The (classical) semisimple rings studied in Chapter 1 are just the artinian J-semisimple rings.

Two famous results involving the use of rad R are the Hopkins-Levitzki Theorem and Nakayama's Lemma. The former implies that left artinian rings are left noetherian, and the latter says that, if $_RM$ is nonzero and finitely generated, then (rad $R)M \neq M$. These results are deservedly among the cornerstones of noncommutative ring theory today.

Of equal importance is the class of von Neumann regular rings, which occur somewhere "between" semisimple rings and J-semisimple rings. By definition, these are the rings R in which every element a can be "factored" into axa for some $x \in R$ (depending on a). Such rings can be characterized by a "weak" semisimplicity property: every *finitely generated* left ideal should be a direct summand of $_RR$. Von Neumann regular rings occur naturally as rings of operators in functional analysis. The theory of such rings is very rich, and constitutes an important field of study.

The exercises in this section begin with an introduction to the Jacobson radical theory for rings possibly without an identity. This is not essential for the problems in the later sections, but it is nice to know how this theory is developed. The other exercises deal with folklore results concerning rad R, for instance lifting invertible matrices from R/I to R where $I \subseteq$ rad R (Exercise 21). The Brown-McCoy radical, not nearly as useful as rad R, receives only a fleeting mention in Exercise 8. Various other kinds of radicals will be dealt with in the Exercises for §10. A few exercises are devoted to von Neumann regular rings and their characterizations: see Exercises 13–15. In Exercise 14B, we introduce the notion of a *unit-regular* ring which has played an increasingly important role in the recent research on von Neumann regular rings. More exercises in this direction can be found in §12, §21, and §22.

Exercises for §4

In *FC*, we deal only with rings with an identity element. In particular, the theory of the Jacobson radical was developed in that text for rings with an identity. However, by doing things a little more carefully, the whole theory can be carried over to rings possibly without an identity. In Exercises 1–7 below, we sketch the steps necessary in developing this more general theory; in these exercises, R denotes a ring possibly without 1 (or a "rng" according to Louis Rowen).

Ex. 4.1. In R, define $a \circ b = a + b - ab$. Show that this binary operation is associative, and that $(R, \circ)$ is a monoid with zero as the identity element.

Solution. For any $a, b, c \in R$, we have

$$(a \circ b) \circ c = (a + b - ab) \circ c = a + b + c - ab - ac - bc + abc,$$
$$= a \circ (b + c - bc) = a \circ (b \circ c).$$

Furthermore, $a \circ 0 = a + 0 - a0 = a$ and $0 \circ a = 0 + a - 0a = 0$, so $(R, \circ)$ is indeed a monoid with zero as the identity element.

Ex. 4.2. An element $a \in R$ is called *left* (resp. *right*) *quasi-regular* if a has a left (resp. right) inverse in the monoid $(R, \circ)$ with identity. If a is both left and right quasi-regular, we say that a is *quasi-regular*.

(1) Show that if ab is left quasi-regular, then so is ba.
(2) Show that any nilpotent element is quasi-regular.
(3) Show that, if R has an identity 1, the map $\varphi : (R, \circ) \to (R, \times)$ sending a to $1 - a$ is a monoid isomorphism. In this case, an element a is left (right) quasi-regular iff $1 - a$ has a left (resp. right) inverse with respect to ring multiplication.

Solution. (1) Say $c \circ ab = 0$, so $c = (ca - a)b$. Left-multiplying by b and right-multiplying by a, we get $bca = b(ca - a)ba$. Therefore,

$$b(ca - a) \circ ba = b(ca - a) + ba - b(ca - a)ba$$
$$= bca - b(ca - a)ba = 0,$$

so ba is also left quasi-regular.

(2) Say $a^{n+1} = 0$. Then

$$a \circ (-a - a^2 - \cdots - a^n) = -a^2 - \cdots - a^n + a(a + a^2 + \cdots + a^n) = 0,$$

and similarly

$$(-a - a^2 - \cdots - a^n) \circ a = 0.$$

(3) Assume now $1 \in R$. For any $a, b \in R$, we have

$$\varphi(a)\varphi(b) = (1 - a)(1 - b) = 1 - (a + b - ab)$$
$$= \varphi(a + b - ab) = \varphi(a \circ b).$$

And of course, $\varphi(0) = 1 - 0 = 1$. Clearly, φ is one-one and onto, so it is a monoid isomorphism. Therefore, $a \in R$ is (say) left quasi-regular with a "left quasi-inverse" b iff $1 - a$ is left invertible in R with a left inverse $1 - b$.

Comment. Any ring R (with or without identity) can always be embedded in a ring S with identity. The map

$$\varphi : a \mapsto 1 - a$$

then defines a monoid embedding of $(R, \circ)$ into $(S, \times)$. This explains why the associative law should hold for $(R, \circ)$. It also explains the construction of a left quasi-inverse for ba in (1) (cf. Exercise 1.6), and the construction of a left quasi-inverse for the nilpotent element a in (2).

Our definition of $a \circ b = a + b - ab$ follows the original one by Jacobson (see his book "Structure of Rings" in the AMS Colloquium Publications). Some later authors found it more convenient to use an alternate definition: $a * b = a + b + ab$. Since

$$(-a) * (-b) = -(a \circ b),$$

we can easily go from one set-up to the other by changing signs. For instance, to say that $a \in R$ is left quasi-regular in one set-up simply amounts to saying that $-a$ is left quasi-regular in the other set-up.

Ex. 4.3. A set $I \subseteq R$ is called quasi-regular (resp. left or right quasi-regular) if every element of I is quasi-regular (resp. left or right quasi-regular). Show that if a left ideal $I \subseteq R$ is left quasi-regular, then it is quasi-regular.

Solution. We must show that every $b \in I$ is *right* quasi-regular. Choose $a \in R$ such that $a \circ b = 0$. Then $a = ab - b \in I$ so there exists $b' \in R$ with $b' \circ a = 0$. But then

$$b = 0 \circ b = (b' \circ a) \circ b = b' \circ (a \circ b) = b',$$

so we have $b \circ a = 0$, as desired.

Ex. 4.4. Define the Jacobson radical of R by

$$\operatorname{rad} R = \{a \in R : Ra \text{ is left quasi-regular}\}.$$

Show that rad R is a quasi-regular ideal which contains every quasi-regular left (resp. right) ideal of R. (In particular, rad R contains every nil left or right ideal of R.) Show that, if R has an identity, the definition of rad R here agrees with the one given in the Introduction to this section.

Solution. To show that rad R *is a left ideal*, it suffices to check that, if $a, b \in \operatorname{rad} R$, then $r(a + b)$ is left quasi-regular for any $r \in R$. For any $c, d \in R$, note that

$$c \circ d \circ r(a + b) = c \circ [d + ra + rb - dr(a + b)]$$
$$= c \circ [d \circ ra + (r - dr)b].$$

Choose d such that $d \circ ra = 0$; then choose c such that $c \circ (r - dr)b = 0$. The above equation shows that $c \circ d$ is a left quasi-inverse of $r(a + b)$, as desired. To show that rad R *is also a right ideal*, we must check that $a \in \operatorname{rad} R$ and $s \in R$ imply $as \in \operatorname{rad} R$. For any $r \in R$, sra is left quasi-regular, so by Exercise 2, $r(as)$ is also left quasi-regular. This shows that $as \in \operatorname{rad} R$.

Next, we show that rad R *is a quasi-regular ideal.* In view of Exercise 3, it suffices to show that every element $a \in \operatorname{rad} R$ is left quasi-regular.[1] Since $a^2 \in Ra$ is left quasi-regular, there exists $b \in R$ such that $b \circ a^2 = 0$. But then

$$(b \circ (-a)) \circ a = b \circ ((-a) \circ a) = b \circ a^2 = 0$$

shows that a is left quasi-regular.

By definition, rad R contains every quasi-regular left ideal. Let I be any quasi-regular *right* ideal. For $a \in I$, aR is right quasi-regular, therefore quasi-regular by the right analogue of Exercise 3. Applying Exercise 2, we see that Ra is (left) quasi-regular, so $I \subseteq \operatorname{rad} R$, as desired.

Finally, assume that $1 \in R$. By Exercise 1, the radical just defined can be described as

$$\{a \in R : 1 - ra \text{ is left-invertible for any } r \in R\}.$$

This is precisely the Jacobson radical for the ring R with identity: cf. *FC-*(4.1).

---

[1] This is non-trivial since a may not lie in Ra!

Comment. If we define the radical to be the intersection of all "modular" maximal left ideals (see Exercise 5), some of the above verifications can be simplified. However, modularity is a rather intricate notion. It is therefore of independent interest to define and characterize the radical (and prove its left-right symmetry) using only the notion of quasi-regularity, without appealing to the notion of modularity.

Ex. 4.5. A left ideal $I \subseteq R$ is said to be *modular* (or *regular*[2]) if there exists $e \in R$ which serves as a "*right* identity mod I"; i.e. $re \equiv r \pmod{I}$ for every $r \in R$.

(a) Show that if $I \subsetneq R$ is a modular left ideal, then I can be embedded in a modular maximal left ideal of R.

(b) Show that rad R is the intersection of all modular maximal left (resp. right) ideals of R.

Solution. (a) Fix the element e above. For any left ideal $I' \supseteq I$, we have $I' \neq R$ iff $e \notin I'$. (For the "only if" part, note that if $e \in I'$, then, for any $r \in R$, we have $r \in re + I \subseteq I'$.) Therefore, we can apply Zorn's Lemma to the family of left ideals

$$\{I' : \ I' \neq R \text{ and } I' \supseteq I\}$$

to prove the existence of a maximal left ideal $\mathfrak{m}$ containing I. Clearly, $\mathfrak{m}$ is also modular, and $e \notin \mathfrak{m}$.

(b) Let J be the intersection of all modular maximal left ideals of R. (If there are no modular maximal left ideals, we define $J = R$.) *First we prove* $J \subseteq \text{rad } R$. Consider any $a \notin \text{rad } R$. Then $e := xa$ is not left quasi-regular for some $x \in R$. It is easy to check that

$$I = \{r - re : r \in R\}$$

is a modular left ideal. Since $e \notin I$, we have $I \neq R$. By (a), $I \subseteq \mathfrak{m}$ for some modular maximal left ideal $\mathfrak{m}$, with $e \notin \mathfrak{m}$. In particular, $e = xa \notin J$, so $a \notin J$. *We now finish by proving that* rad $R \subseteq J$. Assume the contrary. Then rad $R \not\subseteq \mathfrak{m}$ for some modular maximal left ideal $\mathfrak{m}$. In particular, rad $R + \mathfrak{m} = R$. Let $e \in R$ be such that $r \equiv re \pmod{\mathfrak{m}}$ for all $r \in R$. Write $e = a + b$ where $a \in \text{rad } R$ and $b \in \mathfrak{m}$. Then $e - a \in \mathfrak{m}$ so $e - ae \in \mathfrak{m}$. Since rad R is left quasi-regular, there exists $a' \in R$ such that $a' + a - a'a = 0$. But then

$$e = e - (a' + a - a'a)e = (e - ae) - a'(e - ae) \in \mathfrak{m},$$

a contradiction.

---

[2] We mention this alternate term only because it is sometimes used in the literature. Since "regular" has too many meanings, we shall avoid using it altogether.

Ex. 4.6. A left R-module M is said to be *simple* (or *irreducible*) if $R \cdot M \neq 0$ and M has no R-submodules other than (0) and M. Show that $_RM$ is simple iff $M \cong R/\mathfrak{m}$ (as left R-modules) for a suitable modular maximal left ideal $\mathfrak{m} \subset R$. Show that rad R is the intersection A of the annihilators of all simple left R-modules.

Solution. Let M be simple. Then $Rm \neq 0$ for some $m \in M$, and so $Rm = M$. Let $\varphi : R \to M$ be the R-epimorphism defined by $\varphi(r) = rm$, and let $\mathfrak{m} = \ker(\varphi)$. Then $\mathfrak{m}$ is a maximal left ideal of R. Since $m \in M = Rm$, there exists $e \in R$ such that $m = em$. For any $r \in R$:

$$\varphi(r - re) = (r - re)m = rm - rm = 0.$$

Therefore $r \equiv re \pmod{\mathfrak{m}}$, so $\mathfrak{m}$ is modular, with $R/\mathfrak{m} \cong M$.

Conversely, let $M = R/\mathfrak{m}$, where $\mathfrak{m}$ is a modular maximal left ideal. Let $e \in R$ be such that $r \equiv re \pmod{\mathfrak{m}}$ for all $r \in R$. Then $R\bar{e} = M$; in particular, $R \cdot M \neq 0$.

Finally, we show rad $R = A$. For any modular maximal left ideal $\mathfrak{m}$, we have (rad R) $R \subseteq$ rad $R \subseteq \mathfrak{m}$ by Exercise 5. Therefore, (rad R) $\cdot R/\mathfrak{m} = 0$, so rad $R \subseteq A$. Conversely, if $a \in A$, then $aR \subseteq \mathfrak{m}$ for any modular maximal left ideal, so $aR \subseteq$ rad R. Since rad R is quasi-regular, so is aR; hence $a \in$ rad R (cf. Exercise 4).

Comment. For a given maximal left ideal $\mathfrak{m}_o \subset R$, let $M = R/\mathfrak{m}_o$, and consider the following statement:

$(*)$ *M is a simple R-module iff $\mathfrak{m}_o$ is modular.*

We have shown above that the "if" part is true. However, the "only if" part is false in general! Let R be the ring (without identity) of cardinality p^2 constructed in the solution to Exercise 1.10(b). Let

$$\mathfrak{m}_o = \{(d, d) : d \in \mathbb{F}_p\}.$$

This is a left ideal of index p in R, so it is maximal. Let $M = R/\mathfrak{m}_o$. Clearly $RM \neq 0$, so M is a *simple* R-module. For any $(x, y) \in R$, we have

$$(0, 1)(x, y) - (0, 1) = (0, 0) - (0, 1) = (0, -1) \notin \mathfrak{m}_o.$$

Therefore, $\mathfrak{m}_o$ is *not* modular. However, if we carry through the first part of the proof of the Exercise, we will be able to represent M as $R/\mathfrak{m}$ for a suitable *modular* maximal left ideal. Indeed, if we choose $m = \overline{(1,0)}$ to be a cyclic generator of M, we can take $\mathfrak{m}$ to be

$$\{(c, d) : (c, d)(1, 0) \in \mathfrak{m}_o\} = \{(c, d) : (c, 0) \in \mathfrak{m}_o\}$$
$$= \{(0, d) : d \in \mathbb{F}_p\}.$$

Choose $e \in R$ such that $em = m$; for example, $e = (1, 0)$. As predicted by our earlier proof:

$$(a, b)(1, 0) - (a, b) = (a, 0) - (a, b) = (0, -b) \in \mathfrak{m} \qquad (\forall (a, b) \in R),$$

so $\mathfrak{m}$ is indeed modular, with $R/\mathfrak{m} \cong M = R/\mathfrak{m}_o$. In particular, rad $R \subseteq \mathfrak{m}$. On the other hand, $\mathfrak{m}^2 = 0$ implies that $\mathfrak{m} \subseteq$ rad R, so we have in this example rad $R = \mathfrak{m}$.

One final observation: although $(*)$ is false in general, it is true *in case* $\mathfrak{m}_o$ *happens to be an ideal* (e.g. when R is commutative). Indeed, if $M = R/\mathfrak{m}_o$ is simple, then, using the notation in the beginning of the Solution, we have $\mathfrak{m}_o \subseteq \ker \varphi$, so $\mathfrak{m}_o = \ker \varphi$, which is modular!

Ex. 4.7. Show that rad $(R/\text{rad } R) = 0$, and that, if I is an ideal in R, then, viewing I as a ring, rad $I = I \cap$ rad R. This shows, in particular, that a ring R may be equal to its Jacobson radical: if this is the case, R is said to be a *radical ring*. Show that R is a radical ring iff it has no simple left (resp. right) modules.

Solution. Let $\mathfrak{m} \supseteq$ rad R be a typical modular maximal left ideal. Clearly $\mathfrak{m}/\text{rad } R$ is a modular maximal left ideal of $R/\text{rad } R$. Since $\bigcap \mathfrak{m} = \text{rad } R$ by Exercise 5, we see that rad $(R/\text{rad } R) = 0$.

For any ideal $I \subseteq R$, consider $a \in I \cap$ rad R. For any $b \in R$, ba has a left quasi-inverse $r \in R$. But then

$$r = rba - ba \in I,$$

so ba is left quasi-regular *as an element of the ring I*. In particular, this shows that $a \in$ rad I. Conversely, consider any $a \in$ rad I. Since $(Ra)^2 \subseteq Ia$, $(Ra)^2$ is quasi-regular, so $(Ra)^2 \subseteq$ rad R. The image of Ra, being nilpotent in $R/\text{rad } R$, must be zero, since rad $(R/\text{rad } R) = 0$. Therefore, $Ra \subseteq$ rad R. In particular, Ra is quasi-regular, so $a \in I \cap$ rad R.

The last conclusion of the Exercise characterizing radical rings follows easily from Exercise 6.

Comment. As in the text, we may define a ring R (possibly without 1) to be *J-semisimple* if rad $R = 0$. Let $R^* = R \oplus \mathbb{Z}$ be the ring obtained from R by formally adjoining an identity (the 1 in $\mathbb{Z}$). Using the Exercise and the J-semisimplicity of $\mathbb{Z}$, it is easy to show that rad $R = $ rad R^*. In particular, R is J-semisimple iff R^* is J-semisimple.

The notion of a radical ring is a new ingredient in the theory of rings without identity. In simple terms, *a ring R is a radical ring iff every element $a \in R$ has a (say left) quasi-inverse in R*. By the last paragraph, radical rings may be thought of as the radicals of rings with identity. For instance, any nil ring R (every $a \in R$ is nilpotent) is radical. In this connection, it is of interest to mention that Sasiada has constructed an example of a simple radical ring; see E. Sasiada and P.M. Cohn, J. Algebra *5* (1967), 373–377.

More recently, A. Smoktunowicz has constructed an example of a simple nil ring; see Comm. Algebra *30* (2002), 27–59.

In the following problems, we return to our standing assumption that all rings to be considered have an identity element.

Ex. 4.8. An ideal $I \subsetneq R$ is called a *maximal ideal* of R if there is no ideal of R strictly between I and R. Show that any maximal ideal I of R is the annihilator of some simple left R-module, but not conversely. Defining rad$'$ R to be the intersection of all maximal ideals of R, show that rad $R \subseteq$ rad$'R$, and give an example to show that this may be a strict inclusion. (rad$'$ R is called the *Brown-McCoy radical* of R.)

Solution. Let $V = R/\mathfrak{m}$, where $\mathfrak{m}$ is a maximal left ideal of R containing I. Then $IV = 0$, so $I \subseteq \operatorname{ann}(V)$. Since $\operatorname{ann}(V)$ is an ideal of R, the maximality of I implies that $I = \operatorname{ann}(V)$. By *FC*-(4.2), it follows that rad $R \subseteq I$. Therefore, rad $R \subseteq$ rad$'$ R.

Consider $V = \bigoplus_{i=1}^{\infty} e_i k$ where k is any division ring, and let $R = \operatorname{End}(V_k)$. Then V is a simple left R-module. However, $\operatorname{ann}_R(V) = 0$ is *not* a maximal ideal in R (see Exercise 3.15). By *FC*-(4.27), R is von Neumann regular, so by *FC*-(4.24), rad $R = 0$. On the other hand, the only maximal ideal of R is

$$I = \{f \in R : \dim f(V) < \infty\}$$

(by Exercise 3.15), so we have here rad$'$ $R = I \supsetneq$ rad $R = 0$.

Comment. Of course, if R is a commutative ring, then rad$'$ $R =$ rad R. The same holds for simple and semisimple rings (since, in these cases, both radicals are zero).

If R is a ring possibly without identity, the Brown-McCoy radical rad$'$ R may be defined as the intersection of all modular maximal ideals of R (where "modular" is now understood in the 2-sided sense). The radical rad$'$ R is useful in the theory of Banach algebras, under the alternative name "strong radical." In p. 490 of his book on Banach algebras (Cambridge University Press, 1994), T. Palmer observed that "The Jacobson radical of familiar Banach algebras is usually a pathological portion. This is not true of the strong radical." For instance, if H is a separable Hilbert space, then the strong radical of $B(H)$, the ring of all bounded operators, turns out to be the ideal of all compact operators.

Ex. 4.9. Let R be a J-semisimple domain and a be a nonzero central element of R. Show that the intersection of all maximal left ideals not containing a is zero.

Solution. Let x be an element in this intersection. *We claim that* $ax \in$ rad R. Once we have proved this, the hypotheses on R imply that $ax = 0$ and hence $x = 0$. To prove the claim, let us show that, for any maximal left ideal $\mathfrak{m}$, we have $ax \in \mathfrak{m}$. If $a \in \mathfrak{m}$, this is clear since $a \in Z(R)$. If $a \notin \mathfrak{m}$, then by the choice of x we have $x \in \mathfrak{m}$, and hence $ax \in \mathfrak{m}$.

Ex. 4.10. Show that if $f : R \to S$ is a surjective ring homomorphism, then $f(\mathrm{rad}\ R) \subseteq \mathrm{rad}\ S$. Give an example to show that $f(\mathrm{rad}\ R)$ may be smaller than $\mathrm{rad}\ S$.

Solution. The inclusion $f(\mathrm{rad}\ R) \subseteq \mathrm{rad}\ S$ is clear since, for any maximal left ideal $\mathfrak{m}$ of S, the inverse image $f^{-1}(\mathfrak{m})$ is a maximal left ideal of R. (Note that the result $f(\mathrm{rad}\ R) \subseteq \mathrm{rad}\ S$ here may also be viewed as a special case of FC-(5.7).) To give an example of strict inclusion, let $R = \mathbb{Z}$ and let f be the natural projection of R onto $S = \mathbb{Z}/4\mathbb{Z}$. Here,

$$f(\mathrm{rad}\ R) = f(0) = 0,$$

but $\mathrm{rad}\ S = 2\mathbb{Z}/4\mathbb{Z} \neq 0$.

Ex. 4.11. If an ideal $I \subseteq R$ is such that R/I is J-semisimple, show that $I \supseteq \mathrm{rad}\ R$. (Therefore, $\mathrm{rad}\ R$ is the smallest ideal $I \subseteq R$ such that R/I is J-semisimple.)

Solution. The J-semisimplicity of R/I means that the intersection of the maximal left ideals of R containing I is exactly I. It follows that $\mathrm{rad}\ R$, the intersection of all the maximal left ideals of R, is contained in I.

Ex. 4.12A. Let $\mathfrak{A}_i$ $(i \in I)$ be ideals in a ring R, and let $\mathfrak{A} = \bigcap_i \mathfrak{A}_i$. True or False: "If each $R/\mathfrak{A}_i$ is J-semisimple, then so is $R/\mathfrak{A}$"?

Solution. For each i, the J-semisimplicity of $R/\mathfrak{A}_i$ implies that $\mathfrak{A}_i$ is an intersection of maximal left ideals of R. Therefore, so is $\mathfrak{A} = \bigcap_i \mathfrak{A}_i$. This shows the J-semisimplicity of $R/\mathfrak{A}$.

Comment. (1) In the terminology of §12, the affirmative solution to the Exercise implies that *any subdirect product of J-semisimple rings is J-semisimple.*

(2) Using this exercise, we can also give a new solution to the earlier Exercise 3.20. It suffices to show that, if $\mathfrak{A}$, $\mathfrak{B}$ are ideals with $\mathfrak{A} \cap \mathfrak{B} = 0$, and $R/\mathfrak{A}, R/\mathfrak{B}$ are semisimple, then R is semisimple. Now $\mathfrak{B}$ maps injectively into $R/\mathfrak{A}$, so $_R\mathfrak{B}$ is artinian, as is $_R(R/\mathfrak{B})$. This implies that $_RR$ is artinian. Since R is J-semisimple, it must be semisimple, by FC-(4.14).

Ex. 4.12B. Show that, for any direct product of rings $\prod R_i$, $\mathrm{rad}\ (\prod R_i) = \prod \mathrm{rad}\ R_i$.

Solution. Let $y = (y_i) \in \prod R_i$. By FC-(4.1), $y \in \mathrm{rad}\ (\prod R_i)$ amounts to $1 - xy$ being left-invertible for any $x = (x_i) \in \prod R_i$. This, in turn, amounts to $1 - x_i y_i$ being left-invertible in R_i, for any $x_i \in R_i$ (and any i). Therefore, $y \in \mathrm{rad}\ (\prod R_i)$ iff $y_i \in \mathrm{rad}\ R_i$ for all i.

It follows that a direct product of J-semisimple rings is J-semisimple. But this is only a special case of (1) in the *Comment* on the last exercise.

Ex. 4.12C. For a triangular ring $T = \begin{pmatrix} R & M \\ 0 & S \end{pmatrix}$ (where M is an (R, S)-bimodule), show that $\mathrm{rad}(T) = \begin{pmatrix} \mathrm{rad}(R) & M \\ 0 & \mathrm{rad}(S) \end{pmatrix}$. Apply this to compute the radical of the ring $T_n(k)$ of $n \times n$ upper triangular matrices over any ring k.

Solution. Let $J = \mathrm{rad}(R)$, $J' = \mathrm{rad}(S)$, and $I = \begin{pmatrix} J & M \\ 0 & J' \end{pmatrix}$. It is routine to check that I is an ideal of T, with $T/I \cong R/J \times S/J'$. The latter ring is J-semisimple by Ex. 4.12B, so we have $\mathrm{rad}(T) \subseteq I$ by Ex. 4.11. This will be an equality, as asserted, if we can show that $1 + I \subseteq U(T)$. Now any element of $1 + I$ has the form $\begin{pmatrix} u & m \\ 0 & v \end{pmatrix}$, where $m \in M$, $u \in U(R)$, and $v \in U(S)$. This is indeed in $U(T)$, since it has inverse $\begin{pmatrix} u^{-1} & -u^{-1}mv^{-1} \\ 0 & v^{-1} \end{pmatrix}$.

To compute $\mathrm{rad}(T_n(k))$ for any ring k, we can treat $T_n(k)$ as a triangular ring with $R = k$, $S = T_{n-1}(k)$, and $M = k^{n-1}$ (as (R, S)-bimodule). Using the above, and invoking an inductive hypothesis, we see that $\mathrm{rad}(T_n(k))$ consists of $n \times n$ upper triangular matrices with diagonal entries from $\mathrm{rad}(k)$. The same conclusion could also have been obtained, without induction, by a judicious application of the method used in the first paragraph.)

Comment. Analogues of the result in this exercise also hold for other types of radicals; see Ex. 10.23*.

Ex. 4.12D. Redo Ex. 12A with "J-semisimple" replaced by "von Neumann regular".

Solution. We use the notations in Ex. 12A, but surprisingly, the conclusions here are a little different. If the indexing set I is infinite, the answer is "no". For instance, taking $\mathfrak{A}_p = (p)$ for primes p in $R = \mathbb{Z}$, we have $\mathfrak{A} = \bigcap_p \mathfrak{A}_p = (0)$. Here, each $R/\mathfrak{A}_p \cong \mathbb{Z}/p\mathbb{Z}$ is a field and hence von Neumann regular, but $R/\mathfrak{A} \cong \mathbb{Z}$ is *not* von Neumann regular.

To treat the case $|I| < \infty$, let $I = \{1, 2, \ldots, n\}$. We claim that here the answer is "yes". It suffices to prove this for $n = 2$, and we may assume $\mathfrak{A}_1 \cap \mathfrak{A}_2 = (0)$. Consider any $a \in R$. Since $R/\mathfrak{A}_1$ and $R/\mathfrak{A}_2$ are von Neumann regular, there exist $x, y \in R$ such that $(1 - ax)a \in \mathfrak{A}_1$ and $a(1 - ya) \in \mathfrak{A}_2$. Then

$$(1 - ax)a(1 - ya) \in \mathfrak{A}_1 \cap \mathfrak{A}_2 = 0.$$

This yields $a = a(x + y - xay)a$, so R is von Neumann regular.

Comment. Note that the argument above produces an explicit "pseudo-inverse" of a from given pseudo-inverses of a in $R/\mathfrak{A}_1$ and in $R/\mathfrak{A}_2$. As in the *Comment* on Ex. 4.12A, we may conclude that any *finite* subdirect product of von Neumann regular rings is von Neumann regular.

Ex. 4.13. Let R be the ring of all continuous real-valued functions on a topological space A. Show that R is J-semisimple, but "in most cases" not von Neumann regular.

Solution. The following are clearly maximal ideals of R:

$$\mathfrak{m}_a = \{f \in R : f(a) = 0\},$$

where $a \in A$. Therefore,

$$\operatorname{rad} R \subseteq \bigcap_{a \in A} \mathfrak{m}_a = \{f \in R : f(A) = 0\} = 0.$$

To see that in most cases R is not von Neumann regular, consider any non-singleton connected compact Hausdorff space A. Then the only idempotents in R are 0 and 1. Assume R is von Neumann regular. For any nonzero $f \in R$, $fR = eR$ for some idempotent $e \in R$, so we must have $fR = R$, i.e. $f \in U(R)$. Therefore, R is a field. The known classification theorem for maximal ideals of R then implies that $|A| = 1$, a contradiction.

Ex. 4.14. Show that a ring R is von Neumann regular iff $IJ = I \cap J$ for every right ideal I and every left ideal J in R.

Solution. First assume R is von Neumann regular. For I, J as above, it suffices to show that $I \cap J \subseteq IJ$. Let $a \in I \cap J$. There exists $x \in R$ such that $a = axa$. Thus, $a \in (IR)J \subseteq IJ$. Conversely, assume that $IJ = I \cap J$ for any right ideal I and any left ideal J. For any $a \in R$, we have then

$$a \in (aR) \cap (Ra) = (aR)(Ra) = aRa.$$

Comment. What about the equation $IJ = I \cap J$ for left ideals I and right ideals J (or for right ideals I, J)? The answer is a rather pleasant surprise: *the equation holds in these other cases iff R is a von Neumann regular ring without nonzero nilpotent elements* (a so-called *strongly regular* ring). See Exercises 12.6A and 22.4B below!

Ex. 4.14A$_1$. Let $R = \operatorname{End}_k(M)$ where M is a right module over a ring k. Show that an element $f \in R$ is von Neumann regular iff $\ker(f)$ and $\operatorname{im}(f)$ are both direct summands of M.

Solution. In *FC*-(4.27), it was shown that, if M_k is semisimple, then R is von Neumann regular. This exercise may be viewed as a generalization of that result. In fact, the solution here is just a sharpening of the proof of *FC*-(4.27). First, suppose $f \in R$ is such that

$$\ker(f) \oplus P = M = \operatorname{im}(f) \oplus Q,$$

where P, Q are k-submodules of M. Then

$$f \mid P : P \to \mathrm{im}(f)$$

is an isomorphism. Defining $g \in R$ such that $g(Q) = 0^{(*)}$ and $g \mid \mathrm{im}(f)$ is the inverse of $f \mid P$, we get $f = fgf$. Conversely, if $f \in R$ can be written as fgf for some $g \in R$, then the surjection $M \xrightarrow{f} \mathrm{im}(f)$ is split by the map g (since, for every $m \in M$, $fg(f(m)) = f(m)$). This shows that

$$M = \ker(f) \oplus \mathrm{im}(gf).$$

Dually, the injection $\mathrm{im}(f) \to M$ is split by the map fg, so

$$M = \mathrm{im}(f) \oplus \ker(fg).$$

Comment. In the case where M is a projective k-module, we may simplify the "iff" condition to "$\mathrm{im}(f)$ is a direct summand of M." For, if this holds, then $\mathrm{im}(f)$ is also a projective module, and hence the surjection $f : M \to \mathrm{im}(f)$ splits, which implies that $\ker(f)$ is a direct summand of M. Dually, one can show that, in case M is an injective k-module, then the "iff" condition in this exercise can be simplified to "$\ker(f)$ is a direct summand of M." In general, of course, neither simplification is possible.

Ex. 4.14A$_2$. (1) If R is a commutative principal ideal domain, show that the von Neumann regular elements of $\mathbb{M}_n(R)$ are precisely matrices of the form UAV, where $U, V \in \mathrm{GL}_n(R)$, and $A = \mathrm{diag}(1, \ldots, 1, 0, \ldots, 0)$.

(2) Let $B = \begin{pmatrix} 4 & 9 & 2 \\ 3 & 5 & 7 \\ 8 & 1 & 6 \end{pmatrix}$ ("Chinese Magic Square")$^{(\dagger)}$,

or $B = \begin{pmatrix} 1 & 1 & 0 \\ 2 & 5 & 2 \\ -2 & -8 & -4 \end{pmatrix}$. Is B a von Neumann regular element in the ring

$\mathbb{M}_3(\mathbb{Z})$? If it is, find *all* matrices $X \in \mathbb{M}_3(\mathbb{Z})$ such that $B = BXB$.

Solution. (1) First observe that, if u, v are units in any ring S, then *an element $a \in S$ is von Neumann regular iff uav is.* (To see this, it suffices to prove the "only if" part. Suppose $a = asa$ for some $s \in S$; as in Ex. 12D,

---

$^{(*)}$ We take $g(Q) = 0$ only for convenience. In fact, $g \mid Q$ could have been taken to be *any* homomorphism from Q to M.

$^{(\dagger)}$ In the words of A. Bremner (Notices AMS *50* (2003), p. 357), "Magic squares are ancient and common to several civilizations. They were sometimes used as amulets and viewed as possessing talismanic powers. The first references appear to be Chinese, with the familiar square $\begin{pmatrix} 4 & 9 & 2 \\ 3 & 5 & 7 \\ 8 & 1 & 6 \end{pmatrix}$ known as the lo shu."

we call any such s a "*pseudo-inverse*" of a. To find a pseudo-inverse t for uav, we need to solve the equation: $uav = (uav)t(uav)$. This amounts to $a = a(vtu)a$, so it suffices to choose $t = v^{-1}su^{-1}$, and *all pseudo-inverses for uav arise in this way*.) Applying this remark to $M_n(R)$, it follows that, for U, V, A as in (1), UAV is von Neumann regular (since $A = A^2 = A^3$ is). Conversely, let B be any von Neumann regular element in $M_n(R)$. By the Smith Normal Form Theorem, there exist P, $Q \in GL_n(R)$ such that

$$PBQ = \text{diag}(b_1, \ldots, b_n),$$

where $b_{i+1} \in b_i R$ for each $i < n$. By the remark above, $\text{diag}(b_1, \ldots, b_n)$ is von Neumann regular. Thus, upon interpreting $M_n(R)$ as $\text{End}_R(R^n)$,

$$\text{im}(\text{diag}(b_1, \ldots, b_n)) = b_1 R \oplus \cdots \oplus b_n R \subseteq R^n$$

must be a direct summand of R^n (by Ex. 4.14A$_1$). This is possible only if each $b_i R$ is either 0 or R. Since the "elementary divisors" b_i are determined up to units anyway, we may assume that

$$\text{diag}(b_1, \ldots, b_n) = \text{diag}(1, \ldots, 1, 0, \ldots, 0) := A,$$

and so $B = UAV$ for $U = P^{-1}$ and $V = Q^{-1}$ in $GL_n(R)$.

(2) Here, $R = \mathbb{Z}$. If B is the Chinese Magic Square (with magic sum 15), we have $\det(B) = 15 \times 24$. Computing determinants shows that B cannot be written in the form BXB in $M_3(\mathbb{Z})$, so B is *not* von Neumann regular.

   Next, let B be the second matrix in (2). For this choice of B, we offer two different solutions. First, by performing elementary row and column operations on B, we can easily bring it to the Smith Normal Form. Suppressing the calculations, we have, e.g., $PBQ = \text{diag}(1, 1, 0)$, where

$$P = \begin{pmatrix} 1 & 0 & 0 \\ -2 & 1 & 0 \\ -2 & 2 & 1 \end{pmatrix}, \quad \text{and} \quad Q = \begin{pmatrix} 1 & -1 & 2 \\ 0 & 1 & -2 \\ 0 & -1 & 3 \end{pmatrix} \quad \text{(in } GL_3(\mathbb{Z})\text{)}.$$

By (1), B is von Neumann regular, and the pseudo-inverses for B are of the form QYP, where Y ranges over the pseudo-inverses for the idempotent matrix $A = \text{diag}(1, 1, 0)$. By inspection of the equation $AYA = A$, we see easily that Y can be any matrix of the form $\begin{pmatrix} 1 & 0 & c \\ 0 & 1 & d \\ a & b & e \end{pmatrix}$. Thus, the pseudo-inverses of B are expressed in the form of a 5-parameter linear family QYP, with integer parameters a, b, c, d, e. If we set all parameters equal to 0 (that is, taking $Y = A$), we get the following specific pseudo-inverse for B:

$$(*) \quad QYP = QAP = \begin{pmatrix} 1 & -1 & 0 \\ 0 & 1 & 0 \\ 0 & -1 & 0 \end{pmatrix} \begin{pmatrix} 1 & 0 & 0 \\ -2 & 1 & 0 \\ -2 & 2 & 1 \end{pmatrix} = \begin{pmatrix} 3 & -1 & 0 \\ -2 & 1 & 0 \\ 2 & -1 & 0 \end{pmatrix}.$$

For a second solution, we note that the method of proof for Ex. 4.14A$_1$ is in fact good enough for "constructing" all pseudo-inverses for $B \in \text{End}(\mathbb{Z}^3)$, as soon as we know that B is von Neumann regular. Starting from scratch, let $\{e_1, e_2, e_3\}$ be the unit vector basis for $\mathbb{Z}^3$, and let β_1, β_2, β_3 be the three columns of B. To see that $B \in \text{End}(\mathbb{Z}^3)$ is von Neumann regular, we need to determine if the subgroup $\text{im}(B)$ generated by $\{\beta_1, \beta_2, \beta_3\}$ splits in $\mathbb{Z}^3$ (see *Comment* on Ex. 4.14A$_1$). By column transformations, we have

$$
B = \begin{pmatrix} 1 & 1 & 0 \\ 2 & 5 & 2 \\ -2 & -8 & -4 \end{pmatrix} \mapsto \begin{pmatrix} 1 & 0 & 0 \\ 2 & 3 & 2 \\ -2 & -6 & -4 \end{pmatrix} \mapsto \begin{pmatrix} 1 & 0 & 0 \\ 2 & 1 & 2 \\ -2 & -2 & -4 \end{pmatrix}.
$$

Thus, $\text{im}(B)$ is generated by $\beta_1 = (1, 2, -2)^t$ and $\beta := (0, 1, -2)^t = \beta_2 - \beta_1 - \beta_3$. Since $\{\beta_1, \beta, e_3\}$ is (clearly) a basis for $\mathbb{Z}^3$, $\text{im}(B)$ splits in $\mathbb{Z}^3$, so B is von Neumann regular. From the solution to Ex. 4.14A$_1$, it follows that *all* pseudo-inverses X for B can be determined by taking $X \mid \text{im}(B)$ to be any splitting of $B : \mathbb{Z}^3 \to \text{im}(B)$, and taking $X \mid e_3\mathbb{Z}$ to be arbitrary. By an easy computation, we have $\ker(B) = (-2, 2, -3)^t \cdot \mathbb{Z}$. Since

$$
\beta_1 = B(e_1), \quad \text{and} \quad \beta = \beta_2 - \beta_1 - \beta_3 = B(e_2 - e_1 - e_3),
$$

the most general pseudo-inverse $X \in \text{End}(\mathbb{Z}^3)$ is defined by:

$$
X(\beta_1) = \begin{pmatrix} 1 \\ 0 \\ 0 \end{pmatrix} + w \begin{pmatrix} -2 \\ 2 \\ -3 \end{pmatrix}, \quad X(\beta) = \begin{pmatrix} -1 \\ 1 \\ -1 \end{pmatrix} + v \begin{pmatrix} -2 \\ 2 \\ -3 \end{pmatrix} = \begin{pmatrix} -1 - 2v \\ 1 + 2v \\ -1 - 3v \end{pmatrix},
$$

and $X(e_3) = (x, y, z)^t$, for integer parameters v, w, x, y, z. Now $\beta = e_2 - 2e_3$, so the equations above give

$$
X(e_2) = X(\beta + 2e_3) = \begin{pmatrix} -1 - 2v + 2x \\ 1 + 2v + 2y \\ -1 - 3v + 2z \end{pmatrix}.
$$

Finally, $\beta_1 = e_1 + 2e_2 - 2e_3$ yields

$$
X(e_1) = X(\beta_1) - 2X(e_2) + 2X(e_3) = \begin{pmatrix} 3 - 2w + 4v - 2x \\ -2 + 2w - 4v - 2y \\ 2 - 3w + 6v - 2z \end{pmatrix}.
$$

Introducing a new parameter $u = w - 2v$, we see that a pseudo-inverse for B has the following general form:

$$
X = \begin{pmatrix} 3 - 2u - 2x & -1 - 2v + 2x & x \\ -2 + 2u - 2y & 1 + 2v + 2y & y \\ 2 - 3u - 2z & -1 - 3v + 2z & z \end{pmatrix}.
$$

For $u = v = x = y = z = 0$, this retrieves, by a "nice coincidence", the pseudo-inverse obtained earlier in $(*)$!

Ex. 4.14A$_3$. For any ring R, show that the following statements are equivalent:

(1) R is semisimple.
(2) The endomorphism ring of every right R-module M is von Neumann regular.
(3) The endomorphism ring of some infinite-rank free right R-module F is von Neumann regular.
(4) Any countably generated right ideal $A \subseteq R$ splits in R_R.

Solution. (1) $\Longrightarrow$ (2) follows from Ex. 14A$_1$ since M_R is semisimple, and (2) $\Longrightarrow$ (3) is a tautology.

(3) $\Longrightarrow$ (4). Let F and A be as in (3) and (4). Certainly, A is the image of some R-homomorphism from F to R. If we identify R_R with a fixed rank 1 free direct summand of F, then A is the image of some R-endomorphism of F. Since $\text{End}(F_R)$ is von Neumann regular, Ex. 4.14A$_1$ implies that A is a direct summand of F_R, and hence of R.

(4) $\Longrightarrow$ (1). Under assumption (4), R is certainly von Neumann regular. By FC-(4.25), (1) will follow if we can show that R is right noetherian. Assume, instead, that there exists a strictly ascending chain of right ideals in R. From this, we can construct a strictly ascending chain

$$(*) \qquad a_1R \subsetneq a_1R + a_2R \subsetneq a_1R + a_2R + a_3R \subsetneq \cdots \quad \text{(in } R\text{)}.$$

The union A of this chain is a countably generated right ideal, and hence a direct summand of R_R by (4). But this means that $A = eR$ for some $e = e^2 \in R$. Since e belongs to some $a_1R + \cdots + a_nR$, the chain $(*)$ stabilizes after n steps—a contradiction.

Comment. If R is a von Neumann regular ring, it can be shown that the endomorphism ring of a free module of finite rank is also von Neumann regular: see Ex. 6.20 and Ex. 21.10B below. The problem whether the same holds for free modules *of infinite rank* was raised in L.A. Skornyakov's book "Complemented Modular Lattices and Regular Rings," Oliver and Boyd, Edinburgh-London, 1964. (See Problem 16 on p. 165.) The negative answer, together with the main parts of this exercise, appeared in G.M. Cukerman's paper, "Rings of endomorphisms of free modules" (in Russian), Sibersk. Mat. Ž. 7 (1966), 1161–1167. The same results were rediscovered a few years later by R.F. Shanny, in "Regular endomorphism rings of free modules," J. London Math. Soc. 4 (1971), 353–354. I thank K. Goodearl for bringing this exercise to my attention, and for pointing out the relevant literature referenced above.

Ex. 4.14B. For any ring R, show that the following are equivalent:

(1) For any $a \in R$, there exists a unit $u \in \text{U}(R)$ such that $a = aua$.
(2) Every $a \in R$ can be written as a unit times an idempotent.
(2') Every $a \in R$ can be written as an idempotent times a unit.

If R satisfies (1), it is said to be *unit-regular*.

(3) Show that any unit-regular ring R is Dedekind-finite.

Solution. By left-right symmetry, it suffices to prove (1) $\Longleftrightarrow$ (2).

(1) $\Rightarrow$ (2). Write $a = aua$ where $u \in U(R)$. If $e := ua$, then

$$e^2 = uaua = ua = e,$$

and $a = u^{-1}e$, as desired.

(2) $\Rightarrow$ (1). Given $a \in R$, write $a = ve$ where $v \in U(R)$ and $e^2 = e$. The latter implies $v^{-1}a = v^{-1}av^{-1}a$, so $a = av^{-1}a$, as desired.

(3) Suppose $ab = 1 \in R$, where R is unit-regular. Write $a = aua$, where $u \in U(R)$. Then $1 = ab = auab = au$, so $a = u^{-1} \in U(R)$.

Comment. (1) The important notion of a unit-regular ring was introduced by G. Ehrlich: see her paper in Portugal. Math. *27* (1968), 209–212.

(2) Note that, by Exercise 3.10, any semisimple ring is unit-regular.

(3) In general, a von Neumann regular ring R may not be Dedekind-finite. Therefore, R may fail to be unit-regular.

Ex. 4.14C. (Ehrlich, Handelman) Let M be a right module over a ring k such that $R = \operatorname{End}_k(M)$ is von Neumann regular. Show that R is unit-regular iff, whenever $M = K \oplus N = K' \oplus N'$ (in the category of k-modules), $N \cong N'$ implies $K \cong K'$.

Solution. The proof of the sufficiency part is a direct modification of the argument used in Exercise 14A$_1$. For $a \in R$, write

$$M = \ker(a) \oplus P = Q \oplus \operatorname{im}(a)$$

as in that exercise. Since a defines an isomorphism from P to $\operatorname{im}(a)$, the hypothesis implies that $\ker(a) \cong Q(\cong \operatorname{coker}(a))$. Defining $u \in U(R)$ such that u is an isomorphism from Q to $\ker(a)$, and $u : \operatorname{im}(a) \to P$ is the inverse of $a|P : P \to \operatorname{im}(a)$, we have $a = aua \in R$.

For the necessity part, assume R is unit-regular. Suppose

$$M = K \oplus N = K' \oplus N',$$

where $N \cong N'$. Define $a \in R$ such that $a(K) = 0$ and $a|N$ is a fixed isomorphism from N to N'. Write $a = aua$, where $u \in U(R)$. As in Exercise 14A$_1$,

$$(*) \qquad M = \ker(a) \oplus \operatorname{im}(ua) = K \oplus u(N').$$

Since u defines an isomorphism from N' to $u(N')$, it induces an isomorphism from M/N' to $M/u(N')$. Noting that $M/N' \cong K'$ and $M/u(N') \cong K$ (from $(*)$), we conclude that $K \cong K'$.

Comment. Under the assumption that R is von Neumann regular, the above argument also suffices to show that: $a \in R$ can be written in the form aua where $u \in U(R)$ iff $\ker(a) \cong \operatorname{coker}(a)$ as k-modules.

For the relevant literature, see G. Ehrlich, "Units and one-sided units in regular rings," Trans. AMS *216* (1976), 81–90, and D. Handelman, "Perspectivity and cancellation in regular rings", J. Algebra *48* (1977), 1–16.

Ex. 4.14D. Let M be a semisimple right k-module. Show that $R = \operatorname{End}_k(M)$ is unit-regular iff the isotypic components M_i of M (as defined in Exercise 2.8) are all finitely generated.

Solution. First assume that some isotypic component, say M_1, is not finitely generated. Then M_1 is an *infinite* direct sum of a simple k-module, so it is easy to find an epimorphism $f_1 : M_1 \to M_1$ which is not an isomorphism. Extending f_1 by the identity map on the other M_i's, we get an $f : M \to M$ which is an epimorphism but not an isomorphism. Now for any splitting g for f, we have $fg = 1 \neq gf$. By Exercise 14B(3), R cannot be unit-regular.

Next, assume all M_i's are finitely generated. By Exercise 2.8,

$$M = K \oplus N = K' \oplus N' \quad \text{and} \quad N \cong N' \Longrightarrow K \cong K'.$$

By Exercises 14A$_1$ and 14C, we conclude that R is unit-regular. Alternatively, we can give a more direct argument. Since M_i is finitely generated, $R_i := \operatorname{End}_k(M_i)$ is a simple artinian ring, so it is unit-regular by Exercise 3.10(3). It is easy to see that

$$R = \operatorname{End}_k(\oplus_i M_i) \cong \prod_i R_i,$$

so it follows that R is also unit-regular.

Ex. 4.15. For a commutative ring R, show that the following are equivalent:

(1) R has Krull dimension 0.[3]
(2) rad R is nil and $R/\text{rad } R$ is von Neumann regular.
(3) For any $a \in R$, the descending chain $Ra \supseteq Ra^2 \supseteq \dots$ stabilizes.
(4) For any $a \in R$, there exists $n \geq 1$ such that a^n is regular (i.e. such that $a^n \in a^n Ra^n$).

Specializing the above result, show that the following are also equivalent:

(A) R is reduced (no nonzero nilpotents), and K-dim $R = 0$.
(B) R is von Neumann regular.
(C) The localizations of R at its maximal ideals are all fields.

---

[3] Recall that the *Krull dimension* of a commutative ring R is the supremum of the lengths of chains of prime ideals in R. In particular, K-dim $R = 0$ means that all prime ideals in R are maximal ideals.

Solution. (1) $\Rightarrow$ (2). By (1), rad R is the intersection of all prime ideals, so it is nil. For the rest, we may assume that rad $R = 0$. For any $a \in R$, it suffices to show that $Ra = Ra^2$. Let $\mathfrak{p}$ be any prime ideal of R. Since R is reduced, so is $R_{\mathfrak{p}}$. But $\mathfrak{p}R_{\mathfrak{p}}$ is the only prime ideal of $R_{\mathfrak{p}}$, so we have $\mathfrak{p}R_{\mathfrak{p}} = 0$. Therefore, $R_{\mathfrak{p}}$ is a field. In particular, $R_{\mathfrak{p}}a = R_{\mathfrak{p}}a^2$. It follows that $(Ra/Ra^2)_{\mathfrak{p}} = 0$ for every prime ideal $\mathfrak{p} \subset R$; hence $Ra = Ra^2$, as desired.

(2) $\Rightarrow$ (3). Let $a \in R$. By (2), $\bar{a} = \bar{a}^2\bar{b} \in R/\text{rad } R$ for some $b \in R$, so $(a - a^2b)^n = 0$ for some $n \geq 1$. Expanding the LHS and transposing, we get $a^n \in Ra^{n+1}$, and hence $Ra^n = Ra^{n+1} = \cdots$.

(3) $\Rightarrow$ (4) is clear.

(4) $\Rightarrow$ (1). Let $\mathfrak{p}$ be any prime ideal, and $a \notin \mathfrak{p}$. By (4), $a^n = a^{2n}b$ for some $b \in R$ and some $n \geq 1$. Then $a^n(1 - a^nb) = 0$ implies that $1 - a^nb \in \mathfrak{p}$. This shows that $R/\mathfrak{p}$ is a field, so $\mathfrak{p}$ is a maximal ideal.

Upon specializing to reduced rings, (1) becomes (A) and (2) becomes (B), so we have (A) $\Leftrightarrow$ (B). The implication (A) $\Rightarrow$ (C) is already done in (1) $\Rightarrow$ (2) above, and (C) $\Rightarrow$ (A) follows from a similar standard local-global argument in commutative algebra.

Comment. This exercise is related to Exercise 14B, in that *any commutative von Neumann regular ring is unit-regular* in the sense of that exercise. In fact, as long as all idempotents are central in a von Neumann regular ring R, then R is unit-regular: see Exercises 12.6A and 12.6C below.

For more information on the conditions (3) and (4) in the *noncommutative* case, see the *Comments* on Exercises 4.17 and 23.5 below.

Ex. 4.16. (Cf. Exercise 1.12) A left R-module M is said to be *cohopfian* if any injective R-endomorphism of M is an automorphism.

(1) Show that any artinian module M is cohopfian.
(2) Show that the left regular module $_RR$ is cohopfian iff every non right-0-divisor in R is a unit. In this case, show that $_RR$ is also hopfian.

Solution. (1) Let $\alpha : M \to M$ be injective, and M be artinian. The descending chain

$$\text{im}(\alpha) \supseteq \text{im}(\alpha^2) \supseteq \cdots$$

must stabilize, so $\text{im}(\alpha^i) = \text{im}(\alpha^{i+1})$ for some i. For any $m \in M$, we have $\alpha^i(m) = \alpha^{i+1}(m')$ for some $m' \in M$. But then $\alpha^i(m - \alpha(m')) = 0$ implies that $m = \alpha(m')$, so $\alpha \in \text{Aut}_R(M)$.

(2) The first statement is clear since injective endomorphisms of $_RR$ are given by right multiplications by non right-0-divisors, and automorphisms of $_RR$ are given by right multiplications by units. Now suppose non right-0-divisors are units, and suppose $ab = 1$. Then

$$xa = 0 \Longrightarrow xab = 0 \Longrightarrow x = 0,$$

so a is not a right-0-divisor. It follows that $a \in U(R)$, so we have shown that R is Dedekind-finite. By Exercise 1.12(2), $_RR$ is hopfian.

Comment. The fact that $_RR$ is cohopfian $\implies$ $_RR$ is hopfian may be viewed as an analogue of the fact that $_RR$ is artinian $\implies$ $_RR$ is noetherian. It is, however, easy to see that $_RR$ is hopfian need not imply that $_RR$ is cohopfian (e.g. take $R = \mathbb{Z}$). Note also that, for R-modules in general, $_RM$ is cohopfian $\not\Rightarrow$ $_RM$ is hopfian (e.g. take the module

$$M = \{\zeta \in \mathbb{C} : \zeta^{2^i} = 1 \text{ for some } i\}$$

over $\mathbb{Z}$). Finally, one can also show that $_RR$ cohofian $\not\Rightarrow$ R_R cohopfian.

Some classes of rings R have the property that every finitely generated left R-module is cohopfian. By (1) of the Exercise, left artinian rings have this property. Another class of rings with this property is the class of re-duced von Neumann regular rings. (Over such rings, finitely generated left modules are also hopfian. For the proofs, see (3.2) and (6.16) of Goodearl's book on von Neumann regular rings, Krieger Publ. Co., Malabar, Florida, 1991.) For more information on these matters, see the next exercise, as well as Exercise 20.9 and Exercises 23.8–23.11.

Ex. 4.16*. Let $\varphi : R \to S$ be a ring homomorphism such that S is finitely generated when it is viewed as a left R-module via φ. If, over R, all finitely generated left modules are hopfian (resp. cohopfian), show that the same property holds over S.

Solution. Let $f : M \to M$ be a surjective (resp. injective) endomorphism of a finitely generated left S-module M. Via φ, we may view M as a left R-module, and, since $_RS$ is finitely generated, so is $_RM$. Viewing $f : M \to M$ as a surjective (resp. injective) R-homomorphism, we infer from the assumption on R that f is an R-isomorphism, and hence an S-isomorphism.

Ex. 4.17. Let R be a ring in which all descending chains

$$Ra \supseteq Ra^2 \supseteq Ra^3 \supseteq \cdots \qquad \text{(for } a \in R\text{)}$$

stabilize. Show that R is Dedekind-finite, and every non right-0-divisor in R is a unit.

Solution. By (2) of Exercise 4.16, it is sufficient to show that $_RR$ is cohop-fian. Let $\alpha : {}_RR \to {}_RR$ be an injective R-endomorphism. Then α is right multiplication by $a := \alpha(1)$, and α^i is right multiplication by a^i. Therefore, $\operatorname{im}(\alpha^i) = Ra^i$. Since the chain

$$Ra \supseteq Ra^2 \supseteq \cdots$$

stabilizes, the argument in (1) of Exercise 4.16 shows that α is an isomor-phism.

Comment. Rings in which all descending chains

$$Ra \supseteq Ra^2 \supseteq Ra^3 \supseteq \cdots$$

stabilize are known as *strongly π-regular rings*. In Exercise 23.5, it will be shown that this is a left-right symmetric notion. It follows that, *in such a ring, "left-0-divisor" and "right-0-divisor" are both synonymous with "non-unit."* In particular, this is the case for any left artinian ring. For a direct proof of this last fact, see Exercise 21.23. There are, however, no analogues of the above results for (even 2-sided) noetherian rings. For instance, in the noetherian ring

$$R = \begin{pmatrix} \mathbb{Z} & \mathbb{Z}/2\mathbb{Z} \\ 0 & \mathbb{Z} \end{pmatrix},$$

the element $a = \begin{pmatrix} 2 & 0 \\ 0 & 1 \end{pmatrix}$ is not a right 0-divisor. But it is a left 0-divisor, and *à fortiori* a nonunit (see *FC*-p. 3).

Among commutative rings, the strongly π-regular ones are just those with Krull dimension 0, according to Exercise 4.15.

Ex. 4.18. The *socle* soc(M) of a left module M over a ring R is defined to be the sum of all simple submodules of M. Show that

$$\text{soc}(M) \subseteq \{m \in M : (\text{rad } R) \cdot m = 0\},$$

with equality if $R/\text{rad } R$ is an artinian ring.

Solution. The first conclusion follows from the fact that $(\text{rad } R)V = 0$ for any simple left R-module V. Now assume $R/\text{rad } R$ is artinian. Let

$$N = \{m \in M : (\text{rad } R) \cdot m = 0\},$$

which is an R-submodule of M. Viewing N as a module over the semisimple ring $R/\text{rad } R$, we see that $_R N$ is semisimple. Therefore, $N \subseteq \text{soc}(M)$, as desired.

Comment. If $R/\text{rad } R$ is artinian, R is said to be a *semilocal ring*. If R is not semilocal, the submodule N defined above may be *larger* than soc(M). For instance, for $R = \mathbb{Z}$, we have rad $R = 0$, so N is always equal to M. However, if M is not semisimple, we have soc$(M) \subsetneq M$.

Ex. 4.19. Show that for any ring R, soc$(_R R)$ ($=$ sum of all minimal left ideals of R) is an ideal of R. Using this, give a new proof for the fact that if R is a simple ring which has a minimal left ideal, then R is a semisimple ring.

Solution. For any minimal left ideal $I \subseteq R$ and any $r \in R$, Ir is a homomorphic image of $_R I$, so Ir is either 0 or another minimal left ideal. Since soc$(_R R) = \sum I$, we see that soc$(_R R)$ is an ideal. Now suppose R

is a simple ring which has a minimal left ideal. Then $\mathrm{soc}(_R R) \neq 0$ and so $\mathrm{soc}(_R R) = R$. This means that $_R R$ is a semisimple module, so (by FC-(2.5)) R is a semisimple ring.

Ex. 4.20. For any left artinian ring R with Jacobson radical J, show that

$$\mathrm{soc}(_R R) = \{r \in R: \ Jr = 0\} \quad \text{and} \quad \mathrm{soc}(R_R) = \{r \in R: \ rJ = 0\}.$$

Using this, construct an artinian ring R in which $\mathrm{soc}(_R R) \neq \mathrm{soc}(R_R)$.

Solution. Since $R/\mathrm{rad}\ R$ is artinian, the two desired equations follow by applying Exercise 18 (and its right analogue) to the modules $_R R$ and R_R. To construct an artinian ring R for which the two socles differ, take $R = \begin{pmatrix} \mathbb{Q} & \mathbb{Q} \\ 0 & \mathbb{Q} \end{pmatrix}$. For this 3-dimensional $\mathbb{Q}$-algebra, we have $\mathrm{rad}\ R = \begin{pmatrix} 0 & \mathbb{Q} \\ 0 & 0 \end{pmatrix}$, which has right annihilator $\begin{pmatrix} \mathbb{Q} & \mathbb{Q} \\ 0 & 0 \end{pmatrix}$ and left annihilator $\begin{pmatrix} 0 & \mathbb{Q} \\ 0 & \mathbb{Q} \end{pmatrix}$, so

$$\mathrm{soc}(_R R) \neq \mathrm{soc}(R_R).$$

Comment. Actually, another choice of the triangular ring gives a smaller, and more interesting, example. Let R be the triangular ring (of cardinality 2^{k+2}, $k \geq 2$) in Ex. 1.22B. By the solution to that exercise, we have $\mathrm{soc}(_R R) \subseteq \mathrm{soc}(R_R)$, with cardinalities 4 and 8 respectively.

Ex. 4.21. For any ring R, let $\mathrm{GL}_n(R)$ denote the group of units of $\mathbb{M}_n(R)$. Show that for any ideal $I \subseteq \mathrm{rad}\ R$, the natural map $\mathrm{GL}_n(R) \to \mathrm{GL}_n(R/I)$ is surjective.

Solution. First consider the case $n = 1$. For $x \in R$, the argument in FC-(4.8) shows that $x \in \mathrm{U}(R)$ iff $\bar{x} \in \mathrm{U}(R/I)$. Therefore, $\mathrm{U}(R) \to \mathrm{U}(R/I)$ is surjective. Applying this to the matrix ring $\mathbb{M}_n(R)$ and its ideal

$$\mathbb{M}_n(I) \subseteq \mathbb{M}_n(\mathrm{rad}\ R) = \mathrm{rad}\ \mathbb{M}_n(R) \quad (\text{see } FC\text{-p. 57}),$$

we see that

$$\mathrm{U}(\mathbb{M}_n(R)) \to \mathrm{U}(\mathbb{M}_n(R)/\mathbb{M}_n(I)) = \mathrm{U}(\mathbb{M}_n(R/I))$$

is onto; that is, $\mathrm{GL}_n(R) \to \mathrm{GL}_n(R/I)$ is onto.

Comment. The hypothesis $I \subseteq \mathrm{rad}\ R$ in the Exercise is essential. Without this hypothesis, even $\mathrm{U}(R) \to \mathrm{U}(R/I)$ need not be onto. For instance, take $R = \mathbb{Z}$ and $I = p\mathbb{Z}$, where p is any prime number ≥ 5.

Ex. 4.22. Using the definition of $\mathrm{rad}\ R$ as the intersection of the maximal left ideals, show directly that $\mathrm{rad}\ R$ is an ideal.

Solution. For $y \in \mathrm{rad}\ R$, $r \in R$, and $\mathfrak{m}$ any maximal left ideal, we must show that $yr \in \mathfrak{m}$. Assume otherwise; then $Rr + \mathfrak{m} = R$. Consider the left R-module homomorphism $\varphi: R \to R/\mathfrak{m}$ defined by $\varphi(x) = \overline{xr}$. Since

$Rr + \mathfrak{m} = R$, φ is onto. This implies that $\ker(\varphi)$ is a maximal left ideal. Therefore, $y \in \ker(\varphi)$, so we have $0 = \varphi(y) = \overline{y}\overline{r}$, a contradiction.

Comment. With the reader's indulgence, we return for a moment to rings R possibly without an identity. Let J_ℓ (resp. J_r) be the intersection of the maximal left (resp. right) ideals of R. Kaplansky pointed out that J_ℓ (resp. J_r) *is always an ideal of R.* In fact, this follows from the proof above, since the identity element was never used! In view of Exercise 5(b), we have $J_\ell, J_r \subseteq \operatorname{rad} R$, and, if R has a left (resp. right) identity, then

$$J_\ell \subseteq J_r = \operatorname{rad} R \quad (\text{resp. } J_r \subseteq J_\ell = \operatorname{rad} R).$$

However, as also noted by Kaplansky, $J_\ell \neq J_r$ in general. For instance, let R be the "rng" of cardinality p^2 constructed in Exercise 1.10(b). It is easy to check that the only right ideals are (0), R, and

$$\mathfrak{m} = \{(0, d): \ d \in \mathbb{F}_p\}.$$

Since R has a left identity $(1, 0)$, we have $J_r = \operatorname{rad} R = \mathfrak{m}$. But $\mathfrak{m}$ is a maximal left ideal, and

$$\mathfrak{m}_o = \{(d, d): \ d \in \mathbb{F}_p\}$$

is another, so $J_\ell = (0)$. (It follows that $\mathfrak{m}$ is the only *modular* maximal left ideal.)

Ex. 4.23. (Herstein) In commutative algebra, it is well known (as a consequence of Krull's Intersection Theorem) that, for any commutative noetherian R, $\bigcap_{n \geq 1} (\operatorname{rad} R)^n = 0$. Show that this need not be true for noncommutative right noetherian rings.

Solution. Let A be a commutative discrete valuation ring with a uniformizer $\pi (\neq 0)$ and quotient field K. Consider the ring $R = \begin{pmatrix} A & K \\ 0 & K \end{pmatrix}$, which is right noetherian (but not left noetherian) by FC-(1.22). It is easy to check that

$$J: = \begin{pmatrix} \pi A & K \\ 0 & 0 \end{pmatrix}$$

is an ideal of R, and that

$$R/J \cong (A/\pi A) \times K.$$

Since the latter is a semisimple ring, we have $\operatorname{rad} R \subseteq J$ by Exercise 11. On the other hand, $1 + J$ consists of matrices of the form

$$\begin{pmatrix} 1 + \pi a & b \\ 0 & 1 \end{pmatrix} \quad (a \in A,\ b \in K),$$

which are clearly units of R. Therefore, $J \subseteq \operatorname{rad} R$. We have now $J = \operatorname{rad} R$, from which it is easy to see that

$$(\operatorname{rad} R)^n = \begin{pmatrix} \pi^n A & K \\ 0 & 0 \end{pmatrix},$$

for any $n \geq 1$. It follows that

$$\bigcap_{n \geq 1} (\operatorname{rad} R)^n = \begin{pmatrix} 0 & K \\ 0 & 0 \end{pmatrix} \neq 0.$$

Comment. In the above example, $R/\operatorname{rad} R \cong (A/\pi A) \times K$ is semisimple, so R is a *semilocal* ring in the sense of FC-§20. It is also possible to construct examples R which are *local* rings: see Exercise 19.12.

If R is a *left and right* noetherian ring, it has been conjectured that

$$\bigcap_{n \geq 1} (\operatorname{rad} R)^n = 0 \quad (\text{"Jacobson's Conjecture"}).$$

This has been verified for various special classes of noetherian rings, by Lenagan, Jategaonkar, and others. However, the general case has remained open for years.

Ex. 4.24. For any ring R, we know that

$(*)$ $\qquad\qquad \operatorname{rad}(R) \subseteq \{r \in R : r + U(R) \subseteq U(R)\}.$

Give an example to show that this need not be an equality.

Solution. Let $I(R)$ be the set on the *RHS*, and consider the case $R = A[t]$ where A is a commutative domain. It is easy to see that $\operatorname{rad}(R) = 0$. However, $I(R)$ contains $\operatorname{rad}(A)$, since $a \in \operatorname{rad}(A)$ implies that

$$a + U(R) = a + U(A) \subseteq U(A) = U(R).$$

Thus, we see that equality in $(*)$ does not hold for $R = A[t]$ if we choose A to be any commutative domain that is not J-semisimple.

Comment. This exercise might be captioned "How *not* to characterize the Jacobson radical"! I have included this exercise since it has been claimed in a standard ring theory text that equality holds in $(*)$. Of course, equality does hold for *some* classes of rings; see, for instance, Ex. 20.10B.

Ex. 4.25. Let R be the commutative $\mathbb{Q}$-algebra generated by $x_1, x_2, \ldots$ with the relations $x_n^n = 0$ for all n. Show that R does not have a largest nilpotent ideal (so there is no "Wedderburn radical" for R).

Solution. It is not hard to show that each x_n has index of nilpotency exactly equal to n in R. Let I be any nilpotent ideal in R; say $I^n = 0$. Then $x_{n+1} \notin I$, and so $I + x_{n+1}R$ is a nilpotent ideal larger than I.

Ex. 4.26. Let R be a commutative $\mathbb{Q}$-algebra generated by $x_1, x_2, \ldots$ with the relations $x_i x_j = 0$ for all i, j. Show that R is semiprimary (that is, rad R is nilpotent, and $R/\mathrm{rad}\, R$ is semisimple), but neither artinian nor noetherian.

Solution. Clearly, rad R is the ideal generated by the x_i's. We have $(\mathrm{rad}\, R)^2 = 0$, and $R/\mathrm{rad}\, R \cong \mathbb{Q}$, so R is semiprimary. The strictly ascending chain

$$(x_1) \subsetneqq (x_1, x_2) \subsetneqq (x_1, x_2, x_3) \subsetneqq \cdots$$

shows that R is not noetherian, while the strictly descending chain

$$(x_1, x_2, \ldots) \supsetneqq (x_2, x_3, \ldots) \supsetneqq (x_3, x_4, \ldots) \supsetneqq \cdots$$

shows that R is not artinian.

Comment. A nice *noncommutative* semiprimary ring satisfying neither *ACC* nor *DCC* on 1-sided ideals can be found in Ex. 20.5.

Ex. 4.27. Let J be a nilpotent right ideal in a ring R. If I is a subgroup of J such that $I \cdot I \subseteq I$ and $J = I + J^2$, show that $I = J$.

Solution. Since J is nilpotent, it suffices to show that $J = I + J^n$ for all $n \geq 1$. We induct on n, the case $n = 1$ being clear. Assume that $J = I + J^n$ where $n \geq 1$. Then

$$J \cdot I = (I + J^n) \cdot I = I \cdot I + J^n \cdot I \subseteq I + J^{n+1}, \text{and hence}$$
$$J^2 = J \cdot (I + J^n) \subseteq I + J^{n+1} + J^{n+1} = I + J^{n+1}.$$

From this, it follows that $J = I + J^2 \subseteq I + J^{n+1}$, as desired.

Comment. If I was a *right ideal*, we can form the right R-module $M = J/I$, so the conclusion in the exercise would follow from the observation that $MJ = M$. The point of this exercise is that, instead of assuming $I \cdot R \subseteq I$, we need only assume $I \cdot I \subseteq I$ to guarantee the conclusion $J = I$. This exercise is distilled from Lemma (3.1) in I.M. Isaacs' paper "Characters of groups associated with finite algebras," J. Algebra *177* (1995), 708–730.

Ex. 4.28. (1) If R is a commutative ring or a left noetherian ring, show that any finitely generated artinian left R-module M has finite length. (2) Construct an example to show that (1) need not hold over an arbitrary ring R.

Solution. (1) If R is left noetherian, then M is a noetherian (as well as artinian) module, so it has finite length. Now assume R is commutative. Since M is a finite sum of cyclic artinian submodules, we may assume M

itself is cyclic. Represent M in the form R/I, where $I \subseteq R$ is a left ideal. Since R is commutative, I is an ideal, and the fact that $_RM$ is artinian implies that R/I is an artinian ring. By the Hopkins–Levitzki theorem $(FC\text{-}(4.15))$, R/I is also a noetherian ring, so we are back to the case considered above.

(2) We construct here a *cyclic* artinian left module M of infinite length over some (noncommutative, non left-noetherian) ring R. In the triangular ring $R = \begin{pmatrix} \mathbb{Q} & 0 \\ \mathbb{Q} & \mathbb{Z} \end{pmatrix}$, the idempotent $e = \begin{pmatrix} 1 & 0 \\ 0 & 0 \end{pmatrix}$ generates the left ideal $Re = \begin{pmatrix} \mathbb{Q} & 0 \\ \mathbb{Q} & 0 \end{pmatrix}$ (which is in fact an ideal). We express this module in the simpler form $\begin{pmatrix} \mathbb{Q} \\ \mathbb{Q} \end{pmatrix}$, and consider its submodule $\begin{pmatrix} 0 \\ \mathbb{Z}_{(p)} \end{pmatrix}$, where $\mathbb{Z}_{(p)}$ denotes the localization of $\mathbb{Z}$ at a prime ideal (p). Since

$$\begin{pmatrix} a & 0 \\ b & c \end{pmatrix} \begin{pmatrix} 0 & 0 \\ x & 0 \end{pmatrix} = \begin{pmatrix} 0 & 0 \\ cx & 0 \end{pmatrix} \quad (a, b \in \mathbb{Q}; \ c \in \mathbb{Z}),$$

the ideal Re acts trivially on $\begin{pmatrix} 0 \\ \mathbb{Q} \end{pmatrix}$, so the R-submodules of $\begin{pmatrix} 0 \\ \mathbb{Q} \end{pmatrix}$ are just $\begin{pmatrix} 0 \\ G \end{pmatrix}$ where G is any subgroup of $\mathbb{Q}$. Now $\mathbb{Q}/\mathbb{Z}_{(p)}$ is isomorphic to the Prüfer p-group (the group of p^n-th roots of unity for $n \in \mathbb{N}$), which is of infinite length as a $\mathbb{Z}$-module. Therefore, the cyclic R-module

$$M := \begin{pmatrix} \mathbb{Q} \\ \mathbb{Q} \end{pmatrix} \Big/ \begin{pmatrix} 0 \\ \mathbb{Z}_{(p)} \end{pmatrix} \supseteq M' = \begin{pmatrix} 0 \\ \mathbb{Q} \end{pmatrix} \Big/ \begin{pmatrix} 0 \\ \mathbb{Z}_{(p)} \end{pmatrix}$$

is also of infinite length. Now $M/M' \cong \begin{pmatrix} \mathbb{Q} \\ \mathbb{Q} \end{pmatrix} \Big/ \begin{pmatrix} 0 \\ \mathbb{Q} \end{pmatrix}$ is a *simple* R-module, and M' is an *artinian* $\mathbb{Z}$-module (and hence an artinian R-module). It follows that M is also an artinian R-module (of infinite length), as desired.

Comment. Since $\begin{pmatrix} a & 0 \\ b & 0 \end{pmatrix} \begin{pmatrix} y \\ z \end{pmatrix} = \begin{pmatrix} ay \\ by \end{pmatrix}$, it is clear that any submodule of M not contained in M' must be M itself. Therefore, the submodules of M are precisely M and the chain of subgroups of the Prüfer group M'. In particular, the submodules of M form a chain (under inclusion). (A module M with such a property is known as *uniserial*: see FC-§20, Appendix.) Another remarkable property of M is that, *for any submodule $X \subsetneq M$ other than M', the quotient module M/X is isomorphic to M.* Indeed, since X is necessarily of the form $\begin{pmatrix} 0 \\ p^{-n}\mathbb{Z}_{(p)} \end{pmatrix} \Big/ \begin{pmatrix} 0 \\ \mathbb{Z}_{(p)} \end{pmatrix}$ (for some $n \geq 0$), it is precisely the kernel of the surjective endomorphism of M given by multiplication by p^n. From this, it follows that $M/X \cong M$ as R-modules.

Uniserial modules M with this property played an interesting role in the work of Facchini, Salce, and others on the uniqueness of Krull-Schmidt decompositions.

§5. Jacobson Radical Under Change of Rings

Given a ring homomorphism $f : R \to S$, one would like to relate rad R with rad S, if it is at all possible. A few basic results are given in FC-§5. Two important cases are when f is an inclusion, and S is a power series ring or a polynomial ring over R, say in one variable t. The case $S = R[[t]]$ turns out to be very easy, and is covered in Exercise 6. The case $S = R[t]$ is much more difficult, and is not yet fully understood. In this case, a good theorem of Amitsur (FC-(5.10)) guarantees that $J = $ rad $R[t]$ contracts to a nil ideal N of R, and that $J = N[t]$. However, Amitsur's Theorem does not say what the ideal N is. "Köthe's Conjecture" amounts to the hypothetical statement that N is the largest nil ideal of R, but this was never proven. For more information on this, see Exercises 10.24 and 10.25.

The two Exercises 3 and 4 below offer a different proof of Amitsur's Theorem in the case when R is an algebra over a field k. Although this is a somewhat restrictive case, the basic ideas of this alternative proof are simple and direct. Exercise 8, on the other hand, presents a generalization of Amitsur's Theorem to graded rings.

Exercises for §5

Ex. 5.0. (This exercise refines some of the ideas used in the proof of FC-(5.6).) For any subring $R \subseteq S$, consider the following conditions: (1) $_R R$ is a direct summand of $_R S$ and R_R is a direct summand of S_R. (2) R is a *full subring* of S in the sense that $R \cap \mathrm{U}(S) \subseteq \mathrm{U}(R)$. (3) $R \cap$ rad $S \subseteq$ rad R.

(A) Show that $(1) \Rightarrow (2) \Rightarrow (3)$.
(B) Deduce from the above that, if $C = Z(S)$ (the center of S), then $C \cap$ rad $S \subseteq$ rad C.
(C) Does equality hold in general in (B)?

Solution. $(1) \Rightarrow (2)$. First assume $_R S = {}_R R \oplus T$, where T is a suitable R-submodule of $_R S$. If $r \in R$ has a right inverse in S, say $r' + t$ where $r' \in R$ and $t \in T$, then

$$1 = r(r' + t) = rr' + rt$$

implies that $1 = rr'$ (since $rr' \in R$ and $rt \in T$). Therefore, r already has a right inverse in R. Repeating this argument for S_R, we see that, if (1) holds, R must be a full subring of S in the sense of (2).

$(2) \Rightarrow (3)$. Let $x \in R \cap$ rad S. For any $r \in R$, $1 + rx$ is a unit in S, so it is a unit in R. By FC-(4.1), this implies that $x \in$ rad R.

This completes the proof of (A). For (B), it suffices to show that C is a full subring of S. Let $c \in C \cap U(S)$. For any $s \in R$, $sc = cs$ yields $c^{-1}s = sc^{-1}$. This shows that $c^{-1} \in C$, so $c \in U(C)$. For (C), note that, in general, rad C may not lie in rad S. For instance, let C be any commutative domain. Then the free algebra $S = C\langle x, y \rangle$ has center C. It is easy to see that rad $S = 0$, but of course rad C need not be zero.

Comment. There is one important case in which the equality $C \cap \text{rad } S = \text{rad } C$ does hold in (C), namely, in the case when S is left or right artinian. See Exercise 21.25 below. (For an easier case, see Exercise 7.4.)

There are many examples of pairs of rings $R \subseteq S$ where R is full in S. See, for instance, Exercises 1.13 and 6.4.

Ex. 5.1. Let R be a commutative domain and $S^{-1}R$ be the localization of R at a multiplicative set S. Determine if any of the following inclusion relations holds:

(a) rad $R \subseteq R \cap$ rad $S^{-1}R$,
(b) $R \cap$ rad $S^{-1}R \subseteq$ rad R,
(c) rad $S^{-1}R \subseteq S^{-1}(\text{rad } R)$.

Solution. None of these is true in general! For (a), take a commutative local domain $(R, \mathfrak{m})$ with $\mathfrak{m} \neq 0$. Then rad $R = \mathfrak{m}$, but for $S = R \setminus \{0\}$, $S^{-1}R$ is a field, with rad $S^{-1}R = 0$. For (b) and (c), take $R = \mathbb{Z}$ and $S = \mathbb{Z} \setminus 2\mathbb{Z}$. Here, rad $S^{-1}R = 2\mathbb{Z}_{(2)}$ and $R \cap$ rad $S^{-1}R = 2\mathbb{Z}$, but rad $R = 0$.

Ex. 5.1*. Given an element a in a commutative ring R, let S be the multiplicative set $1 + aR \subseteq R$.

(1) Show that $a \in \text{rad}(S^{-1}R)$ (where a means $a/1 \in S^{-1}R$).
(2) Does (1) still hold if $S \subseteq R$ is taken to be only a multiplicative set containing $1 + aR$?

Solution. (1) It suffices to show that $1 + R'a \subseteq U(R')$ where $R' = S^{-1}R$. A typical element of $1 + R'a$ has the form

$$b = 1 + (s^{-1}r)a = s^{-1}(s + ra), \quad \text{where } r \in R \text{ and } s \in S.$$

Writing s in the form $1 + r_1 a$ $(r_1 \in R)$, we have

$$b = s^{-1}(1 + (r_1 + r)a) \in s^{-1}S \subseteq U(R').$$

(2) If we assume only $S \supseteq 1 + aR$, we may no longer have $a \in \text{rad}(S^{-1}R)$. For instance, take a to be a nonzero element in a commutative domain R, and take $S = R \setminus \{0\}$ so $S^{-1}R$ is the quotient field of R. Then $\text{rad}(S^{-1}R) = 0$ does not contain a.

Ex. 5.2. Give an example of a ring R with rad $R \neq 0$ but rad $R[t] = 0$.

Solution. Again, take $(R, \mathfrak{m})$ to be a reduced commutative local ring with $0 \neq \mathfrak{m} = \text{rad } R$. By Amitsur's Theorem FC-(5.10) (or Snapper's Theorem FC-(5.1)), the fact that R is reduced implies that rad $R[t] = 0$.

In the following two Exercises, we sketch another proof for Amitsur's Theorem on rad $R[t]$ (FC-(5.10C)), *in the special case when R is an algebra over a field k.* As in FC-(5.10C), we let $S = R[t]$, $J = \operatorname{rad} S$, and $N = R \cap J$. The fact that N is a nil ideal is easy to see (cf. FC-(5.10A)). The main job is that of proving the equation $J = N[t]$.

Ex. 5.3. Assume k is an infinite field. Show that $J = N[t]$.

Solution. We first show that $J \neq 0 \Rightarrow N \neq 0$. Let

$$f(t) = a_o + a_1 t + \cdots + a_n t^n \in J \backslash \{0\}$$

with n chosen minimal. For any constant $\alpha \in k$, $t \mapsto t + \alpha$ defines an R-automorphism of S, so we have $f(t + \alpha) - f(t) \in J$. Since this polynomial has degree less than n, we must have $f(t + \alpha) = f(t)$. Setting $t = 0$ gives $a_n \alpha^n + \cdots + a_1 \alpha = 0$ for any $\alpha \in k$. Since k is infinite, the usual Vandermonde matrix argument shows that

$$a_n = \cdots = a_1 = 0,$$

so $f(t) = a_o$ gives a nonzero element in $R \cap J = N$.

In the general case, consider the ideal $N[t] \subseteq J = \operatorname{rad} R[t]$. For $\overline{R} = R/N$, we have by FC-(4.6):

$$\operatorname{rad} \overline{R}[t] = \operatorname{rad}(R[t]/N[t]) = J/N[t].$$

Since $J/N[t]$ contracts to zero in $\overline{R}$, the case we dealt with in the first paragraph implies that $\operatorname{rad} \overline{R}[t] = 0$, that is, $J = N[t]$.

Ex. 5.4. Assume k is a finite field. Show again that $J = N[t]$.

Solution. Let $\tilde{k}$ be the algebraic closure of k, and

$$\tilde{R} = R \otimes_k \tilde{k}, \quad \tilde{J} = \operatorname{rad}(\tilde{R}[t]).$$

By the first part of FC-(5.14), $J = R[t] \cap \tilde{J}$. Let $f(t) = \sum a_i t^i \in J$. Then $f(t) \in \tilde{J}$, and, since $\tilde{k}$ is infinite, Exercise 3 yields $a_i \in \tilde{J}$ for all i. But then

$$a_i \in R \cap \tilde{J} = R \cap (R[t] \cap \tilde{J}) = R \cap J = N,$$

so $J = N[t]$.

Ex. 5.5. Let R be any ring whose additive group is torsion-free. Show (without using Amitsur's Theorem) that $J = \operatorname{rad} R[t] \neq 0$ implies that $R \cap J \neq 0$.

Solution. Let $f(t) \in J \backslash \{0\}$ be of minimal degree, say

$$f(t) = a_n t^n + \cdots + a_o,$$

where $a_i \in R$, $a_n \neq 0$. We are done if we can show that $n = 0$. Assume, instead, $n \geq 1$. As in the solution to Exercise 3, we must have $f(t + 1) = f(t)$. Comparing the coefficients of t^{n-1}, we see that $n a_n = 0$ and hence $a_n = 0$, a contradiction.

Ex. 5.6. For any ring R with (Jacobson) radical J, show that the power series ring $A = R[[x_i : i \in I]]$ (in a nonempty set of, say, commuting variables $\{x_i\}$) has radical $J + \sum_{i\in I} Ax_i$. (In particular, A can never be J-semisimple if $R \neq 0$.)

Solution. Clearly, $J' := J + \sum_i Ax_i$ is an ideal of A. Since $A/J' \cong R/J$ is J-semisimple, Ex. 4.11 yields $\mathrm{rad}(A) \subseteq J'$. This will be an equality, as asserted, if we can show that $1 + J' \subseteq U(A)$. Now any element of $1 + J'$ has the form $f_0 + f_1 + f_2 + \cdots$ where $f_0 \in U(R)$, and f_n ($n \geq 1$) is a homogeneous polynomial (over R) of degree n in $\{x_i\}$. According to FC-(1.5), this power series is invertible in A.

Ex. 5.7. For any k-algebra R and any finite field extension K/k, show that $\mathrm{rad}\, R$ is nilpotent iff $\mathrm{rad}\, R^K$ is nilpotent.

Solution. By FC-(5.14), we have

$$\mathrm{rad}\, R = R \cap \mathrm{rad}\, R^K.$$

In particular, $\mathrm{rad}\, R \subseteq \mathrm{rad}\, R^K$. Thus, if $\mathrm{rad}\, R^K$ is nilpotent, so is $\mathrm{rad}\, R$. Conversely, assume that $\mathrm{rad}\, R$ is nilpotent, say $(\mathrm{rad}\, R)^m = 0$. Then, clearly, $[(\mathrm{rad}\, R)^K]^m = 0$. But if $[K : k] = n$, FC-(5.14) also gives $(\mathrm{rad}\, R^K)^n \subseteq (\mathrm{rad}\, R)^K$. Therefore, $(\mathrm{rad}\, R^K)^{nm} = 0$.

Ex. 5.8. (This problem, due to G. Bergman, is the origin of the proof of FC-(5.10B).) Let R be a graded ring, i.e.

$$R = R_0 \oplus R_1 \oplus \cdots$$

where the R_i's are additive subgroups of R such that $R_i R_j \subseteq R_{i+j}$ (for all i, j) and $1 \in R_0$. Show that $J = \mathrm{rad}\, R$ is a *graded ideal* of R, in the sense that J has a decomposition

$$J = J_0 \oplus J_1 \oplus \cdots,$$

where $J_i = J \cap R_i$.

Solution. Let

$$\alpha = \alpha_n + \alpha_{n+1} + \cdots + \alpha_m \in J,$$

where $\alpha_i \in R_i$, and $\alpha_n \neq 0 \neq \alpha_m$. We shall show that $\alpha_i \in J$ by induction on the number of nonzero homogeneous components in α (for all rings R). If this number is 1, there is clearly no problem. For the inductive step, we consider a prime p (to be specified) and the ring

$$(*) \qquad S = \frac{R[\zeta]}{(1 + \zeta + \cdots + \zeta^{p-1})} = R \oplus R\zeta \oplus \cdots \oplus R\zeta^{p-2}.$$

Note that $\zeta^p = 1$, and if we view ζ as an element of degree 0, S is also a graded ring, with

$$S_i = R_i \oplus R_i\zeta \oplus \cdots \oplus R_i\zeta^{p-2}.$$

On the ring S, we have a special automorphism φ defined on homogeneous elements $s_i \in S_i$ by $\varphi(s_i) = \zeta^i s_i$. From $(*)$, we have $J = R \cap \operatorname{rad} S$ (by FC-(5.6), (5.7)). Thus, $\alpha \in \operatorname{rad} S$, and so $\operatorname{rad} S$ contains

$$\varphi(\alpha) - \zeta^m \alpha = (\zeta^n \alpha_n + \cdots + \zeta^m \alpha_m) - (\zeta^m \alpha_n + \cdots + \zeta^m \alpha_m)$$
$$= (\zeta^n - \zeta^m)\alpha_n + \cdots + (\zeta^{m-1} - \zeta^m)\alpha_{m-1}.$$

By the inductive hypothesis, we have

$$(\zeta^n - \zeta^m)\alpha_n \in \operatorname{rad} S, \quad \text{and so} \quad (1 - \zeta^{m-n})\alpha_n \in \operatorname{rad} S.$$

Now assume $p > m - n$. Then $p \in (1 - \zeta^{m-n})S$. (Modulo $1 - \zeta^{m-n}$, we have $\overline{\zeta}^{m-n} = 1$, $\overline{\zeta}^p = 1$, so $\overline{\zeta} = 1$ and $\overline{p} = \overline{1} + \overline{\zeta} + \cdots + \overline{\zeta}^{p-1} = 0$.) From this, we conclude that

$$p\alpha_n \in R \cap \operatorname{rad} S = J.$$

Since this holds for any prime $p > m - n$, it follows that $\alpha_n \in J$, and by induction, $\alpha_i \in J$ for all i.

Ex. 5.9. Let $A = R[T]$, where T is an infinite set of commuting indeterminates. Show that $\operatorname{rad} A$ is a nil ideal.

Solution. Consider any $\alpha \in \operatorname{rad} A$. Pick a finite subset $S \subseteq T$ such that $\alpha \in R' := R[S]$, and let $T' = T \setminus S \neq \emptyset$. Then

$$A = R[S][T'] = R'[T'],$$

and $\alpha \in R' \cap \operatorname{rad} R'[T']$. By Amitsur's Theorem FC-(5.10), $R' \cap \operatorname{rad} R'[T']$ is a nil ideal, so α is nilpotent.

Comment. One may ask if the conclusion of the exercise remains true if T is a *finite* (nonempty) set of commuting indeterminates. I do not know the answer.

Ex. 5.10. Let us call a ring R "rad-nil" if its Jacobson radical is a nil ideal. (Examples include: J-semisimple rings, algebraic algebras, and commutative affine algebras over fields, etc.) Show that:

(1) a commutative ring is Hilbert iff all of its quotients are rad-nil;
(2) any commutative artinian ring is Hilbert;
(3) any commutative ring is a quotient of a commutative rad-nil ring.
(4) Construct a commutative noetherian rad-nil ring that is not Hilbert.

Solution. (1) Recall that a commutative ring R is called Hilbert if every prime ideal in R is an intersection of maximal ideals. Note that this property is inherited by all quotients. If R is Hilbert, then $\operatorname{Nil}(R)$, being (always) the intersection of prime ideals in R, is also an intersection of maximal ideals in R. This shows that $\operatorname{Nil}(R) = \operatorname{rad}(R)$, and the same equation also holds for all quotients of R. This proves the "only if" part, and the "if" part is clear.

(2) Let R be a commutative artinian ring, and consider any quotient $\overline{R}$ of R. Then $\overline{R}$ is also artinian, and so $\mathrm{rad}(\overline{R}) = \mathrm{Nil}(\overline{R})$ (by FC-(4.12)). By (1), R is Hilbert. [Alternatively, we can use the well-known fact that R, being artinian, has Krull-dimension 0. (This follows from Ex. 4.15.) This means that any prime ideal of R is maximal, so R is clearly Hilbert.]

(3) For any commutative ring A, let $R = A[t]$. By Snapper's Theorem FC-(5.1), R is a rad-nil ring, and we have $A \cong R/(t)$.

(4) Let A be a commutative noetherian ring that is not rad-nil (e.g. the localization of $\mathbb{Z}$ at any maximal ideal). Then $R = A[t]$ is noetherian (by the Hilbert Basis Theorem), rad-nil (by Snapper's Theorem), but its quotient $R/(t) \cong A$ is *not* rad-nil. Thus, by (1), R is not Hilbert.

Ex. 5.11. Let J be an ideal in any ring R.

(1) If $J \subseteq \mathrm{rad}(R)$, show that, for any $i \geq 1$, the multiplicative group $(1 + J^i)/(1 + J^{i+1})$ is isomorphic to the additive group J^i/J^{i+1}.

(2) Suppose $J^{n+1} = 0$ (so in particular, $J \subseteq \mathrm{rad}(R)$). Show that $1 + J$ is a nilpotent group of class $\leq n$; that is, $1 + J$ has a central series of length $\leq n$. If, moreover, the group J/J^2 has a finite exponent m, show that $(1 + J)^{m^n} = 1$.

(3) If J is a nil ideal and the ring R/J has prime characteristic p, show that $1 + J$ is a p-group.

Solution. (1) First note that, for $i \geq 1$, $1 + J^i$ is a normal subgroup of $U(R)$, as it is the kernel of the natural group homomorphism $U(R) \longrightarrow U(R/J^i)$. We define a map $\sigma : J^i \longrightarrow (1 + J^i)/(1 + J^{i+1})$ by

$$\sigma(x) = (1 + x) \cdot (1 + J^{i+1}), \quad \text{for every } x \in J^i.$$

This is a group homomorphism since, for $x, y \in J^i$:

$$\begin{aligned}
\sigma(x)\sigma(y) &= (1+x)(1+y) \cdot (1 + J^{i+1}) \\
&= (1 + x + y + xy) \cdot (1 + J^{i+1}) \\
&= (1 + x + y)[1 + (1 + x + y)^{-1}xy] \cdot (1 + J^{i+1}) \\
&= \sigma(x + y).
\end{aligned}$$

Since σ is surjective and has kernel J^{i+1}, it induces a group isomorphism $J^i/J^{i+1} \cong (1 + J^i)/(1 + J^{i+1})$.

(2) If $J^{n+1} = 0$, the subgroups $\{1 + J^i \colon 1 \leq i \leq n+1\}$ give a normal series in the group $1 + J$. Moreover, the same calculation as in (1) above, with $x \in J^i$ and $y \in J$, shows that $(1 + J^i)/(1 + J^{i+1})$ is in the center of $(1 + J)/(1 + J^{i+1})$, for each $i \geq 1$. Therefore, $\{1 + J^i \colon 1 \leq i \leq n+1\}$ is a *central* series in $1 + J$ of length $\leq n$, so by definition, $1 + J$ is a nilpotent group of

class $\leq n$. If, in addition, $mJ \subseteq J^2$ for some $m \geq 1$, then $m \cdot J^i/J^{i+1} = 0$ for all $i \geq 1$, so (1) implies that $(1 + J^i)^m \subseteq 1 + J^{i+1}$. From this, we see by induction that $(1 + J)^{m^n} \subseteq 1 + J^{n+1} = 1$.

(3) Here, we have $p \in J$ and $p^n = 0 \in R$ for some $n \geq 1$. For any $a \in J$, pick a large integer r such that $a^{2^r} = 0$. Applying Ex. 1.34 with $m = p^n$, we see that $(1 + a)^{p^{nr}} = 1$, so every element in $1 + J$ has a p-power order.

Comment. This exercise is surely part of the folklore in noncommutative ring theory, although a complete statement for all three parts of the exercise is not easy to find in the literature. A good approximation appeared in Lemma 8.12 in the paper of H. Bass and M.P. Murthy, "Grothendieck groups and Picard groups of abelian group rings," Annals of Math. *86* (1967), 16–73. This lemma states that, if J is an ideal in a ring R with $mJ = 0 = J^n$, then $(1 + a)^{m^n} = 1$ for all $a \in J$. Part (2) of this exercise gives a sharper result, with the hypothesis $mJ = 0$ weakened to $mJ \subseteq J^2$. Also, part (3) here applies to nil (instead of nilpotent) ideals J, thanks to the earlier Ex. 1.34.

Finally, we note that, for the ring R of $(n + 1) \times (n + 1)$ upper triangular matrices over a field k, and $J = \text{rad}(R)$ (with $J^{n+1} = 0$), $1 + J$ is the *unitriangular group* $\text{UT}_{n+1}(k)$ consisting of upper triangular matrices with diagonal entries 1. This exercise shows that $\text{UT}_{n+1}(k)$ is nilpotent of class $\leq n$. For a proof that equality actually holds here, see, e.g. p. 107 of "Fundamentals of the Theory of Groups" by Kargapolov and Merzljakov, GTM Vol. 62, Springer, 1972. In the case $k = \mathbb{R}$, $1 + J$ is also called the *Heisenberg group* (at least when $n = 3$). For more relevant information about the groups $1 + J$ (for nil ideals J in k-algebras), see Exercises 9.4A–C.

Ex. 5.12. (Suggested by H.W. Lenstra) For any ring R with Jacobson radical J, we have an exact sequence of groups

$(*)$ $1 \to 1 + J \to U(R) \to U(\overline{R}) \to 1$ (where $\overline{R} = R/J$),

induced by the projection map $\pi : R \to \overline{R}$ (see *FC*-(4.5)). Show that this sequence splits if

(1) R is a commutative rad-nil $\mathbb{Q}$-algebra, or

(2) R is a commutative artinian ring. (To prove this part, you may use the Akizuki–Cohen Theorem in *FC*-(23.12).)

Solution. (1) Under the hypothesis of (1), the groups in $(*)$ are all abelian, and $1 + J$ is a *divisible* group since, for any $x \in J$ (a nilpotent element), we can use the binomial expansion

$$(1 + x)^{1/n} = \sum_{i=0}^{\infty} \binom{1/n}{i} x^i \in 1 + J \quad \text{(a \textit{finite} sum)}$$

to get an nth root of $1 + x$ in $1 + J$. Thus, $1 + J$ is an injective $\mathbb{Z}$-module, and so $(*)$ splits.

(2) We assume here that R is *commutative artinian*. By the theorem of Ak-izuki and Cohen, R is isomorphic to a finite direct product of commutative local artinian rings. Thus, we need only treat the case where R is *local*. Let k be the residue class field R/J.

Case A. $\text{char}(k) = 0$. In this case, $r \cdot 1 \notin J$ for any positive integer r, so $r \in U(R)$. This implies that $\mathbb{Q} \subseteq R$. Thus, R is a (rad-nil) $\mathbb{Q}$-algebra, and we are done by (1).

Case B. $\text{char}(k) = p > 0$. Here, a basic case to keep in mind is $R = \mathbb{Z}/p^{n+1}\mathbb{Z}$ where p is a prime. In this special case, $J = p\mathbb{Z}/p^{n+1}\mathbb{Z}$ so $|1 + J| = |J| = p^n$. Since $|U(R)| = p^n(p-1)$, G is precisely the p-Sylow subgroup of the finite abelian group $U(R)$; of course it splits in $U(R)$.

We now try to "generalize" the above to our local ring (R, J). By FC-(4.12), $J^{n+1} = 0$ for some integer $n \geq 0$. Then, since $p \cdot J/J^2 = 0$, Ex. (5.11)(2) gives $(1 + J)^{p^n} = 1$. Also, in the field k of characteristic p, $x^p = 1 \Longrightarrow x = 1$ so $U(k)$ is p-torsionfree. Our conclusion, therefore, is im-mediate from the following general result in the theory of abelian groups.

Proposition. *Let $0 \to A \to B \to C \to 0$ be an exact sequence of (additive) abelian groups. If, for some prime p, C is p-torsionfree, and $p^n A = 0$ for some n, then the exact sequence splits.*

Proof. We induct on n. For $n = 1$, $pA = 0$ implies that $A \cap pB = 0$. The induced inclusion map $A \to B/pB$ splits since both groups are vector spaces over $\mathbb{Z}_p$. This implies that $A \to B$ also splits. For $n \geq 2$, let $D = p^{n-1}A$. Applying the inductive hypothesis to the inclusion $A/D \to B/D$, we have a subgroup $V \subseteq B$ such that $B = A + V$ and $A \cap V = D$. By the $n = 1$ case, the inclusion $D \to V$ also splits (since $V/D \hookrightarrow B/A$ is p-torsionfree). Thus, a direct complement to D in V provides a direct complement to A in B.

Second Proof. Readers familiar with basic homological algebra may pre-fer the following proof of the Proposition using the Ext-functor over $\mathbb{Z}$. We need to prove $\text{Ext}^1(C, A) = 0$, and proceed by induction on n. The case $n = 1$ can be handled as before. For $n \geq 2$, let $D = p^{n-1}A$ and use the exact sequence

$$\text{Ext}^1(C, D) \to \text{Ext}^1(C, A) \to \text{Ext}^1(C, A/D).$$

By the $n = 1$ case, $\text{Ext}^1(C, D) = 0$, and by the inductive hypothesis, $\text{Ext}^1(C, A/D) = 0$. Therefore, $\text{Ext}^1(C, A) = 0$, as desired.

Comment. The Proposition above can also be deduced easily from a more general group-theoretic theorem: *if an abelian group A of finite exponent is a pure subgroup of an abelian group B (that is, $A \cap mB = mA$ for any integer m), then A splits in B.* For a proof of this theorem, see (4.3.8) in Robinson's book, "A Course in the Theory of Groups," GTM Vol. 80, Springer, 1972. We note, however, that the Proposition no longer holds if

A is only assumed to be a p-group. An example is given by taking A to be the torsion subgroup of $B := \mathbb{Z}_p \times \mathbb{Z}_{p^2} \times \mathbb{Z}_{p^3} \times \cdots$. Here, A is a p-group and B/A is torsionfree, but A *does not* split in B: a short proof for this can be found in (4.3.10) in Robinson's book.

It is also worth pointing out that the exact sequence $(*)$ is known to split for a certain large class of finite-dimensional (not necessarily commutative) algebras. In fact, if R is such an algebra over a field k, and if the factor algebra $\overline{R} = R/J$ $(J = \mathrm{rad}(R))$ is *separable* over k, the Principal Theorem of Wedderburn-Mal'cev[*] says that the projection may $\pi : R \to \overline{R}$ splits by a suitable k-algebra homomorphism $\overline{R} \to R$. This implies, of course, that $\pi : U(R) \to U(\overline{R})$ also splits.

Ex. 5.13. Let R be a commutative domain that is not a field.

(1) If R is semilocal (that is, R has only finitely many maximal ideals), show that R is not J-semisimple.

(2) Show that the converse holds in (1) if R is a 1-dimensional noetherian domain, but not in general for noetherian domains.

Solution. (1) If $\mathfrak{m}_1, \ldots, \mathfrak{m}_n$ are the maximal ideals of R (each nonzero since R is not a field), then $0 \neq \mathfrak{m}_1 \cdots \mathfrak{m}_n \subseteq \mathrm{rad}(R)$.

(2) Let R be a 1-dimensional noetherian domain. For convenience, we'll assume the commutative version of Ex. 10.15 below. If $\mathrm{rad}(R) \neq 0$, any maximimal ideal of R is a minimal prime over $\mathrm{rad}(R)$. By Ex. 10.15, there are only finitely many such primes, so R is semilocal.

For higher dimensions, the converse to (1) fails in general. For instance, the 2-dimensional noetherian domain $R = \mathbb{Z}[[x]]$ is not J-semisimple by Ex. 5.6 above, but has infinitely many maximal ideals: (p, x) for $p = 2, 3, 5, \ldots$.

Comment. The converse to (1) also fails if R is a 1-dimensional non-noetherian domain, though we will not produce an example here.

§6. Group Rings and the J-Semisimplicity Problem

It all started in 1898 when H. Maschke proved that finite group algebras over fields of characteristic zero are semisimple rings. Maschke's Chicago colleague L.E. Dickson subsequently pointed out that the theorem already holds when the characteristic of the field does not divide the order of the group. Certainly, Maschke's Theorem played a key role in Noether's reformulation of the representation theory of finite groups, now enshrined in van

---

[*] See (72.19) in "Representation Theory of Finite Groups and Associative Algebras" by Curtis and Reiner, J. Wiley Interscience, 1962.

der Waerden's "Modern Algebra." We shall return to this matter shortly in the next chapter.

With Jacobson's discovery of the notion of J-semisimple rings in the 1940's, it was only natural to search for analogues of Maschke's Theorem, with J-semisimplicity replacing classical semisimplicity. Rickart's 1950 result, to the effect that $\mathbb{C}G$ and $\mathbb{R}G$ are both J-semisimple for any group G, was a stunning testament to the prowess and efficacy of Banach algebra methods. But of course the algebraists were not to be outdone. Amitsur showed more generally that, for any group G, KG is J-semisimple if K is a nonalgebraic field extension of $\mathbb{Q}$, and Passman proved the analogue of this for p'-groups over fields of characteristic p. Surprisingly, the toughest case with the "J-semisimplicity Problem" for group rings occurs when the ground field is an *algebraic* extension of the prime field. In this case, the J-semisimplicity Problem for group rings has apparently remained unsolved to this date.

There are other problems and conjectures too, mostly concerning the structure of kG when k is a domain and G is a torsion-free group. *Is kG a domain, or is it at least reduced* (no nonzero nilpotent elements)? *What is* $U(kG)$, *and, if* $G \neq \{1\}$, *is kG always J-semisimple?* None of these questions has been fully answered, but a large number of partial results have been obtained. A short exposition for some of these is given in FC-§6, along with a quick glimpse into Passman's Δ-group methods.

The exercises in this section are a mixed bag, ranging from variations and applications of Maschke's Theorem, to the relationship between rad (kH) and rad (kG) for H normal in G, to explicit computations of rad (kG) for specific groups, and to purely group-theoretic results such as Hölder's Theorem (Exercise 11), Neumann's Theorem (Exercise 17), and the delightful Dietzmann's Lemma (Exercise 15). The overall theme behind all of these exercises is simply that group theory and ring theory sometimes parade as inseparable twins. In the study of group rings, the two theories are merged together so harmoniously that it seems no longer possible to tell where one ends and the other begins.

Exercises for §6

In the following exercises, k denotes a field and G denotes a group, unless otherwise specified.

Ex. 6.1. Let V be a kG-module and H be a subgroup in G of finite index n not divisible by char k. Modify the proof of Maschke's Theorem to show the following: If V is semisimple as a kH-module, then V is semisimple as a kG-module.

Solution. Fix a coset decomposition $G = \bigcup_{i=1}^{n} H\sigma_i$. Let W be any kG-submodule of V. Following the notations in the proof of FC-(6.1), let $f :$

$V \to W$ be a kH-homomorphism with $f|W = 1_W$. Here we try to modify f into a kG-homomorphism $g : V \to W$ with $g|W = 1_W$. Define

$$g(v) = n^{-1} \sum_{i=1}^{n} \sigma_i^{-1} f(\sigma_i v) \quad (v \in V).$$

It is clear that $g|W = 1_W$, so we only need to prove $g(\tau v) = \tau g(v)$ for every $\tau \in G$. Write $\sigma_i \tau = \theta_i \sigma_{\alpha(i)}$ where $\theta_i \in H$ and α is a permutation of $\{1, \ldots, n\}$. Then

$$g(\tau v) = n^{-1} \sum \sigma_i^{-1} f(\sigma_i \tau v) = n^{-1} \sum \sigma_i^{-1} f(\theta_i \sigma_{\alpha(i)} v)$$

$$= n^{-1} \sum \sigma_i^{-1} \theta_i f(\sigma_{\alpha(i)} v) = n^{-1} \sum \tau \sigma_{\alpha(i)}^{-1} f(\sigma_{\alpha(i)} v)$$

$$= \tau \left(n^{-1} \sum \sigma_{\alpha(i)}^{-1} f(\sigma_{\alpha(i)} v) \right) = \tau g(v).$$

Ex. 6.2. Let A be a normal elementary p-subgroup of a finite group G such that the index of the centralizer $C_G(A)$ is prime to p. Show that for any normal subgroup B of G lying in A, there exists another normal subgroup C of G lying in A such that $A = B \times C$.

Solution. Let $k = \mathbb{F}_p$ and let G act on A by conjugation. The subgroup $C_G(A)$ is also normal in G, and acts trivially on A. Writing $\overline{G} = G/C_G(A)$, we may therefore view A as a $k\overline{G}$-module. Since $|\overline{G}| = [G : C_G(A)]$ is prime to $p = \operatorname{char} k$, $k\overline{G}$ is a semisimple ring. The assumption that $B \lhd G$ implies that B is a $k\overline{G}$-submodule of A. Therefore, $A = B \oplus C$ for a suitable $k\overline{G}$-submodule $C \subseteq A$. Going back to the multiplicative notation, we have $A = B \times C$, and $C \lhd G$.

Ex. 6.3. Let G be a finite group whose order is a unit in a ring k, and let $W \subseteq V$ be left kG-modules.

(1) If W is a direct summand of V as k-modules, then W is a direct summand of V as kG-modules.
(2) If V is projective as a k-module, then V is projective as a kG-module.

Solution. (1) Fixing a k-homomorphism $f : V \to W$ such that $f \mid W$ is the identity, we can define $g : V \to V$ as in the solution to Ex. 1 by the "averaging" device:

$$g(v) = |G|^{-1} \sum_{\sigma \in G} \sigma^{-1} f(\sigma v), \quad \text{for } v \in V.$$

Again, we check easily that g is a kG-homomorphism with $g \mid W = \operatorname{Id}_W$, and so $V = W \oplus \ker(g)$.

(2) Take a kG-epimorphism $\varphi : F \to V$, where F is a suitable free kG-module, and let $E = \ker(\varphi)$. Since V is projective as a k-module, E is a

direct summand of F as k-modules. By (1), E is a direct summand of F as kG-modules. Thus, V is isomorphic to a direct kG-complement of E in F, so V is a projective kG-module.

Ex. 6.4. (This exercise is valid for any ring k.) For any subgroup H of a group G, show that

(∗) $kH \cap U(kG) \subseteq U(kH)$ and $kH \cap \operatorname{rad} kG \subseteq \operatorname{rad} kH$.

Deduce that, if kH is J-semisimple for any finitely generated subgroup H of G, then kG itself is J-semisimple.

Solution. Let $R = kH \subseteq S = kG$, and fix a coset decomposition

$$G = \bigcup_{i \in I} H\sigma_i.$$

Then we have $S = \bigoplus_i R\sigma_i$. We may assume that some $\sigma_{i_0} = 1$. Therefore, $_R R = R\sigma_{i_0}$ is a direct summand of $_R S$. Similarly, R_R is a direct summand of S_R. Applying Exercise 5.0, we get the two conclusions in (∗).

Assume now $\operatorname{rad} kH = 0$ for any finitely generated subgroup $H \subseteq G$. For any $\alpha \in \operatorname{rad} kG$, we have $\alpha \in kH$ for some such H, so

$$\alpha \in kH \cap \operatorname{rad} kG \subseteq \operatorname{rad} kH = 0.$$

This implies that $\operatorname{rad} kG = 0$, as desired.

Ex. 6.5. (Amitsur, Herstein) If k is an uncountable field, show that, for any group G, $\operatorname{rad} kG$ is a nil ideal.

Solution. Let $\alpha \in \operatorname{rad} kG$. Then

$$\alpha \in kH \cap \operatorname{rad} kG \subseteq \operatorname{rad} kH$$

for some finitely generated subgroup $H \subseteq G$. Now $\dim_k kH = |H|$ is countable, and k is uncountable. Therefore, FC-(4.20) gives the desired conclusion that $\alpha^n = 0$ for some $n \geq 1$.

Ex. 6.6. Let H be a normal subgroup of G. Show that $I = kG \cdot \operatorname{rad} kH$ is an ideal of kG. If $\operatorname{rad} kH$ is nilpotent, show that I is also nilpotent. (In particular, if H is finite, I is always nilpotent.)

Solution. Any $\sigma \in G$ defines a conjugation automorphism on the subring $kH \subseteq kG$, and this automorphism must take $\operatorname{rad} kH$ to $\operatorname{rad} kH$. Therefore,

$$(\operatorname{rad} kH)\sigma \subseteq \sigma \cdot \operatorname{rad} kH \subseteq I,$$

which shows that I is an ideal of kG. This method also shows that

$$I^n = kG \cdot (\operatorname{rad} kH)^n$$

for any $n \geq 1$, so the rest of the Exercise follows.

Ex. 6.7. (For this Exercise, we assume Wedderburn's Theorem that finite division rings are commutative. A proof of this theorem can be found in FC-(13.1).) Show that if k_0 is any finite field and G is any finite group, then $(k_0G/\operatorname{rad} k_0G) \otimes_{k_0} K$ is semisimple for any field extension $K \supseteq k_0$.

Solution. Let $A = k_0G/\operatorname{rad} k_0G$. In fact, A can be any finite-dimensional semisimple k_0-algebra below. By Wedderburn's Theorem, $A \cong \prod_i \mathbb{M}_{n_i}(D_i)$, where the D_i's are finite-dimensional k_0-division algebras. By the other theorem of Wedderburn quoted in the statement of the Exercise, each D_i is just a finite field extension of k_0. Now

$$A \otimes_{k_0} K \cong \left(\prod_i \mathbb{M}_{n_i}(D_i) \right) \otimes_{k_0} K \cong \prod_i \left(\mathbb{M}_{n_i}(D_i) \otimes_{k_0} K \right)$$

$$\cong \prod_i \mathbb{M}_{n_i} \left(D_i \otimes_{k_0} K \right).$$

We are done if we can show that each $D_i \otimes_{k_0} K$ is a finite direct product of fields. To simplify the notation, write D for D_i. The crux of the matter is that D/k_0 is a (finite) *separable* extension. (The finite field k_0 is a perfect field!) Say $D = k_0(\alpha)$, and let $f(x)$ be the minimal polynomial of α over k_0. Then $f(x)$ is a *separable* polynomial over k_0. Over $K[x]$, we have a factorization $f(x) = f_1(x) \cdots f_m(x)$ where the f_i's are nonassociate irreducible polynomials in $K[x]$. By the Chinese Remainder Theorem:

$$D \otimes_{k_0} K \cong \frac{k_0[x]}{(f(x))} \otimes_{k_0} K \cong \frac{K[x]}{(f_1(x) \cdots f_m(x))} \cong \prod_j \frac{K[x]}{(f_j(x))}.$$

This is a finite direct product of finite (separable) field extensions of K, as desired.

Ex. 6.8. Let $k \subseteq K$ be two fields and G be a finite group. Show that

$$\operatorname{rad}(KG) = (\operatorname{rad} kG) \otimes_k K.$$

Solution. We may assume that char $k = p > 0$, for otherwise both sides of the equation are zero. Let $\mathbb{F}_p$ denote the prime field of k. We shall first prove the following special case of the Exercise:

$$(*) \qquad\qquad \operatorname{rad}(kG) = (\operatorname{rad} \mathbb{F}_p G) \otimes_{\mathbb{F}_p} k.$$

By Exercise 7,

$$(\mathbb{F}_p G/\operatorname{rad} \mathbb{F}_p G) \otimes_{\mathbb{F}_p} k \cong kG/ \left((\operatorname{rad} \mathbb{F}_p G) \otimes_{\mathbb{F}_p} k \right)$$

is semisimple. Therefore, by Exercise 4.11, we must have

$$\operatorname{rad} kG \subseteq (\operatorname{rad} \mathbb{F}_p G) \otimes_{\mathbb{F}_p} k.$$

Since the latter is a nilpotent ideal in kG, equality holds. This proves $(*)$, and we deduce immediately that

$$\operatorname{rad}(KG) = (\operatorname{rad} \mathbb{F}_p G) \otimes_{\mathbb{F}_p} K = (\operatorname{rad} \mathbb{F}_p G) \otimes_{\mathbb{F}_p} k \otimes_k K$$
$$= (\operatorname{rad} kG) \otimes_k K.$$

Ex. 6.9. Let $k \subseteq K$ and G be as above. Show that a kG-module M is semisimple iff the KG-module $M^K = M \otimes_k K$ is semisimple.

Solution. By Exercise 8, $\operatorname{rad} KG = (\operatorname{rad} kG) \otimes_k K$. Therefore,

$$(\operatorname{rad} KG) \cdot M^K = (\operatorname{rad} kG \otimes_k K) \cdot (M \otimes_k K)$$
$$= (\operatorname{rad} kG \cdot M) \otimes_k K.$$

It follows that:

$$M \text{ is semisimple} \Longleftrightarrow \operatorname{rad} kG \cdot M = 0$$
$$\Longleftrightarrow (\operatorname{rad} KG) \cdot M^K = 0$$
$$\Longleftrightarrow M^K \text{ is semisimple.}$$

Ex. 6.10. Let k be a commutative ring and G be any group. If kG is left noetherian (resp. left artinian), show that kG is right noetherian (resp. right artinian).

Solution. Define a map $\varepsilon : kG \to kG$ by

$$\varepsilon \left(\sum a_g g \right) = \sum a_g g^{-1}.$$

Since $(gh)^{-1} = h^{-1} g^{-1}$, and k is commutative, we can show that $\varepsilon(\alpha\beta) = \varepsilon(\beta)\varepsilon(\alpha)$. Of course ε is one-one, onto, and an additive homomorphism. Since we also have $\varepsilon^2 = 1$, ε is an involution on kG. (This is a special case of the discussion in FC-p. 83.) If

$$I_1 \subsetneq I_2 \subsetneq \cdots$$

was an ascending chain of right ideals in kG,

$$\varepsilon(I_1) \subsetneq \varepsilon(I_2) \subsetneq \cdots$$

would have given an ascending chain of left ideals in kG. This gives the desired conclusion in the noetherian case, and the artinian case follows similarly.

Comment. The existence of the involution ε above implies that kG is isomorphic to its opposite ring. For a ring R to be isomorphic to its opposite ring, what we need exactly is an "anti-isomorphism" $\varepsilon : R \to R$ ("anti" referring to the property $\varepsilon(\alpha\beta) = \varepsilon(\beta)\varepsilon(\alpha)$): we *do not* need to have $\varepsilon^2 = \text{Id}$. If R is any ring that is isomorphic to its opposite ring, our argument above shows that R is left noetherian (resp. artinian) iff it is right noetherian (resp. artinian). But of course, if k is not commutative, the Exercise doesn't work: just take $G = \{1\}$.

Ex. 6.11. (Hölder's Theorem) An ordered group $(G, <)$ is said to be *archimedean* if, for any $a, b > 1$ in G, we have $a < b^n$ for some integer $n \geq 1$. Show that if $(G, <)$ is archimedean, then G is commutative and $(G, <)$ is order-isomorphic to an additive subgroup of $\mathbb{R}$ with the usual ordering.

Solution. It is more convenient to present the proof here by switching to the *additive* notation for the ordered group $(G, <)$. Thus, we write "$+$" for the operation on G, and 0 for its identity element. The positive cone for the ordered group $(G, <)$ is therefore

$$P = \{a \in G : a > 0\}.$$

We just have to work a bit more carefully since it is *not* assumed that $a + b = b + a$.

The argument is divided into the following two cases.

Case 1. The set P admits a least element, say a. In this case, we prove that $(G, <)$ is order-isomorphic to $(\mathbb{Z}, <)$ with $a \leftrightarrow 1$. It suffices to show that every $b > a$ is an integer multiple of a. Let n be the least positive integer such that $b < na$. Then $n \geq 2$ and we have $(n-1)a \leq b < na$. Left-adding $(1-n)a$, we get

$$0 \leq (1-n)a + b < a,$$

and therefore $b = -(1-n)a = (n-1)a$.

Case 2. P has no least element. In this case, first note that

(A)   *For any $c > 0$, there exists $d > 0$ such that $2d \leq c$.*

To prove this, fix any positive $e < c$. We may assume that $c < 2e$. Adding $-e$ from both left and right, we get $-e + c - e < 0$. Adding c from the left, we get $2d < c$ with $d := c - e > 0$. Next, we claim that:

(B)   *For any $a, b > 0$, $a + b = b + a$ in G.*

Indeed, if otherwise, there exist $a, b > 0$ with $a + b > b + a$. Let

$$c = (a + b) - (b + a) > 0.$$

By (A), $0 < 2d \leq c$ for some d. Pick integers $n, m \geq 0$ such that

$$md \leq a < (m+1)d, \quad nd \leq b < (n+1)d.$$

Then $a + b < (m + n + 2)d$, and $b + a \geq (m + n)d$. Right-adding

$$-(b + a) \leq -(m + n)d$$

to the former, we get $c < 2d$, a contradiction. From (B), we deduce easily that G *is commutative*. (From here on, we can afford to be a bit less cautious with the notation, for now we have $n(a + b) = na + nb$ for $n \in \mathbb{Z}$, etc.) *Fix any element $\omega > 0$.* We shall construct an order-embedding $\varphi : G \to \mathbb{R}$ with $\varphi(\omega) = 1$. The idea is: for any rational number $r \in \mathbb{Q}$ and any $a \in G$, we can define what is meant by "$r\omega < a$". Namely, if $r = m/n$ where $m, n \in \mathbb{Z}$ and $n > 0$, we take $r\omega < a$ to mean $m\omega < na$. (It is easy to see that this depends only on r, and not on the choice of m, n.) For any $a \in G$, define now

$$L_a = \{r \in \mathbb{Q} : \quad r\omega < a\}.$$

Since $(G, <)$ is archimedean, we see that L_a is nonempty and bounded from above. Therefore, we can define $\varphi : G \to \mathbb{R}$ by

$$\varphi(a) = \sup(L_a) \in \mathbb{R} \quad (\text{with} \quad \varphi(\omega) = 1).$$

Using the commutativity of G, it can be checked that $L_a + L_b = L_{a+b}$. (We omit the routine proof of this equality.) Therefore, we have

$$\varphi(a + b) = \sup(L_{a+b}) = \sup(L_a + L_b)$$
$$= \sup(L_a) + \sup(L_b) = \varphi(a) + \varphi(b).$$

It only remains to show that φ is an *order-embedding*. Say $a < b$ in G. Then $c := b - a > 0$ and $\omega < nc$ for some integer $n > 0$. But then $n^{-1} \in L_c$ and

$$\varphi(b) = \varphi(a) + \varphi(b - a) \geq \varphi(a) + n^{-1} > \varphi(a),$$

as desired.

Comment. (1) It can be shown that, for the given element $\omega > 0$ in G, the map φ constructed above is the *only* order-embedding of $(G, <)$ into $(\mathbb{R}, <)$ taking ω to 1.

(2) A result similar to that in Exercise 11 also holds for ordered rings: *If $(R, <)$ is an archimedean ordered ring, then R is commutative, and order-isomorphic to a subring of $\mathbb{R}$ with the usual ordering.* For this result, see *FC*-(17.21). The proof there uses exactly the same ideas. However, the work is significantly easier, since $(R, +)$ is already commutative, and there is a natural choice for the element $\omega > 0$ used in the construction of the order-embedding, namely, $\omega = 1$! The general proof for the ordered group case, as we saw above, is quite a bit more subtle.

Hölder's Theorem is as old as the last century. It appeared in Hölder's paper with the somewhat unlikely title "Die Axiome der Quantität und die Lehre vom Mass," Ber. Verk. Sächs. Wiss. Leipzig, Math. Phys. Cl. *53* (1901), 1–64.

Ex. 6.12. Assume $\text{char}(k) = 3$, and let $G = S_3$ (symmetric group on three letters).

(1) Compute the Jacobson radical $J = \text{rad}(kG)$, and the factor ring kG/J.
(2) Determine the index of nilpotency for J, and find a k-basis for J^i for each i.

Solution. Consider the homomorphism $\varphi : kG \to kS_2$ induced by the group surjection $G \to G/\langle(123)\rangle = S_2$. It is easy to see that $\ker(\varphi) = kG \cdot \alpha_1$, where $\alpha_1 = (123) - 1$. Since

$$kS_2 \cong k[t]/(t^2 - 1) \cong k \times k$$

is semisimple, we have $J \subseteq \ker(\varphi)$ (by Ex. 4.11). *We claim that* $J = \ker(\varphi)$ (whereby $kG/J \cong k \times k$). For this, it suffices to check that (123) acts as the identity on any left simple kG-module V. Now

$$V_0 = \{v \in V : (123)v = v\} \neq 0$$

since $\alpha_1^3 = (123)^3 - 1 = 0 \in kG$ implies that α_1 acts as a nilpotent transformation on V. It is easy to check that V_0 is a kG-submodule of V, so indeed $V_0 = V$. We have thus shown that

$$J = \ker(\varphi)$$
$$= \{x + y(123) + z(132) + u(12) + v(13) + w(23) :$$
$$x + y + z = u + v + w = 0\}.$$

A k-basis for J is therefore given by

$$\alpha_1 = (123) - 1, \quad \alpha_2 = (132) - 1, \quad \alpha_3 = (12) - (13), \quad \alpha_4 = (12) - (23).$$

A simple computation shows that each product $\alpha_i \alpha_j$ is of the form $\pm\beta_1$ or $\pm\beta_2$, where

$$\beta_1 = 1 + (123) + (132), \quad \text{and} \quad \beta_2 = (12) + (13) + (23).$$

Thus, J^2 has k-basis $\{\beta_1, \beta_2\}$. By inspection, we see that $\alpha_i \beta_j = 0$ for all i, j. Therefore, $J^3 = 0$, so the index of nilpotency for J is 3.

Comment. The computation of $\text{rad}(kG)$ here is a special case of Wallace's result in *FC*-(8.7). Another case of interest is when $\text{char}(k) = 2$: this is covered by the first part of the next exercise, as well as by Ex. 8.30(2) later.

Ex. 6.13. (Passman) Assume char $k = 2$. Let A be an abelian $2'$-group and let G be the semidirect product of A and a cyclic group $\langle x \rangle$ of order 2, where x acts on A by $a \mapsto a^{-1}$.

(1) If $|A| < \infty$, show that $\text{rad } kG = k \cdot \sum_{g \in G} g$, and $(\text{rad } kG)^2 = 0$.
(2) If A is infinite, show that kG has no nonzero nil ideals.

Solution. First note that, by FC-(6.13), kA has no nonzero nil ideals. Since kA is commutative, this simply means that the only nilpotent element of kA is zero. For

$$\alpha = \sum_{a \in A} \alpha_a a \in kA,$$

let $\alpha^* = \sum \alpha_a a^{-1}$ (cf. FC-p. 86). This defines an involution on kA, with $x\alpha = \alpha^* x$ for any $\alpha \in kA$. Any element $\sigma \in kG$ can be expressed uniquely in the form $\alpha + \beta x$, with $\alpha, \beta \in kA$. Let I be any nil ideal in kG, and let $\sigma \in I$. Then

$$(\alpha + \beta x)(\alpha^* + x\beta^*) = \alpha\alpha^* + \beta\beta^* + (\beta\alpha + \alpha\beta)x$$
$$= \alpha\alpha^* + \beta\beta^* \in kA.$$

Since this element is nilpotent, we must have $\alpha\alpha^* = \beta\beta^*$. Therefore,

$$(\alpha + \beta x)^2 = \alpha^2 + \beta x\beta x + \alpha\beta x + \beta x\alpha$$
$$= \alpha^2 + \beta\beta^* + (\alpha\beta + \beta\alpha^*)x$$
$$= \alpha^2 + \alpha\alpha^* + (\alpha\beta + \alpha^*\beta)x$$
$$= (\alpha + \alpha^*)(\alpha + \beta x).$$

Say $(\alpha + \beta x)^n = 0$. Then we have $0 = (\alpha + \alpha^*)^{n-1}(\alpha + \beta x)$, so that $(\alpha + \alpha^*)^{n-1}\alpha = 0$. But then

$$0 = x[(\alpha + \alpha^*)^{n-1}\alpha]x = (\alpha + \alpha^*)^{n-1}\alpha^*.$$

Therefore, by addition, $(\alpha + \alpha^*)^n = 0$, so $\alpha = \alpha^*$. Now consider any $b \in A$. Then $b(\alpha + \beta x) \in I$ implies $b\alpha = (b\alpha)^*$. Suppose $\alpha \neq 0$; say α involves some group element $b^{-1} \in A$. Then $1 \in \operatorname{supp}(b\alpha)$, and $b\alpha = (b\alpha)^*$ implies that $|\operatorname{supp}(\alpha)| = |\operatorname{supp}(b\alpha)|$ is odd, since A has no element of order 2. But if $\operatorname{supp}(\alpha)$ *misses* some element $c^{-1} \in A$, then $1 \notin \operatorname{supp}(c\alpha)$, and $c\alpha = (c\alpha)^*$ would imply that $|\operatorname{supp}(\alpha)| = |\operatorname{supp}(c\alpha)|$ is even. Therefore, we must have $\operatorname{supp}(\alpha) = A$. If A is infinite, this is impossible. In this case, we conclude that $\alpha = 0$, and since $\sigma x = \beta x^2 = \beta$ is nilpotent, $\beta = 0$ too, so $\sigma = 0$. This completes the proof in Case (2).

Next, assume $|A| < \infty$. Then $|A|$ is odd, so every element of A is a square. We continue to work with the element $\sigma = \alpha + \beta x \in I$. Write $\alpha = \sum_{a \in A} \alpha_a a$. For any $a_1, a_2 \in A$, choose $d \in A$ such that $d^2 = a_1^{-1}a_2^{-1}$. Then $da_1 = (da_2)^{-1}$, and $d\alpha = (d\alpha)^*$ implies that $\alpha_{a_1} = \alpha_{a_2}$. Therefore, $\alpha = \varepsilon \sum_{a \in A} a$ for some $\varepsilon \in k$. Since

$$(\alpha + \beta x)x = \beta + \alpha x \in I,$$

we have similarly $\beta = \varepsilon' \sum_{a \in A} a$. Now let $\tau = \sum_{g \in G} g \in Z(kG)$. We have

$$\tau^2 = |G|\tau = 2|A|\tau = 0,$$

so $k\tau$ is an ideal with $(k\tau)^2 = 0$. This implies that $k\tau \subseteq \operatorname{rad} kG$. Applying the above analysis to $\sigma \in I : = \operatorname{rad} kG$, we have now

$$\sigma + \varepsilon'\tau = (\varepsilon + \varepsilon') \sum_{a \in A} a \in I.$$

Since this is a nilpotent element in kA, we conclude that $\varepsilon = \varepsilon'$, so $\sigma = \varepsilon\tau$. This shows that $I = k\tau$, with $I^2 = 0$.

Comment. The above clever argument is taken from Passman's article "Nil ideals in group rings," in Mich. Math. J. *9* (1962), 375–384.

Ex. 6.14. (Wallace) Assume char $k = 2$, and let $G = A \cdot \langle x \rangle$ as in Exercise 13, where A is the infinite cyclic group $\langle y \rangle$. (G is the infinite dihedral group.) Show that $R = kG$ is J-semisimple (even though G has an element of order 2).

Solution. Let $H_r = \langle y^{3^r} \rangle \lhd G \ (1 \le r < \infty)$ and let $G_r = G/H_r$, which is a dihedral group of order $2 \cdot 3^r$. We first show that the natural ring homomorphism

$(*)$ $$\varphi = (\varphi_r): \quad kG \longrightarrow \prod_r kG_r$$

is *injective*. Indeed, consider a nonzero element

$$\alpha = a_1 g_1 + \cdots + a_n g_n \in R,$$

where $a_i \ne 0$ for all i, and $g_1, \ldots, g_n$ are distinct elements of G. Since $H_1 \supseteq H_2 \supseteq \cdots$ and $\bigcap H_r = \{1\}$, there exists an s such that $g_i^{-1} g_j \notin H_s$ for all $i \ne j$. But then $g_i H_s \ne g_j H_s$, and we have

$$\varphi_s(\alpha) = a_1 g_1 H_s + \cdots + a_n g_n H_s \ne 0 \in kG_s.$$

Now, by part (1) of Exercise 13, $(\operatorname{rad} kG_r)^2 = 0$ for all r. From

$$\varphi_r(\operatorname{rad} kG) \subseteq \operatorname{rad} kG_r$$

(cf. Exercise 4.10), we have $\varphi_r((\operatorname{rad} kG)^2) = 0$. Since this holds for all r, the injectivity of φ yields $(\operatorname{rad} kG)^2 = 0$. But then part (2) of Exercise 13 gives $\operatorname{rad} kG = 0$, as desired.

Comment. Let $\{H_r : r \in I\}$ be a family of normal subgroups of a group G such that $\bigcap H_r = \{1\}$, and for every r, r', there exists s with $H_s \subseteq H_r \cap H_{r'}$. Let $G_r = G/H_r$. A slight generalization of the argument used above shows that the map $\varphi = (\varphi_r)$ in $(*)$ is injective.

   Therefore, one may derive information on $\operatorname{rad} kG$ (e.g. nilpotency properties) via information on $\{\operatorname{rad} kG_r\}$. In the most basic case, if each kG_r is J-semisimple, kG is also J-semisimple. Wallace has used this method to show, for instance, that kG is J-semisimple for certain residually finite groups. For this and much more information on $\operatorname{rad} kG$, see Wallace's article in Math. Zeit. *100* (1967), 282–294.

Ex. 6.15. (Dietzmann's Lemma) Let G be a group generated by $x_1, \ldots, x_n$ where each x_i has finite order and has only finitely many conjugates in G. Show that G is a finite group.

Solution. After enlarging the generating set, we may assume that

$$X : = \{x_1, \ldots, x_n\}$$

is closed under conjugation. Every element of G is a product of the x_i's. (We don't need their inverses since each x_i has finite order.) Fix a large number e such that $x_i^e = 1$ for all i. We finish by showing that every $a \in G$ is a product of no more than $n(e-1)$ elements from X. (This yields an explicit bound $|G| \leq n^{n(e-1)}$.)

Suppose $a = a_1 \cdots a_m$ where $a_i \in X$ and $m > n(e-1)$. It suffices to show that we can "reexpress" a with fewer factors. Since $m > n(e-1)$, some $x \in X$ must appear at least e times in the above expression. Say a_i is the first factor equal to x. Let $a_j' = x^{-1} a_j x \in X$ for $j < i$. Then

$$a = x a_1' \cdots a_{i-1}' a_{i+1} \cdots a_m.$$

Repeating this process, we can move e factors of x to the left, to get

$$a = x^e a_1^* \cdots a_{i-1}^* \cdots a_{m-e}^*,$$

where $a_j^* \in X$. Since $x^e = 1$, we have now $a = a_1^* a_2^* \cdots a_{m-e}^*$, as desired.

Ex. 6.16. (1) Let G be a group such that $[G : Z(G)] < \infty$. Show that the commutator subgroup $[G, G]$ is finite.

(2) Let G be an f. c. group, i.e. each $g \in G$ has only finitely many conjugates in G. Show that $[G, G]$ is torsion. If, moreover, G is finitely generated, show that $[G, G]$ is finite.

Solution. (1) Let $n = [G : Z(G)]$. By FC-(6.22), G is n-abelian, i.e. $(xy)^n = x^n y^n$ for all $x, y \in G$. In particular, for any commutator $[a, b]$, we have $[a, b]^n = [a^n, b^n] = 1$ (since $a^n \in Z(G)$). Writing

$$G = \bigcup_{i=1}^{n} Z(G) a_i,$$

we see that $[G, G]$ is generated by the finite set $\{[a_i, a_j]\}$. By Exercise 15, $|[G, G]| < \infty$.

(2) We first handle the case where the f.c. group G is finitely generated, say by $\{g_1, \ldots, g_m\}$. Here,

$$Z(G) = \bigcap_{i=1}^{m} C_G(g_i)$$

is of finite index, so $\|[G, G]\| < \infty$ by (1). If G is just f. c., consider any $g \in [G, G]$. There exists a finitely generated subgroup $H \subseteq G$ such that $g \in [H, H]$. Since H is also f. c., the order of g divides $\|[H, H]\| < \infty$.

Ex. 6.17. For any group G, let

$$\Delta(G) = \{g \in G : \ [G : C_G(g)] < \infty\}, \quad \text{and}$$
$$\Delta^+(G) = \{g \in \Delta(G) : \ g \text{ has finite order}\}.$$

(1) Show that $\Delta^+(G)$ is a characteristic subgroup of G and that $\Delta^+(G)$ is the union of all finite normal subgroups of G.
(2) (B. H. Neumann) Show that $\Delta(G)/\Delta^+(G)$ is torsion-free abelian.

Solution. (1) $\Delta^+(G)$ is clearly closed under inverses. To see that it is closed under multiplication, consider $a_1, a_2 \in \Delta^+(G)$, and the subgroup H they generate in G. Since each a_i has finite order and finitely many conjugates in G, Exercise 15 implies that H is finite. Therefore $a_1 a_2 \in H$ also has finite order, and we have $a_1 a_2 \in \Delta^+(G)$. Any automorphism of G induces an automorphism of $\Delta(G)$, which in turn induces an automorphism of $\Delta^+(G)$. Therefore, $\Delta^+(G)$ is a characteristic subgroup of G. If K is any finite normal subgroup of G, we have clearly $K \subseteq \Delta^+(G)$. Conversely, if $a \in \Delta^+(G)$, let $A = \{a_1, \ldots, a_n\}$ be the set of all conjugates of a. Then $A \subseteq \Delta^+(G)$, and by Exercise 15 again, A generates a finite normal subgroup of $\Delta^+(G)$ containing a.

(2) Let $a \in \Delta(G)$ be such that $\bar{a}$ has finite order in the quotient group $Q = \Delta(G)/\Delta^+(G)$. Then $a^n \in \Delta^+(G)$ for some $n \geq 1$, and hence $(a^n)^m = 1$ for some $m \geq 1$. But then $a \in \Delta^+(G)$, and so $\bar{a} = 1$. This shows that Q is torsion-free. Since $\Delta(G)$ is an f.c. group, so is Q. But then by FC-(6.24), Q must be abelian.

Ex. 6.18. A total ordering "$<$" of the elements of a group is said to be a *right ordering* of G if $x < y \Longrightarrow xz < yz$ for any $x, y, z \in G$. Show that, if G can be right ordered, then, for any domain k, $A = kG$ has only trivial units and is a domain. Moreover, if $G \neq \{1\}$, show that A is J-semisimple.

Solution. In FC-(6.29), this was proved in the case where "$<$" is an ordering. Here, we need only re-examine the proof given before, and extend it to the case where "$<$" is a *right ordering*. Consider a product $\alpha\beta$ where

$$\alpha = a_1 g_1 + \cdots + a_m g_m, \quad g_1 < \cdots < g_m, \quad a_i \neq 0 \quad (1 \leq i \leq m),$$
$$\beta = b_1 h_1 + \cdots + b_n h_n, \quad h_1 < \cdots < h_n, \quad b_j \neq 0 \quad (1 \leq j \leq n).$$

Choose i_0, j_0 such that $g_{i_0} h_{j_0}$ is *least* among $\{g_i h_j\}$. Then $i_0 = 1$ (for otherwise $g_1 < g_{i_0} \Longrightarrow g_1 h_{j_0} < g_{i_0} h_{j_0}$). In particular, $g_{i_0} h_{j_0} = g_i h_j$ implies $i = i_0 = 1$ and hence $j = j_0$. This shows that, in the product $\alpha\beta$, $a_1 b_{j_0} g_1 h_{j_0}$ cannot be "canceled out" by any other term, so $\alpha\beta \neq 0$. To compute $U(A)$, suppose $\alpha\beta = 1$. By the above consideration, we see that there is also a *largest* product among the $g_i h_j$'s, which cannot be "canceled out" by other terms in the expansion of $\alpha\beta$. Thus, the only way for $\alpha\beta = 1$ to be possible is when $m = n = 1$, so A has only trivial units. The argument given before (see *FC*-(6.21)) for the J-semisimplicity of A (when $G \neq \{1\}$) carries over verbatim.

Ex. 6.19. For any von Neumann regular ring k, show that any finitely generated submodule M of a projective k-module P is a direct summand of P (and hence also a projective k-module).

Solution. Say $M \subseteq P$ are right k-modules. Clearly, it suffices to handle the case where P is free with a *finite* basis $e_1, \ldots, e_n$. In this new situation, we carry out the proof by induction on n, the case $n = 1$ being covered by *FC*-(4.23).

For $n \geq 2$, let $P_0 = e_1 k \oplus \cdots \oplus e_{n-1} k$, and $M_0 = M \cap P_0$. By taking the projection $\pi : P \to e_n k$, we get a short exact sequence

$$(*) \qquad 0 \to M_0 \to M \xrightarrow{\pi} I \to 0, \text{ where } I = \pi(M).$$

By the beginning case of the induction (applied to the finitely generated submodule $I \subseteq e_n k$), we have $e_n k = I \oplus J$ for some k-submodule $J \subseteq e_n k$. Thus, I is projective. Hence $(*)$ splits, and M_0 is finitely generated. By the inductive hypothesis, $P_0 = M_0 \oplus N$ for some k-module N. Since $P_0 + M = P_0 + I$, we have now

$$P = P_0 \oplus I \oplus J = (P_0 + M) \oplus J = M \oplus (N \oplus J),$$

as desired.

Comment. As far as I can ascertain, the conclusion of this exercise first appeared in I. Kaplansky's paper, "Projective modules," Annals of Math. *68* (1958), 372–377.

Note that the proof above can be used to show that M is isomorphic to a finite sum of right ideals of the form $e \cdot k$ where the e's are idempotents of k. In fact, in Theorem 4 of the above paper, Kaplansky showed that *any* projective right k-module (finitely generated or otherwise) is always a direct sum of right ideals of the form $e \cdot k$ where the e's are idempotents of k.

Ex. 6.20. Show that the conclusion of the last exercise is equivalent to the fact that, if k is a von Neumann regular ring, then so is $\mathbb{M}_n(k)$ for any $n \geq 1$.

Solution. For any endomorphism $f : k^n \to k^n$, $\mathrm{im}(f)$ is finitely generated. By the last exercise, $\mathrm{im}(f)$ is a direct summand of k^n, and hence a projective k-module. This implies that $\ker(f)$ is finitely generated, and hence also a direct summand of k^n. By Ex. $4.14A_1$, it follows that $\mathbb{M}_n(k) \cong \text{End}_k(k^n)$ is von Neumann regular.

Conversely, suppose $\mathbb{M}_n(k)$ is von Neumann regular for all $n \geq 1$. To solve the last exercise, we may again restrict ourselves to the case where M is a finitely generated (right) k-submodule of $P = k^n$. By enlarging n if necessary, we may assume that M can be generated by n elements. Thus $M = \text{im}(f)$ for some $f \in \text{End}_k(k^n) \cong \mathbb{M}_n(k)$. Since $\mathbb{M}_n(k)$ is von Neumann regular, Ex. $4.14A_1$ implies that M is a direct summand of k^n.

Comment. Of course, it is also possible to give a proof for the fact that $\mathbb{M}_n(k)$ is von Neumann regular if k is by using direct matrix computations. See the solution to Ex. 21.10B.

It is natural to ask what happens if $\mathbb{M}_n(k)$ is replaced by $\mathbb{M}_\mathbb{N}(k)$, the ring of column-finite $\mathbb{N} \times \mathbb{N}$-matrices over k (which is the endomorphism ring of the infinite-rank free module $k \oplus k \oplus \cdots$). Such a replacement brings a radical change: according to Ex. $4.14A_3$, $\mathbb{M}_\mathbb{N}(k)$ *is von Neumann regular iff k is a semisimple ring*!

Ex. 6.21. (Auslander, McLaughlin, Connell) For any nonzero ring k and any group G, show that the group ring kG is von Neumann regular iff k is von Neumann regular, G is locally finite (that is, any finite subset in G generates a finite subgroup), and the order of any finite subgroup of G is a unit in k.

Solution. First suppose $R = kG$ is von Neumann regular. As a quotient ring of R, k is certainly von Neumann regular. Consider any finite subgroup $E \subseteq G$, say of order m. It suffices to show that any prime $p \mid m$ is a unit in k. To see this, we fix an element $\sigma \in E$ of order p (which exists by Cauchy's Theorem). Taking an element $\alpha \in R$ such that $1 - \sigma = (1 - \sigma)\alpha(1 - \sigma)$, we can argue as in the proof of FC-(6.1) that $p \in U(k)$. Finally, let H be a subgroup of G generated by a finite set of elements, say $h_1, \ldots, h_n$. It is easy to see that

$$I := \sum_{h \in H} R \cdot (h - 1) = \sum_{i=1}^{n} R \cdot (h_i - 1)$$

(using facts such as $h_i^{-1} - 1 = h_i^{-1}(1 - h_i)$, and $h_i h_j - 1 = h_i(h_j - 1) + h_i - 1$). Since R is von Neumann regular, $I = Re$ for some idempotent e. Now I is contained in the augmentation ideal of R, so $e \neq 1$ (since $k \neq 0$). Thus, $f := 1 - e \neq 0$. But for any $h \in H$,

$$(h - 1)f \in I \cdot f = R \cdot ef = 0 \Longrightarrow f = hf.$$

Since f involves only finitely many elements of G, this implies that $|H| < \infty$.

For the converse, let us assume the given conditions on k and G. Consider any element $\alpha = a_1 h_1 + \cdots + a_n h_n \in R$ $(a_i \in k, h_i \in G)$, for which we want to show $\alpha \in \alpha R \alpha$. Let H be the (finite) subgroup of G generated by $h_1, \ldots, h_n$. Since $\alpha \in kH$, we are done if we can show that $S := kH$ is von Neumann regular. Consider any principal left ideal, say $S \cdot \beta$, where $\beta \in S$. Viewing $S \cdot \beta \subseteq S$ as k-modules, we have

$$S \cong k^{|H|} \quad \text{and} \quad S \cdot \beta = \sum_{h \in H} k \cdot (h\beta).$$

Since k is von Neumann regular, $S \cdot \beta$ is a direct summand of S as k-*modules* (by Ex. 19). From Ex. 3 (which applies since $|H| \in U(k)$), it follows that $S \cdot \beta$ is a direct summand of S as S-modules. This checks that S is von Neumann regular, as desired.

Chapter 3

Introduction to Representation Theory

§7. Modules over Finite-Dimensional Algebras

In this new chapter, we shift our main emphasis from rings to modules. More specifically, we study modules over finite-dimensional k-algebras, where k is a field. For such an algebra R, $R/\mathrm{rad}\, R$ is a finite-dimensional semisimple k-algebra, whose structure is completely determined by Wedderburn's Theorem. Thus, the study of simple (and semisimple) left R-modules is well under control.

In studying finite-dimensional k-algebras R, it is often essential to consider base field extensions K/k, and the associated algebra scalar extensions $R^K = R \otimes_k K$. This leads quickly to the notion of an absolutely irreducible module: an irreducible (left) R-module M is said to be *absolutely irreducible* if $M^K = M \otimes_k K$ remains irreducible over R^K for any field extension K/k. (An equivalent condition is that End $(_R M) = k$: see FC-(7.5).) We say that a field K/k is a splitting field for R (or that R splits over k) if every (left) irreducible R^K-module is absolutely irreducible. (An equivalent condition is that $R^K/\mathrm{rad}\,(R^K)$ is a finite direct product of matrix algebras over K: see FC-(7.7).) Roughly speaking, if K/k is a splitting field for R, the classification of simple modules reaches a "stable state" when we go to the scalar extension algebra R^K. The precise meaning of this is given in FC-(7.14); see also Exercise 8 below.

Another advantage of the k-algebra setting is that we can define *characters* of finite-dimensional R-modules. In characteristic 0, the characters of such a module determines its composition factors (FC-(7.19)). In the case of a split algebra, independently of the field characteristic, characters determine the *simple* modules (*FC*-(7.21)).

The exercises in this section mostly deal with the concept of splitting, and with the behavior of finite-dimensional modules under scalar extensions of the ground field.

Exercises for §7

In the following exercises, k denotes a field.

Ex. 7.1. Let M, N be finite-dimensional modules over a finite-dimensional k-algebra R. For any field $K \supseteq k$, show that M^K and N^K have a common composition factor as R^K-modules iff M and N have a common composition factor as R-modules.

Solution. Fix a composition series

$$0 = M_0 \subsetneq M_1 \subsetneq \cdots \subsetneq M_n = M$$

for M. Then

$$0 = M_0^K \subsetneq M_1^K \subsetneq \cdots \subsetneq M_n^K = M^K$$

can be refined into a composition series for M^K. Therefore, if $\{V_i\}$ are the composition factors of M, then those of M^K are the composition factors of $\{V_i^K\}$. This gives immediately the "if" part of the Exercise, and the "only if" part also follows by an appeal to *FC*-(7.13)(2).

Ex. 7.2. Let R be a finite-dimensional k-algebra which splits over k. Show that, for any field $K \supseteq k$, rad $(R^K) = (\text{rad } R)^K$.

Solution. Without any assumptions on k, the nilpotency of $(\text{rad } R)^K$ implies that $(\text{rad } R)^K \subseteq \text{rad}(R^K)$ (see *FC*-(5.14), and also Exercise 7 below). Now assume R splits over k. The quotient $_R(R/\text{rad } R)$ is a semisimple R-module. Since every simple R-module remains a simple R^K-module upon scalar extension to K, $(R/\text{rad } R)^K$ remains a semisimple R^K-module. Therefore, $\text{rad}(R^K)$ annihilates

$$(R/\text{rad } R)^K = R^K/(\text{rad } R)^K,$$

which amounts precisely to rad $(R^K) \subseteq (\text{rad } R)^K$.

Ex. 7.3. Let R be a finite-dimensional k-algebra, M be an R-module and $E = \text{End}_R M$. Show that if $f \in E$ is such that $f(M) \subseteq (\text{rad } R)M$, then $f \in \text{rad } E$.

Solution. Let

$$I = \{f \in E : \quad f(M) \subseteq (\text{rad } R)M\}.$$

It is routine to check that I is an ideal in the endomorphism ring E. For this ideal I, we have $I^n M \subseteq (\text{rad } R)^n M$. Since rad R is nilpotent, $I^n M = 0$ for a sufficiently large n, so $I^n = 0$. This implies that $I \subseteq \text{rad } E$, as desired.

Ex. 7.4. Let R be a left artinian ring and C be a subring in the center $Z(R)$ of R. Show that Nil $C = C \cap \text{rad } R$. If R is a finite-dimensional algebra over a subfield $k \subseteq C$, show that rad $C = C \cap \text{rad } R$.

Solution. Let $a \in \text{Nil } C$. Then aR is a nil ideal of R, so $aR \subseteq \text{rad } R$. This shows that Nil $C \subseteq C \cap \text{rad } R$. For the converse, note that rad R is nil (in fact nilpotent, since R is left artinian). Therefore, $C \cap \text{rad } R \subseteq \text{Nil } C$. Now assume $\dim_k R < \infty$, where k is a field $\subseteq C$. Then we also have $\dim_k C < \infty$, so rad C is nil(potent). This implies that rad $C = \text{Nil } C = C \cap \text{rad } R$, as desired.

Comment. In the case when $C = Z(R)$ where R is any left artinian ring, the equality rad $C = C \cap \text{rad } R$ always holds. The proof is considerably harder: see Exercise 21.25 below.

Ex. 7.5. Let R be a finite-dimensional k-algebra which splits over k. Show that any k-subalgebra $C \subseteq Z(R)$ also splits over k.

Solution. By Exercise 4, rad $C = C \cap \text{rad } R$. Therefore, we have a k-embedding $\varphi : C/\text{rad } C \to R/\text{rad } R$. Since R splits over k,

$$R/\text{rad } R \cong \prod^r \mathbb{M}_{n_i}(k)$$

for suitable $n_1, \ldots, n_r$ (by FC-(7.7)). We have then

$$\varphi(C/\text{rad } C) \subseteq Z(R/\text{rad } R) \cong \prod^r k.$$

We claim that *every k-subalgebra of $B := \prod^r k$ is k-isomorphic to $\prod^s k$ for some $s \leq r$.* Assuming this claim, we will have

$$C/\text{rad } C \cong \varphi(C/\text{rad } C) \cong \prod^s k,$$

so by FC-(7.7), C splits over k. To prove our claim, consider a k-subalgebra $A \subseteq B$. Since A is commutative, reduced, and artinian,

$$A = K_1 \times \cdots \times K_s$$

for suitable finite field extensions K_i/k (e.g. use the Wedderburn-Artin Theorem!). We finish by showing that $K_i = k$ for all i. Let e_i be the identity of K_i. For a suitable coordinate projection π of $B = \prod^r k$ onto k, we have $\pi(e_i) \neq 0$. Since $\pi(e_i)$ is an idempotent, we must have $\pi(e_i) = 1$. Thus, π defines a k-algebra homomorphism from K_i to k, and it follows that $\pi : K_i \cong k$, as desired.

Comment. If C is not assumed to be in the center $Z(R)$ of R, the conclusion in the Exercise may no longer hold. For instance, take $k = \mathbb{R}$ and $R = M_2(\mathbb{R})$. Then R splits over $\mathbb{R}$. However, the $\mathbb{R}$-subalgebra

$$C = \left\{ \begin{pmatrix} a & -b \\ b & a \end{pmatrix} : \quad a, b \in \mathbb{R} \right\} \subseteq R$$

is $\mathbb{R}$-isomorphic to $\mathbb{C}$, which *does not* split over $\mathbb{R}$. Of course, here

$$C \not\subseteq Z(R) = \mathbb{R}.$$

Alternatively, we can also take $R = M_4(\mathbb{R})$, and C to be an $\mathbb{R}$-subalgebra of R isomorphic to the division ring of real quaternions: see *FC*-p. 15. Many other counterexamples can be constructed using group algebras, say over $\mathbb{Q}$. For "most" finite groups G, $C = \mathbb{Q}G$ fails to split over $\mathbb{Q}$. However, by Cayley's Theorem, G can be embedded in the symmetric group S_n for some n, and a well-known theorem in the representation theory of groups says that $R = \mathbb{Q}S_n$ (which contains C) always splits over $\mathbb{Q}$.

Ex. 7.6. For a finite-dimensional k-algebra R, let $T(R) = \operatorname{rad} R + [R, R]$, where $[R, R]$ denotes the subgroup of R generated by $ab - ba$ for all $a, b \in R$. Assume that k has characteristic $p > 0$. Show that

$$T(R) \subseteq \{a \in R : \quad a^{p^m} \in [R, R] \quad \text{for some } m \geq 1\},$$

with equality if k is a splitting field for R.

Solution. For $a \in T(R)$, write $a = b + c$ where $b \in \operatorname{rad} R$ and $c \in [R, R]$. By *FC*-(7.15), we have for every m:

$$b^{p^m} = (a - c)^{p^m} \equiv a^{p^m} - c^{p^m} \equiv a^{p^m} \quad (\text{mod } [R, R]).$$

Choosing m to be large enough, we have $b^{p^m} = 0$ (since $\operatorname{rad} R$ is nil). Therefore, the above congruence shows that $a^{p^m} \in [R, R]$. Now assume k is a splitting field for R, and let $a \in R$ be such that $a^{p^m} \in [R, R]$ for some m. Let

$$\overline{R} = R/\operatorname{rad} R \cong \prod_i A_i$$

where $A_i = M_{n_i}(k)$. Our job is to show that $\bar{a} = a + \operatorname{rad} R$ belongs to $[\overline{R}, \overline{R}]$. Using the direct product decomposition above, we are reduced to showing that, for any i, the image $\bar{a}_i \in A_i$ of $\bar{a}$ belongs to $[A_i, A_i]$ (given that $\bar{a}_i^{p^m} \in [A_i, A_i]$ for some m). Therefore, we may as well assume that $R = M_n(k)$. Here, let us compare

$$T'(R) : = \{a \in R : \quad a^{p^m} \in [R, R] \quad \text{for some } m \geq 1\}$$

with $T(R) = [R, R]$. By *FC*-(7.15) again, $T'(R)$ is a k-subspace of R containing $T(R)$. But by *FC*-(7.16), $T(R)$ has codimension 1 in R. Since

$$a = \operatorname{diag}(1, 0, \ldots, 0) \notin T'(R)$$

(also by *FC*-(7.16)), we must have $T(R) = T'(R)$.

Ex. 7.7. Using FC-(4.1) and FC-(7.13), give another proof for the fact (already established in FC-(5.14)) that for any finite-dimensional k-algebra R and any field extension $K \supseteq k$, $(\operatorname{rad} R)^K \subseteq \operatorname{rad}(R^K)$.

Solution. By FC-(4.1), it suffices to show that $(\operatorname{rad} R)^K$ annihilates any simple left R^K-module V. But by FC-(7.13)(1), V is a composition factor of M^K for some simple left R-module M. Since $(\operatorname{rad} R)M = 0$, it follows that $(\operatorname{rad} R)^K M^K = 0$. But then $(\operatorname{rad} R)^K V = 0$, as desired.

Ex. 7.8. Let R be a finite-dimensional k-algebra and let $L \supseteq K \supseteq k$ be fields. Assume that L is a splitting field for R. Show that K is a splitting field for R iff, for every simple left R^L-module M, there exists a (simple) left R^K-module U such that $U^L \cong M$.

Solution. The "only if" part follows directly from FC-(7.14). For the "if" part, assume every simple (left) R^L-module "comes from" a simple R^K-module. Let $M_1, \ldots, M_n$ be all the distinct simple R^L-modules, and let $M_i = U_i^L$, where $U_1, \ldots, U_n$ are (necessarily distinct) simple R^K-modules. If there is another simple R^K-module V not isomorphic to any U_i, then by FC-(7.13)(2), a composition factor of V^L would give a simple R^L-module not isomorphic to any M_i, a contradiction. Thus, $\{U_1, \ldots, U_n\}$ give all the distinct simple R^K-modules. Each U_i remains irreducible over L and hence over the algebraic closure of L. Therefore, each U_i is *absolutely* irreducible (by FC-(7.5)), and we have proved that K is a splitting field for R.

Ex. 7.9. If $K \supseteq k$ is a splitting field for a finite-dimensional k-algebra R, does it follow that K is also a splitting field for any quotient algebra $\overline{R}$ of R?

Solution. Since this fails for subalgebras (see *Comment* following Exercise 5), an impulsive guess might be that it also fails for quotient algebras. However, the statement turns out to be true for quotient algebras. To prove this, we may assume that $K = k$. Consider the natural surjection $\varphi: R \to \overline{R}$. Any simple left $\overline{R}$-module V may be viewed as a simple left R-module via φ. For any field extension $L \supseteq k$, V^L remains simple as a left R^L-module. Therefore, V^L is also a simple left $\overline{R}^L$-module. This shows that V is absolutely irreducible, so k is indeed a splitting field for $\overline{R}$.

Ex. 7.10. Let R be a k-algebra where k is a field, and M, N be left R-modules, with $\dim_k M < \infty$. In FC-(7.4), it was shown that, for any field extension $K \supseteq k$, the natural map

$$\theta: (\operatorname{Hom}_R(M, N))^K \longrightarrow \operatorname{Hom}_{R^K}(M^K, N^K)$$

is an isomorphism of K-vector spaces. Replacing the hypothesis $\dim_k M < \infty$ by "M is a finitely presented R-module," give a basis-free proof for the fact that θ is a K-isomorphism.

Solution. First consider the special case when $M \cong R^n$. In this case, we have the canonical isomorphisms

$$(\operatorname{Hom}_R(M, N))^K \cong (\operatorname{Hom}_R(R^n, N))^K \cong (nN)^K \cong n \cdot N^K,$$
$$\operatorname{Hom}_{R^K}(M^K, N^K) \cong \operatorname{Hom}_{R^K}((R^K)^n, N^K) \cong n \cdot N^K.$$

From this, we can safely conclude that θ is an isomorphism. (Some commutative diagrams must be checked, but it is mostly routine work.) Next we assume M is a *finitely presented* R-module, which means that there exists an exact sequence of R-modules

$$M_1 \to M_2 \to M \to 0$$

where $M_1 = R^n$ and $M_2 = R^m$. Applying the left-exact Hom-functors into N and into N^K, we have the following commutative diagram:

$$
\begin{array}{ccccccc}
0 \to & (\operatorname{Hom}_R(M, N))^K & \to & (\operatorname{Hom}_R(M_2, N))^K & \to & (\operatorname{Hom}_R(M_1, N))^K \\
& \downarrow \theta & & \downarrow \theta_2 & & \downarrow \theta_1 \\
0 \to & \operatorname{Hom}_{R^K}(M^K, N^K) & \to & \operatorname{Hom}_{R^K}(M_2^K, N^K) & \to & \operatorname{Hom}_{R^K}(M_1^K, N^K)
\end{array}
$$

Since θ_1, θ_2 are both isomorphisms, an easy diagram chase shows that θ is also an isomorphism.

§8. Representations of Groups

In this section, we specialize our finite-dimensional k-algebras to group algebras $R = kG$, where G is any finite group. Here, the module theory over R becomes the theory of representations of the group G. In this beautiful theory, groups, rings, and modules find a common melting pot, and group characters emerge as a central notion unifying the three.

In studying group representations, one needs to distinguish "ordinary representations" (char $k \nmid |G|$) from "modular representations" (char $k \mid |G|$). The theory of modular representations is quite deep, and requires various sophisticated tools. Here, one has to deal with indecomposable modules as well as irreducible ones. In a beginning treatment, one usually focuses on ordinary representation theory, and postpones the modular theory in order to avoid excessive technical difficulties. Our approach to basic representation theory in this section is certainly guided by this piece of traditional wisdom.

The theory of group characters works particularly well in characteristic zero over a splitting field. In this case, the character table of the group G encodes much of the key information about G itself (although it is not sufficient to determine G). The characters of a group are usually eminently computable, since so many precise facts are known about them. Quite often, information about the characters of a group G can be used to prove purely group-theoretic facts about G. Burnside's Theorem on the solvability of groups of order $p^a q^b$ has remained the quickest and one of the most beautiful applications of the theory of group characters. The Feit-Thompson Theorem on the solvability of groups of odd order went much further in combining character-theoretic methods with purely group-theoretic methods. With the proof of the odd-order theorem, the theory of finite group representations was elevated to a new plateau.

The many exercises in this section deal with examples of representations and their characters, explicit computations of the decompositions of semisimple group algebras, units of integral group rings, and the isomorphism problem for group algebras. Most problems are in the framework of ordinary representations, but a couple of elementary problems on modular representations (cf. Exs. 11, 12, 26, 30, 31) offer a quick glimpse into what happens in the world of characteristic $p \mid |G|$. Some familiarity with basic finite group theory on the part of the reader will be assumed throughout these exercises.

Exercises for §8

Ex. 8.1. Give an example of a pair of finite groups G, G' such that, for some field k, $kG \cong kG'$ as k-algebras, but $G \not\cong G'$ as groups.

Solution. Let G, G' be abelian groups of order n, and let $k = \mathbb{C}$. Since k has no proper algebraic extensions, Wedderburn's Theorem implies that kG, kG' are both isomorphic (as k-algebras) to $k \times \cdots \times k$ (n factors). But of course, G and G' need not be isomorphic!

Comment. The solution to the above problem was easy since we were allowed to pick the field k. (See also Exercise 15). If the field k is specified, the problem of deciding whether $kG \cong kG'$ implies $G \cong G'$ can be much harder. There are various affirmative results. For instance, a theorem of Perlis and Walker states that, for finite abelian groups G and G', $\mathbb{Q}G \cong \mathbb{Q}G'$ (as $\mathbb{Q}$-algebras) does imply $G \cong G'$. (See their article in Trans. Amer. Math. Soc. *68* (1950), 420–426.) But if p is a prime > 2 and G, G' are the two nonisomorphic nonabelian groups of order p^3, then $\mathbb{Q}G \cong \mathbb{Q}G'$. In characteristic p, Deskins has shown that, for finite abelian p-groups G, G', and any field k of characteristic p, $kG \cong kG'$ implies $G \cong G'$. (See Duke Math. J. *23* (1956), 35–40.) Most remarkably, Dade has constructed a pair of nonisomorphic finite groups G, G' such that $kG \cong kG'$ (as k-algebras) for *all* fields k. (See Math. Zeit. *119* (1971), 345–348.) For an exposition of the work done on this kind of "Isomorphism Problem" over fields, see Chapter 14 of Passman's book "The Algebraic Structure of Group Rings", J. Wiley Interscience, 1977.

The case of *integral* group rings (whether $\mathbb{Z}G \cong \mathbb{Z}G'$ implies $G \cong G'$) presents yet another facet of the Isomorphism Problem: we have only barely touched the "tip of the iceberg" of this problem in *FC*-(8.25).

For *nilpotent* (finite) groups, an affirmative answer to this problem has been given by K. Roggenkamp and L. Scott (Ann. Math. *126* (1987), 593–647). However, M. Hertweck has recently constructed two *nonisomorphic* solvable groups G and G' of order $2^{21} \cdot 97^{28}$ with $\mathbb{Z}G \cong \mathbb{Z}G'$; see his paper in Ann. Math. *154* (2001), 115–138. One remaining problem is whether such examples can be found for *odd-order* groups. For an excellent survey on the group ring Isomorphism Problem (and its interesting relations to the various conjectures of Zassenhaus), see D. Passman's feature review of Hertweck's article in MR 2002e:20010.

Ex. 8.2. Let k be a field whose characteristic is prime to the order of a finite group G. Show that the following two statements are equivalent:

(a) each irreducible kG-module has k-dimension 1;
(b) G is abelian, and k is a splitting field for G.

Solution. For the proof, it is convenient to use the notations set up in *FC*-(8.1). Assume (a). Then we have $1 = \dim_k M_i = n_i \dim_k D_i$, which implies that $n_i = 1$ and $\dim_k D_i = 1$. Therefore, $D_i = k$, and we have

$$kG \cong k \times \cdots \times k$$

(since $\operatorname{rad} kG = 0$ here). Clearly, this gives (b). Conversely, if (b) holds, then $n_i = \dim_k D_i = 1$ for all i, and we have $\dim_k M_i = n_i \dim_k D_i = 1$, as desired.

Ex. 8.3. Let $G = S_3$. Show that $\mathbb{Q}G \cong \mathbb{Q} \times \mathbb{Q} \times \mathrm{M}_2(\mathbb{Q})$ and compute the central idempotents of $\mathbb{Q}G$ that give this decomposition of $\mathbb{Q}G$ into its simple components. Compute, similarly, the decompositions of $\mathbb{Q}G_1, \mathbb{Q}G_2$, where G_1 is the Klein 4-group, and G_2 is the quaternion group of order 8.

Solution. Since the three irreducible complex representations of $G = S_3$ can be defined over $\mathbb{Q}$, we know that $\mathbb{Q}$ is a splitting field for G. The dimension equation $|G| = \sum \chi_i(1)^2$ here is $6 = 1^2 + 1^2 + 2^2$, so we have

$$\mathbb{Q}G \cong \mathbb{Q} \times \mathbb{Q} \times \mathrm{M}_2(\mathbb{Q}).$$

The three central idempotents of $\mathbb{Q}G$ giving such a decomposition can be read off from *FC*-(8.15)(1), by using the character table of S_3. They are:

$$e_1 = \tfrac{1}{6}\{(1) + (123) + (132) + (12) + (23) + (13)\},$$
$$e_2 = \tfrac{1}{6}\{(1) + (123) + (132) - (12) - (23) - (13)\},$$
$$e_3 = 1 - e_1 - e_2 = \tfrac{2}{3}(1) - \tfrac{1}{3}(123) - \tfrac{1}{3}(132).$$

Next, let G_1 be the Klein 4-group $\langle a \rangle \times \langle b \rangle$. Here,

$$\mathbb{Q}G_1 \cong \mathbb{Q} \times \mathbb{Q} \times \mathbb{Q} \times \mathbb{Q},$$

and the (central) idempotents giving this decomposition are determined again from the irreducible characters:

$$e_1 = \tfrac{1}{4}(1 + a + b + ab), \qquad e_2 = \tfrac{1}{4}(1 + a - b - ab),$$

$$e_3 = \tfrac{1}{4}(1 - a + b - ab), \qquad e_4 = \tfrac{1}{4}(1 - a - b + ab).$$

Finally, let G_2 be the quaternion group $\{1, i, j, k, \varepsilon, \varepsilon i, \varepsilon j, \varepsilon k\}$. Since there is an irreducible $\mathbb{Q}G_2$-module given by the rational quaternions $\mathbb{H}_0$, and four 1-dimensional $\mathbb{Q}G_2$-modules, we have

$$\mathbb{Q}G_2 \cong \mathbb{Q} \times \mathbb{Q} \times \mathbb{Q} \times \mathbb{Q} \times \mathbb{H}_0.$$

For the corresponding central idempotents, we can first construct the ones associated with the linear characters (cf. Exercise 18 below):

$$e_1 = \tfrac{1}{8}(1 + \varepsilon + i + \varepsilon i + j + \varepsilon j + k + \varepsilon k),$$

$$e_2 = \tfrac{1}{8}(1 + \varepsilon - i - \varepsilon i + j + \varepsilon j - k - \varepsilon k),$$

$$e_3 = \tfrac{1}{8}(1 + \varepsilon + i + \varepsilon i - j - \varepsilon j - k - \varepsilon k),$$

$$e_4 = \tfrac{1}{8}(1 + \varepsilon - i - \varepsilon i - j - \varepsilon j + k + \varepsilon k).$$

The last central idempotent can then be obtained as:

$$e_5 = 1 - e_1 - e_2 - e_3 - e_4 = \tfrac{1}{2}(1 - \varepsilon).$$

(This is indeed an idempotent, as $e_5^2 = \tfrac{1}{4}(1 - 2\varepsilon + \varepsilon^2) = \tfrac{1}{2}(1 - \varepsilon) = e_5$.)

Ex. 8.4. Let $R = kG$ where k is any field and G is any group. Let I be the ideal of R generated by $ab - ba$ for all $a, b \in R$. Show that $R/I \cong k[G/G']$ as k-algebras, where G' denotes the commutator subgroup of G. Moreover, show that $I = \sum_{a \in G'} (a - 1)kG$.

Solution. Let $\overline{G}$ be the abelian group G/G'. Then $k\overline{G}$ is a commutative ring, so the natural ring homomorphism $\varphi_0 : R \to k\overline{G}$ sends $ab - ba$ to zero, for all $a, b \in R$. This shows that $\varphi_0(I) = 0$, so φ_0 induces a k-algebra homomorphism $\varphi : R/I \to k\overline{G}$. Next, we shall try to construct a k-algebra homomorphism $\psi : k\overline{G} \to R/I$. Consider the group homomorphism $\psi_0 : G \to \mathrm{U}(R/I)$ defined by $\psi_0(g) = g + I$. Since

$$\psi_0(gh) = gh + I = hg + I = \psi_0(hg) \quad (\forall g, h \in G),$$

ψ_0 induces a group homomorphism $\overline{G} \to \mathrm{U}(R/I)$, which in turn induces a k-algebra homomorphism $\psi : k\overline{G} \to R/I$. It is easy to check that ψ and φ are mutually inverse maps, so we have $R/I \cong k\overline{G}$.

To prove the last part of the Exercise, let

$$J = \sum_{a \in G'} (a - 1)kG.$$

Since G' is normal in G, we have for any $g \in G$ and $a \in G'$:

$$g(a-1)kG = (gag^{-1} - 1)g \cdot kG \subseteq J,$$

so J is an ideal in kG. The same method above can be used to show that $R/J \cong k\overline{G}$. From this, we conclude easily that $I = J$.

Ex. 8.5. For any field k and for any normal subgroup H of a group G, show that $kH \cap \mathrm{rad}\, kG = \mathrm{rad}\, kH$.

Solution. Even without H being normal in G, we have in general

$$kH \cap \mathrm{rad}\, kG \subseteq \mathrm{rad}\, kH,$$

by Exercise 6.4. Therefore, it suffices to prove that, if H is normal in G, then $\mathrm{rad}\, kH \subseteq \mathrm{rad}\, kG$. Consider any simple left kG-module V. By Clifford's Theorem FC-(8.5), $_{kH}V$ is a semisimple kH-module. Therefore, $(\mathrm{rad}\, kH)V = 0$. By FC-(4.1), this implies that $\mathrm{rad}\, kH \subseteq \mathrm{rad}\, kG$.

Comment. If H is not normal in G, the inclusion $\mathrm{rad}\, kH \subseteq \mathrm{rad}\, kG$ may not be true, even when H, G are finite groups. To construct an explicit example, let k be any field of characteristic 2, $G = S_3$, and $H = \langle (12) \rangle$. Then $y = 1 + (12) \in \mathrm{rad}\, kH$, but we claim that $y \notin \mathrm{rad}\, kG$. Indeed, if $y \in \mathrm{rad}\, kG$, then

$$1 + (23)y = 1 + (23) + (132) \in \mathrm{U}(kG).$$

Suppose $a + b(12) + c(13) + d(23) + e(123) + f(132)$ is its inverse. A quick calculation leads to six incompatible linear equations in $a, b, \dots, f$.

Ex. 8.6. In the above Exercise, assume further that $[G : H]$ is finite and prime to char k. Let V be a kG-module and W be a kH-module. Show that

(1) V is a semisimple kG-module iff $_{kH}V$ is a semisimple kH-module.
(2) W is a semisimple kH-module iff the induced module $kG \otimes_{kH} W$ is a semisimple kG-module.

Solution. (1) The "only if" part follows from Clifford's Theorem FC-(8.5), since a semisimple kG-module is a direct sum of simple kG-modules. The "if" part is the earlier Exercise 6.1.

For (2), let $V = kG \otimes_{kH} W$, and let $G = \bigcup_i x_i H$ be the coset decomposition with respect to H. Then $kG = \bigoplus x_i kH$, and so $V = \bigoplus x_i \otimes_{kH} W$. Here, each $x_i \otimes_{kH} W$ is a kH-submodule of V. Indeed, for any $h \in H$, we have $h' := x_i^{-1} h x_i \in H$, so for any $w \in W$:

$$h(x_i \otimes w) = h x_i \otimes w = x_i h' \otimes w = x_i \otimes h'w.$$

This shows, in fact, that the kH-module $x_i \otimes_{kH} W$ is just a "twist" of $_{kH}W$ (using the automorphism of H induced by the conjugation by x_i). Assume that $_{kH}W$ is semisimple. It is easy to see that the "twisted" kH-modules $x_i \otimes_{kH} W$ are also semisimple. Therefore, $_{kH}V$ is semisimple, and part (1) above implies that $_{kG}V$ is semisimple. Conversely, if $_{kG}V$ is semisimple, then by part (1) so is $_{kH}V$. Since $W = 1 \otimes_{kH} W$ is a kH-submodule of $_{kH}V$, it is also semisimple.

Ex. 8.7. (Villamayor, Green-Stonehewer, Willems) Let k be a field, and H be a normal subgroup of a finite group G. Show that rad $kG = kG \cdot$ rad kH iff char $k \nmid [G : H]$.

Solution. Write $G = \bigcup_i x_i H$ as in the solution of Exercise 6. First assume char $k \nmid [G : H]$. Since $kG = \bigoplus_i x_i kH$, any $\alpha \in$ rad kG can be written in the form $\sum x_i \alpha_i$ where $\alpha_i \in kH$. We claim that each $\alpha_i \in$ rad kH. Indeed, consider any simple kH-module W. By Exercise 6, $kG \otimes_{kH} W$ is a semisimple kG-module, so

$$0 = \alpha \cdot kG \otimes_{kH} W \supseteq \sum x_i \alpha_i \otimes_{kH} W = \sum x_i \otimes \alpha_i W.$$

This implies that $\alpha_i W = 0$ for all i, so $\alpha_i \in$ rad kH. We have now shown that rad $kG \subseteq kG \cdot$ rad kH, and the reverse inclusion follows from Exercise 5. (Note that the work above actually yields an explicit dimension equation: $\dim_k$ rad $kG = [G : H] \dim_k$ rad kH.)

For the converse, let us now assume that

$$\text{rad } kG = kG \cdot \text{rad } kH.$$

Consider the group algebra $R = k[G/H]$ as a left kG-module (via the action of G on G/H). Since H acts as the identity on R, the augmentation ideal of kH acts as zero, and therefore so do rad kH and rad $kG = kG \cdot$ rad kH. We can then view R as a $kG/\text{rad } kG$-module. Since $kG/\text{rad } kG$ is a semisimple ring, R is a semisimple $kG/\text{rad } kG$-module. Therefore, R is also a semisimple module over kG and over R, so R is itself a semisimple ring. Now $FC\text{-}(6.1)$ implies that char $k \nmid [G : H]$.

Ex. 8.8. Let G be a finite group such that, for some field k, kG is a finite direct product of k-division algebras. Show that any subgroup $H \subseteq G$ is normal.

Solution. Since kG is semisimple, $|G|$ is a unit in k, by $FC\text{-}(6.1)$. Therefore, for any given $H \subseteq G$, we can form the element

$$e = |H|^{-1} \sum_{h \in H} h \in kH \subseteq kG.$$

We have

$$e^2 = |H|^{-2} \cdot |H| \sum_{h \in H} h = e,$$

so e is an idempotent. Writing $kG = D_1 \times \cdots \times D_r$ where the D_i's are division rings, we see that any idempotent in kG is central. Therefore, for any $g \in G$, we have $geg^{-1} = e$. This implies that $gHg^{-1} = H$, so $H \triangleleft G$.

Comment. A group G is said to be *Hamiltonian* if any subgroup $H \subseteq G$ is normal. (In some books, G is also required to be nonabelian.) The structure of (finite or infinite) Hamiltonian groups is completely known: see Theorem 12.5.4 in Marshall Hall's book "The Theory of Groups," Macmillan, 1959.

Ex. 8.9. Show that the First Orthogonality Relation in *FC*-(8.16)(A) can be generalized to

$$\sum_{g \in G} \chi_i(g^{-1})\chi_j(hg) = \delta_{ij}|G|\chi_i(h)/n_i,$$

where h is any element in G, and $n_i = \chi_i(1)$. (*FC*-(8.16)(A) was the special case of this formula for $h = 1$.)

Solution. Here χ_i are the characters of the simple left kG-modules, where k is a splitting field for G of characteristic not dividing $|G|$. The proof depends on the expressions for the central idempotents in kG giving the Wedderburn decomposition of kG into its simple components. According to *FC*-(8.15)(1), these central idempotents are given by

$$e_i = |G|^{-1} n_i \sum_{g \in G} \chi_i(g^{-1}) g.$$

Now use the equations $e_i e_j = \delta_{ij} e_i$, where the δ_{ij}'s are the Kronecker deltas. The coefficient of h^{-1} on the RHS is $\delta_{ij}|G|^{-1} n_i \chi_i(h)$, and the coefficient of h^{-1} on the LHS is

$$n_i n_j |G|^{-2} \sum_{g \in G} \chi_i(g^{-1})\chi_j(hg).$$

Therefore, the desired formula follows.

Ex. 8.10. Under the same assumptions on kG as in Exercise 9, let $F_k(G)$ be the k-space of class functions on G, given the inner product

$$[\mu, \nu] = \frac{1}{|G|} \sum_g \mu(g^{-1})\nu(g).$$

Show that, for any class function $f \in F_k(G)$, there is a "Fourier expansion" $f = \sum_i [f, \chi_i] \chi_i$, and that, for any two class functions $f, f' \in F_k(G)$, there is a "Plancherel formula"

$$[f, f'] = \sum_i [f, \chi_i] [f', \chi_i].$$

Assuming that char $k = 0$, show that $f = \chi_M$ for some kG-module M iff $[f, \chi_i]$ is a nonnegative integer for all i, and that M is irreducible iff $[\chi_M, \chi_M] = 1$.

Solution. We know that the irreducible characters $\{\chi_i\}$ form an orthonormal basis for $F_k(G)$ under $[\,,\,]$. If $f = \sum_i a_i \chi_i$, we see immediately that

$$[f, \chi_j] = [\sum_i a_i \chi_i, \chi_j] = a_j.$$

If similarly, $f' = \sum_i a'_i \chi_i$, then

$$[f, f'] = [\sum_i a_i \chi_i, \sum_j a'_j \chi_j] = \sum_i a_i a'_i$$

$$= \sum_i [f, \chi_i] \, [f', \chi_i].$$

If $f = \chi_M$, we can resolve χ_M into $\sum s_i \chi_i$ where the s_i's are nonnegative integers. In this case $[f, \chi_i] = s_i$, as desired. Conversely, if $s_i := [f, \chi_i]$ are nonnegative integers, we can form the module $M := \bigoplus_i s_i M_i$ where M_i is the simple kG-module affording the character χ_i. Then

$$f = \sum_i [f, \chi_i] \, \chi_i = \sum_i s_i \chi_i = \chi_M,$$

as desired. If M is an irreducible kG-module, say $M \cong M_i$, then $\chi_M = \chi_i$ and of course $[\chi_M, \chi_M] = [\chi_i, \chi_i] = 1$. Conversely, let M be a kG-module with $[\chi_M, \chi_M] = 1$. Write $\chi_M = \sum s_i \chi_i$ as before, where $s_i \geq 0$. Then $\sum s_i^2 = 1$. Since char $k = 0$, we must have some $s_i = 1$ and other s_j's zero. Therefore $\chi_M = \chi_i$, and $M \cong M_i$ by FC-(7.19).

Ex. 8.11. Let k be the algebraic closure of $\mathbb{F}_p$ and $K = k(t)$, where t is an indeterminate. Let G be an elementary p-group of order p^2 generated by a, b. Show that

$$(*) \qquad a \mapsto A = \begin{pmatrix} 1 & 1 \\ 0 & 1 \end{pmatrix}, \qquad b \mapsto B = \begin{pmatrix} 1 & t \\ 0 & 1 \end{pmatrix}$$

defines a representation of G over K which is not equivalent to any representation of G over k.

Solution. Since $A^p = B^p = I$ and $AB = BA$, $(*)$ defines a representation $D : G \to \mathrm{GL}_2(K)$. We shall prove a slightly stronger statement: For any extension field $L \supseteq K$, D cannot be equivalent over L to any representation of G over k. Indeed, assume otherwise. Then there exists

$$U = \begin{pmatrix} a & b \\ c & d \end{pmatrix} \in \mathrm{GL}_2(L)$$

which conjugates both A and B into $GL_2(k)$. By explicit matrix multiplication,

$$U^{-1}BU = e^{-1} \begin{pmatrix} * & td^2 \\ -tc^2 & * \end{pmatrix}, \qquad U^{-1}AU = e^{-1} \begin{pmatrix} * & d^2 \\ -c^2 & * \end{pmatrix},$$

where $e = \det U = ad - bc \in L^*$. Therefore,

$$td^2e^{-1}, \quad d^2e^{-1}, \quad tc^2e^{-1}, \quad c^2e^{-1}$$

all belong to k. Now c, d cannot both be zero, since $U \in \mathrm{GL}_2(L)$. If $c \neq 0$, we have $t = tc^2e^{-1}/c^2e^{-1} \in k$. If $d \neq 0$, we have similarly

$$t = td^2e^{-1}/d^2e^{-1} \in k.$$

This gives the desired contradiction.

Ex. 8.12. Let k be any field of characteristic 2, and let $G = S_4$. Let M be the kG-module given by

$$ke_1 \oplus \cdots \oplus ke_4/k(e_1 + \cdots + e_4),$$

on which G acts by permuting the e_i's. Compute the kG-composition factors of M.

Solution. We think of V as a k-space with basis e_1, e_2, e_3, and an element $e_4 = e_1 + e_2 + e_3$. Let

$$W = k(e_1 + e_2) \oplus k(e_2 + e_3).$$

Since W also contains $(e_1 + e_2) + (e_2 + e_3) = e_1 + e_3$ and $e_4 + e_i$ for $1 \leq i \leq 3$, we see that W is a kG-submodule of V. Clearly, V/W is the 1-dimensional trivial kG-module. We finish by showing that W is *irreducible*. (If so, the composition factors of V are W and V/W.) Assume, on the contrary, that W is reducible. Then there exists a "line" kv in W stabilized by G. Let

$$v = a(e_1 + e_2) + b(e_2 + e_3) \neq 0.$$

Since (123) is a commutator in $S_3 \subset S_4$, we must have

$$a(e_1 + e_2) + b(e_2 + e_3) = (123)v = a(e_2 + e_3) + b(e_3 + e_1).$$

Comparing the coefficients of e_1, e_2, e_3, we get $a = b = 0$, a contradiction.

Ex. 8.13. Let k be a field of algebraic numbers, and A be its ring of algebraic integers. Let G be any finite group. Using the character of the regular representation of G, give another proof for the fact $(FC\text{-}(8.26))$ that AG has no idempotents except 0 and 1.

Solution. Let $e = \sum a_g g$ and $e' = 1 - e$ be complementary idempotents in AG, and let $R = kG$. We write χ_{reg} for the character associated with the left regular module $_RR$. Then $\chi_{\mathrm{reg}}(e)$ is the trace of the linear transformation $E : R \to R$ defined by left multiplication by e. Decomposing R into $eR \oplus e'R$, we observe that E is zero on $e'R$, and the identity on eR. Therefore, $\chi_{\mathrm{reg}}(e) = \dim_k(eR)$. On the other hand, $\chi_{\mathrm{reg}}(e) = a_1 \cdot |G|$. Therefore,

$$a_1 = |G|^{-1}\dim_k(eR) \in \mathbb{Q} \cap A = \mathbb{Z}.$$

Since we also have $0 \leq a_1 \leq 1$, we see that $a_1 = 0$ or 1. If $a_1 = 0$, we have $eR = 0$, so $e = 0$. If $a_1 = 1$, then $|G| = \dim_k(eR)$, and we have $e' = 0$; that is, $e = 1$.

Comment. The proof given above is due to D. Coleman; see his article in Proc. Amer. Math. Soc. *17* (1966), p. 961. The idea of computing the trace of the linear map E by using a decomposition of $R = kG$ is rather akin to that used for the proof of Littlewood's Formula in Exercise 25 below.

Ex. 8.14. (Kaplansky)   Let $G = \langle x \rangle$ be a cyclic group of order 5. Show that $u = 1 - x^2 - x^3$ is a unit of infinite order in $\mathbb{Z}G$, with inverse $v = 1 - x - x^4$. Then show that

$$U(\mathbb{Z}G) = \langle u \rangle \times (\pm G) \cong \mathbb{Z} \oplus \mathbb{Z}_2 \oplus \mathbb{Z}_5.$$

For the more computationally inclined reader, show that $a = 2x^4 - x^3 - 3x^2 - x + 2$ is a unit of infinite order in $\mathbb{Z}G$, with inverse $b = 2x^4 - 3x^3 + 2x^2 - x - 1$.

Solution. Let $R = \mathbb{Z}[\zeta]$ where ζ is a primitive 5th root of unity. By Dirichlet's Unit Theorem, $U(R)$ has rank 1. To shorten the proof, we shall assume the number-theoretic fact that $1 + \zeta$ is a fundamental unit in R, that is,

$$U(R) = \langle 1 + \zeta \rangle \cdot \{\pm \zeta^i\}.$$

(Of course, $1 + \zeta$ has infinite order. Its inverse is $-(\zeta + \zeta^3)$.) Now consider the projection $\varphi : \mathbb{Z}G \to R$ defined by $\varphi(x) = \zeta$. *We claim that* $\varphi : U(\mathbb{Z}G) \to U(R)$ *is an injection.* To see this, suppose

$$\varphi(\alpha_0 + \alpha_1 x + \cdots + \alpha_4 x^4) = 1,$$

where $\alpha_i \in \mathbb{Z}$. This implies that

$$\begin{aligned}
1 &= \alpha_0 + \alpha_1 \zeta + \alpha_2 \zeta^2 + \alpha_3 \zeta^3 + \alpha_4(-1 - \zeta - \zeta^2 - \zeta^3) \\
&= (\alpha_0 - \alpha_4) + (\alpha_1 - \alpha_4)\zeta + (\alpha_2 - \alpha_4)\zeta^2 + (\alpha_3 - \alpha_4)\zeta^3,
\end{aligned}$$

so we have $\alpha_1 = \alpha_2 = \alpha_3 = \alpha_4 = \alpha$ (say), and $\alpha_0 = 1 + \alpha$. On the other hand, the "augmentation"

$$\alpha_0 + \alpha_1 + \cdots + \alpha_4 = (1 + \alpha) + 4\alpha = 1 + 5\alpha$$

must be ± 1, since $\sum \alpha_i x^i \in U(\mathbb{Z}G)$. Therefore, we must have $\alpha = 0$, and $\sum \alpha_i x^i = 1$, which proves our claim. *We can now determine* $U(\mathbb{Z}G)$ *by computing* $\varphi(U(\mathbb{Z}G))$. Of course, $\varphi(\pm G) = \{\pm \zeta^i\}$. A simple calculation using the same ideas as above shows that

$$1 + \zeta \notin \varphi(U(\mathbb{Z}G)).$$

On the other hand, $uv = 1$ (by direct calculation), and

$$\begin{aligned}
\varphi(xu) &= \zeta(1 - \zeta^2 - \zeta^3) = \zeta - \zeta^3 + (1 + \zeta + \zeta^2 + \zeta^3) \\
&= 1 + 2\zeta + \zeta^2 = (1 + \zeta)^2.
\end{aligned}$$

This, together with $\varphi(\pm G) = \{\pm \zeta^i\}$, imply that $[U(R) : \varphi(U(\mathbb{Z}G))] \leq 2$. Since $1 + \zeta \notin \varphi(U(\mathbb{Z}G))$, equality must hold, and we must have

$$\varphi(U(\mathbb{Z}G)) = \langle (1 + \zeta)^2 \rangle \cdot \{\pm \zeta^i\}, \quad \text{and therefore}$$

$$U(\mathbb{Z}G) = \langle xu \rangle \times (\pm G) = \langle u \rangle \times (\pm G) \cong \mathbb{Z} \oplus \mathbb{Z}_2 \oplus \mathbb{Z}_5.$$

We finish with a sketch of the proof that a is a unit of infinite order in $\mathbb{Z}G$. Again, we exploit the map φ. Note that

$$\varphi(a) = -2(1 + \zeta + \zeta^2 + \zeta^3) - \zeta^3 - 3\zeta^2 - \zeta + 2$$
$$= -\zeta(3 + 5\zeta + 3\zeta^2).$$

On the other hand,

$$(1 + \zeta)^4 = 1 + 4\zeta + 6\zeta^2 + 4\zeta^3 - (1 + \zeta + \zeta^2 + \zeta^3)$$
$$= \zeta(3 + 5\zeta + 3\zeta^2).$$

Therefore, $\varphi(-a(xu)^{-2}) = 1$. Since $-a(xu)^{-2}$ has clearly augmentation 1, our earlier argument implies that $-a(xu)^{-2} = 1$. It follows that $a = -(xu)^2$ is a unit of infinite order in $\mathbb{Z}G$, with

$$a^{-1} = -(xu)^{-2} = -x^3 v^2$$
$$= -x^3(1 - x - x^4)^2$$
$$= -x^3(1 + x^2 + x^8 - 2x - 2x^4 + 2x^5)$$
$$= 2x^4 - 3x^3 + 2x^2 - x - 1.$$

Comment. In general, for any finite abelian group G,

$$U: = U(\mathbb{Z}G)$$

is known to be a finitely generated abelian group. The rank r of U can be explicitly computed, and the torsion subgroup of U is $\pm G$ according to a theorem of G. Higman. Therefore, $U \cong \mathbb{Z}^r \times (\pm G)$. If G is only finite but not necessarily abelian, U is still a finitely generated group, and the torsion subgroup of the center of U is $\pm Z(G)$. This generalization of Higman's Theorem is proved in FC-(8.21). However, U may have elements of finite order outside of $\pm G$.

Ex. 8.15. For finite abelian groups G and H, show that $\mathbb{R}G \cong \mathbb{R}H$ as $\mathbb{R}$-algebras iff $|G| = |H|$ and $|G/G^2| = |H/H^2|$.

Solution. Since G is abelian, Wedderburn's Theorem gives:

$$\mathbb{R}G \cong \mathbb{R} \times \cdots \times \mathbb{R} \times \mathbb{C} \times \cdots \times \mathbb{C}.$$

Suppose there are s factors of $\mathbb{R}$, and t factors of $\mathbb{C}$, so that $|G| = s + 2t$. The number s is the number of 1-dimensional real representations of G. This is the number of group homomorphisms from G to $\{\pm 1\} \subseteq \mathbb{R}^*$, so $s = |G/G^2|$. Therefore, the isomorphism type of $\mathbb{R}G$ (as an $\mathbb{R}$-algebra) is uniquely determined by $|G|$ and $|G/G^2|$. The conclusion of the Exercise follows immediately from this.

Comment. For instance, if G, H are abelian groups of odd order $2t + 1$, then $\mathbb{R}G \cong \mathbb{R}H$ as $\mathbb{R}$-algebras. In fact, the analysis above shows that both group algebras are $\cong \mathbb{R} \times \mathbb{C} \times \cdots \times \mathbb{C}$ with t factors of $\mathbb{C}$.

Ex. 8.16. Show that, for any two groups G, H, there exists a (nonzero) ring R such that $RG \cong RH$ as rings.

Solution. We claim that there exists a group K such that $K \times G \cong K \times H$ as groups. First let us assume this claim. Then $\mathbb{Z}[K \times G] \cong \mathbb{Z}[K \times H]$ as rings. It is easy to see that

$$\mathbb{Z}[K \times G] \cong (\mathbb{Z}[K])[G], \quad \text{and similarly} \quad \mathbb{Z}[K \times H] \cong (\mathbb{Z}[K])[H].$$

Therefore, for the integral group ring $R := \mathbb{Z}[K]$, we have $RG \cong RH$ as rings.

To prove our claim, take the group

$$K = (G \times H) \times (G \times H) \times \cdots.$$

We have

$$\begin{aligned}
G \times K &\cong G \times (H \times G) \times (H \times G) \times \cdots \\
&\cong (G \times H) \times (G \times H) \times \cdots \cong K, \quad \text{and} \\
H \times K &\cong H \times (G \times H) \times (G \times H) \times \cdots \\
&\cong (H \times G) \times (H \times G) \times \cdots \cong K.
\end{aligned}$$

In particular, $G \times K \cong H \times K$, as desired.

Comment. The above kind of infinite constructions is sometimes referred to as "Eilenberg's trick." Note that the construction also works if direct sums are used in lieu of direct products. This enables us to use a "smaller" group K: for instance, if G, H are countable, we can also take K to be countable.

This exercise suggests that, for *arbitrary* rings R, we should not expect the group ring RG to capture much of the information about the group G. One very interesting case is, however, where $R = \mathbb{Z}$ and $|G| < \infty$, as we have seen in the *Comment* on Ex. 1.

Ex. 8.17. Using the theory of group representations, show that for any prime p, a group G of order p^2 must be abelian.

Solution. Say $\mathbb{C}G \cong M_{n_1}(\mathbb{C}) \times \cdots \times M_{n_r}(\mathbb{C})$, with $n_1 = 1$. Then

$$n_1^2 + \cdots + n_r^2 = |G| = p^2.$$

But by FC-p. 129 (bottom of the page), each $n_i | p^2$. This clearly implies that each $n_i = 1$. In particular,

$$\mathbb{C}G \cong \mathbb{C} \times \cdots \times \mathbb{C},$$

so G is abelian.

Ex. 8.18. Let $G = \{\pm 1, \pm i, \pm j, \pm k\}$ be the quaternion group of order 8. It is known that, over $\mathbb{C}, G$ has four 1-dimensional representations, and a unique irreducible 2-dimensional representation D (FC-p. 120). Construct D explicitly, and compute the character table for G.

Solution. Let $\mathbb{H}$ be the division ring of real quaternions. A well-known model of $\mathbb{H}$ is the $\mathbb{R}$-subalgebra

$$\left\{ \begin{pmatrix} \alpha & \beta \\ -\bar{\beta} & \bar{\alpha} \end{pmatrix} : \alpha, \beta \in \mathbb{C} \right\} \subseteq M_2(\mathbb{C}).$$

Here, a real quaternion $w + xi + yj + zk \in \mathbb{H}$ is "identified" with

$$\begin{pmatrix} w + xi & y + zi \\ -y + zi & w - xi \end{pmatrix} = wI + x \begin{pmatrix} i & 0 \\ 0 & -i \end{pmatrix} + y \begin{pmatrix} 0 & 1 \\ -1 & 0 \end{pmatrix} + z \begin{pmatrix} 0 & i \\ i & 0 \end{pmatrix}.$$

In particular, we arrive at a 2-dimensional representation $D : G \to GL_2(\mathbb{C})$ with

$$(*) \quad D(1) = I, \ D(i) = \begin{pmatrix} i & 0 \\ 0 & -i \end{pmatrix}, \ D(j) = \begin{pmatrix} 0 & 1 \\ -1 & 0 \end{pmatrix}, \ D(k) = \begin{pmatrix} 0 & i \\ i & 0 \end{pmatrix}.$$

These matrices are easily checked to be $\mathbb{C}$-linearly independent, so they span $M_2(\mathbb{C})$. This implies that D is irreducible, so D is the representation we want.

Another way to construct D is to embed $\mathbb{C} = \mathbb{R} \oplus \mathbb{R}i \subset \mathbb{H}$, and view $_{\mathbb{C}}\mathbb{H}$ as an irreducible $\mathbb{C}G$-module via the G-action

$$g * h = hg^{-1} \quad (g \in G, \ h \in \mathbb{H}).$$

Note that this action is $\mathbb{C}$-linear by the associative law on $\mathbb{H}$. Using the left $\mathbb{C}$-basis $\{j, -1\}$ on $\mathbb{H}$, and noting that

$$i * j = ij, \qquad\qquad i * (-1) = i = (-i)(-1),$$
$$j * j = 1 = (-1)(-1), \qquad j * (-1) = j,$$

we see that, under this "new" representation:

$$i \mapsto \begin{pmatrix} i & 0 \\ 0 & -i \end{pmatrix}, \quad j \mapsto \begin{pmatrix} 0 & 1 \\ -1 & 0 \end{pmatrix}.$$

Thus, we have retrieved the representation D.

Note that the $\mathbb{C}$-independence of the matrices in $(*)$ implies that we can identify $\mathbb{C} \otimes_{\mathbb{R}} \mathbb{H}$ with $M_2(\mathbb{C})$ as $\mathbb{C}$-algebras. Via this identification, the $\mathbb{C}G$-module affording the representation D is just the "canonical module" $\left\{ \begin{pmatrix} a \\ b \end{pmatrix} : a, b \in \mathbb{C} \right\}$ over $M_2(\mathbb{C})$. In particular, if D_0 denotes the irreducible $\mathbb{R}$-representation given by the $\mathbb{R}G$-module $\mathbb{H}$ (see FC-p. 120), we have a decomposition $\mathbb{C} \otimes_{\mathbb{R}} D_0 \cong D \oplus D$.

Using $1, i, j, k, -1$ as the conjugacy class representatives for G (and noting that $1^2 + 1^2 + 1^2 + 1^2 + 2^2 = 8 = |G|$), we arrive at the following character table:

	1	i	j	k	-1
χ_1	1	1	1	1	1
χ_2	1	1	-1	-1	1
χ_3	1	-1	1	-1	1
χ_4	1	-1	-1	1	1
χ_5	2	0	0	0	-2

Comment. Note that the matrices in $(*)$ are *unitary* matrices, so D is a unitary representation. If we multiply $D(i), D(j), D(k)$ by $-i$, we get the following three *Hermitian* matrices

$$\sigma_z = \begin{pmatrix} 1 & 0 \\ 0 & -1 \end{pmatrix}, \quad \sigma_y = \begin{pmatrix} 0 & -i \\ i & 0 \end{pmatrix}, \quad \sigma_x = \begin{pmatrix} 0 & 1 \\ 1 & 0 \end{pmatrix}.$$

These are the famous *Pauli spin matrices*, which are used by physicists in the study of the quantum mechanical motion of a spinning electron. The following relations among the Pauli spin matrices are familiar to all quantum physicists:

$$\sigma_x \sigma_y \sigma_z = iI, \quad \sigma_x^2 = \sigma_y^2 = \sigma_z^2 = I,$$

$$\sigma_x \sigma_y = -\sigma_y \sigma_x = i\sigma_z, \quad \sigma_y \sigma_z = -\sigma_z \sigma_y = i\sigma_x, \quad \sigma_z \sigma_x = -\sigma_x \sigma_z = i\sigma_y.$$

Ex. 8.19. Let G be the dihedral group of order $2n$ generated by two elements r, s with relations $r^n = 1$, $s^2 = 1$ and $srs^{-1} = r^{-1}$. Let $\theta = 2\pi/n$.

(1) For any integer h $(0 \leq h \leq n)$, show that

$$D_h(r) = \begin{pmatrix} \cos h\theta & -\sin h\theta \\ \sin h\theta & \cos h\theta \end{pmatrix}, \quad D_h(s) = \begin{pmatrix} 1 & 0 \\ 0 & -1 \end{pmatrix}$$

defines a real representation of G.

(2) Show that, over $\mathbb{C}$, D_h is equivalent to the representation D'_h defined by

$$D'_h(r) = \begin{pmatrix} e^{-ih\theta} & 0 \\ 0 & e^{ih\theta} \end{pmatrix}, \quad D'_h(s) = \begin{pmatrix} 0 & 1 \\ 1 & 0 \end{pmatrix}.$$

(3) For $n = 2m + 1$, show that $D_1, \ldots, D_m$ give all irreducible representations of G (over $\mathbb{R}$ or over $\mathbb{C}$) with dimensions > 1. For $n = 2m$, show the same for $D_1, \ldots, D_{m-1}$.

(4) Construct the character table for G.

(5) Verify that the two nonabelian groups of order 8 (the dihedral group and the quaternion group) have the same character table (upon a suitable enumeration of the characters and the conjugacy classes of the two groups).

Solution. (1) We omit the routine check that $D_h(r)$ and $D_h(s)$ satisfy the relations between r and s.

(2) Extending scalars to $\mathbb{C}$, we can take in $\mathbb{C}^2$ the new basis

$$\left\{ \begin{pmatrix} 1 \\ i \end{pmatrix}, \begin{pmatrix} 1 \\ -i \end{pmatrix} \right\}:$$

these are the eigenvectors of $D_h(r)$, with corresponding eigenvalues $e^{-ih\theta}$ and $e^{ih\theta}$. By using the new basis, we obtain the following diagonalization of the rotation matrix $D_h(r)$:

$$\begin{pmatrix} 1 & 1 \\ i & -i \end{pmatrix}^{-1} \begin{pmatrix} \cos h\theta & -\sin h\theta \\ \sin h\theta & \cos h\theta \end{pmatrix} \begin{pmatrix} 1 & 1 \\ i & -i \end{pmatrix} = \begin{pmatrix} e^{-ih\theta} & 0 \\ 0 & e^{ih\theta} \end{pmatrix} = D'_h(r).$$

By a simple calculation,

$$\begin{pmatrix} 1 & 1 \\ i & -i \end{pmatrix}^{-1} \begin{pmatrix} 1 & 0 \\ 0 & -1 \end{pmatrix} \begin{pmatrix} 1 & 1 \\ i & -i \end{pmatrix} = \begin{pmatrix} 0 & 1 \\ 1 & 0 \end{pmatrix} = D'_h(s).$$

(3) We first determine the 1-dimensional representations of G. Noting that

$$r^{-2} = (srs^{-1})r^{-1} \in [G, G],$$

we see that $[G, G]$ is $\langle r \rangle$ when $n = 2m + 1$, and $\langle r^2 \rangle$ when $n = 2m$. The commutator quotient group $G/[G, G]$ is therefore $\langle \bar{s} \rangle$ and $\langle \bar{r} \rangle \times \langle \bar{s} \rangle$ accordingly. In particular, G has 1-dimensional characters χ_1, χ_2 in the first case, and $\chi_1, \chi_2, \chi_3, \chi_4$ in the second case. These will appear in the character tables below. Let $n = 2m + 1$, and let $\alpha_h = \chi_{D_h}$ for $1 \leq h \leq m$. Since $\alpha_h(r) = 2\cos h\theta$ and $0 < h\theta < \pi$, we have $\alpha_h \neq \alpha_{h'}$ for $h \neq h'$. Also, since χ_1, χ_2 take only values ± 1, α_h cannot decompose into a sum of the χ_i's. Therefore, D_h ($1 \leq h \leq m$) are irreducible over $\mathbb{C}$. Since

$$\chi_1(1)^2 + \chi_2(1)^2 + \sum_{h=1}^{m} \alpha_h(1)^2 = 2 + 4m = 2n = |G|,$$

we have now found all the irreducible $\mathbb{C}$-representations. The argument for the case $n = 2m$ is the same, by using $\alpha_h = \chi_{D_h}$ for $1 \le h \le m - 1$. (It is easy to see that $D_h \cong D_{n-h}$ in general, so it is not surprising that we use only a subset of the D_h's. Note also that, in the case $n = 2m$, the representation D_m is useless, since it is obviously reducible.)

(4) For $n = 2m + 1$, the conjugacy class representatives can be taken to be

$$\{1, s, r, \ldots, r^j, \ldots, r^m\}.$$

We have the character table (where $1 \le h \le m$):

	1	s	r	$\cdots$	r^j	$\cdots$	r^m
x_1	1	1	1	$\cdots$	1	$\cdots$	1
x_2	1	-1	1	$\cdots$	1	$\cdots$	1
α_h	2	0	$2\cos h\theta$	$\cdots$	$2\cos hj\theta$	$\cdots$	$2\cos hm\theta$

For $n = 2m$, the class representatives can be taken to be

$$\{1, s, sr, r, \ldots, r^j, \ldots, r^m\}.$$

Here we have the character table (where $1 \le h \le m - 1$):

	1	s	sr	r	$\cdots$	$r^j (j < m)$	$\cdots$	r^m
χ_1	1	1	1	1	$\cdots$	1	$\cdots$	1
χ_2	1	1	-1	-1	$\cdots$	$(-1)^j$	$\cdots$	$(-1)^m$
χ_3	1	-1	1	-1	$\cdots$	$(-1)^j$	$\cdots$	$(-1)^m$
χ_4	1	-1	-1	1	$\cdots$	1	$\cdots$	1
α_h	2	0	0	$2\cos h\theta$	$\cdots$	$2\cos hj\theta$	$\cdots$	$2(-1)^h$

(5) For $m = 2$, we have $|G| = 8$. Taking $h = 1$ and $\theta = \pi/2$, the second character table above (for the dihedral group of order 8) is seen to be the same as that in Exercise 18 (for the quaternion group of order 8).

Ex. 8.20. Let $G = S_4$, which acts irreducibly on

$$M = \mathbb{Q}e_1 \oplus \mathbb{Q}e_2 \oplus \mathbb{Q}e_3 \oplus \mathbb{Q}e_4/\mathbb{Q}(e_1 + e_2 + e_3 + e_4).$$

Let $M' = \sigma \otimes M$, where σ denotes the sign representation of G. Show that M' is equivalent to the representation D of G as the group of rotational symmetries of the cube (or of the octahedron).

Solution. We shall show that $\sigma \otimes D$ is equivalent to the representation afforded by M. One way to show this is by using characters. On the five conjugacy classes of G represented by

$$(1), \ (12), \ (123), \ (1234) \ \text{and} \ (12)(34),$$

χ_M takes the values $(3, 1, 0, -1, -1)$ (cf. FC-p. 132). By writing down the matrices $D((1)), D((12)), \ldots$, etc., we see that χ_D takes the values $(3, -1, 0, 1, -1)$. Since $(12), (1234)$ are odd and the other class representatives are even, we have $\chi_{\sigma \otimes D} = \chi_M$, so $\sigma \otimes D$ is equivalent to M by FC-(7.19). However, this formal computation of characters is not really very illuminating. In the following, we shall give a more geometric proof for the equivalence of $\sigma \otimes D$ with M.

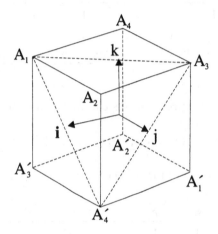

We label the vertices of the cube as in the above picture, so that the four long diagonals are $A_i A_i'$ ($1 \leq i \leq 4$). Under the representation D, the elements of G realize all the 24 permutations of the four long diagonals. We now let G act on $\mathbb{R}^3$ by the twisted representation $\sigma \otimes D$. To see that this is "the same" as the representation M, we choose in $\mathbb{R}^3$ the four vectors

$$e_1 = \vec{OA_1}, \quad e_2 = \vec{OA_2'}, \quad e_3 = \vec{OA_3}, \quad e_4 = \vec{OA_4'}.$$

Since $\vec{OA_1} + \vec{OA_3} = 2\vec{k}$ and $\vec{OA_2'} + \vec{OA_4'} = -2\vec{k}$, we have

$$e_1 + e_2 + e_3 + e_4 = 0.$$

We finish by showing that, under $\sigma \otimes D$, the action of any $g \in G = S_4$ is exactly the permutation action of g on $\{e_1, e_2, e_3, e_4\}$. Since G is generated by (143) and (1234), it suffices to check the desired conclusion for $g_1 = (143)$ and $g_2 = (1234)$. For the even permutation g_1,

$$(\sigma \otimes D)(g_1) = D(g_1)$$

is a 120°-rotation about the axis $A_2 A_2'$, under which

$$e_1 \to e_4, \quad e_2 \to e_2, \quad e_3 \to e_1, \quad e_4 \to e_3,$$

as desired. For the odd permutation g_2,

$$(\sigma \otimes D)(g_2) = -D(g_2)$$

is the counterclockwise 90°-rotation about the z-axis followed by $-I$. Under this 2-stage action:

$$e_1 \to \vec{OA_2} \to \vec{OA_2'} = e_2, \quad e_2 \to \vec{OA_3'} \to \vec{OA_3} = e_3,$$
$$e_3 \to \vec{OA_4} \to \vec{OA_4'} = e_4, \quad e_4 \to \vec{OA_1'} \to \vec{OA_1} = e_1.$$

Again, this is the permutation action of g_2 on $\{e_1, e_2, e_3, e_4\}$.

Ex. 8.21. Show that, over $\mathbb{Q}$, $G = A_5$ has four irreducible representations, of dimensions 1, 4, 5, 6 respectively.

Solution. Let $R = \mathbb{Q}G$. It will be convenient to assume the knowledge of the irreducible $\mathbb{C}G$-modules constructed in FC-p.133. The 4-dimensional and 5-dimensional irreducible $\mathbb{C}$-representations both come from irreducible R-modules. Therefore, we have

$$R \cong \mathbb{Q} \times \mathbb{M}_4(\mathbb{Q}) \times \mathbb{M}_5(\mathbb{Q}) \times S$$

where S is the product of the remaining simple components, with

$$\dim_{\mathbb{Q}} S = 60 - (1^2 + 4^2 + 5^2) = 18.$$

Let $\mathbb{M}_r(D)$ be a simple component of S, and let V be the corresponding simple left R-module. By Exercise 7.1, $V^{\mathbb{C}}$ must break into a sum of 3-dimensional irreducible $\mathbb{C}G$-modules. If $V^{\mathbb{C}}$ does not involve *both* of these modules, it would follow that one of the 3-dimensional irreducible characters of G is rational-valued, which is not the case. Therefore, $V^{\mathbb{C}}$ must involve both of the 3-dimensional irreducible $\mathbb{C}$-representations, and, by Exercise 7.1 again, there are no more simple R-modules (beyond what we have constructed). In particular, we must have

$$r^2 \dim_{\mathbb{Q}} D = 18 \quad \text{and} \quad r \dim_{\mathbb{Q}} D = \dim_{\mathbb{Q}} V \geq 6.$$

From these, we conclude that $r = 3$, $\dim_{\mathbb{Q}} D = 2$, and $\dim_{\mathbb{Q}} V = 6$. (A little further work will show that, in the Wedderburn decomposition

$$R \cong \mathbb{Q} \times \mathbb{M}_4(\mathbb{Q}) \times \mathbb{M}_5(\mathbb{Q}) \times \mathbb{M}_3(D),$$

we have $D \cong \mathbb{Q}(\sqrt{5})$.)

Comment. It is of interest to give a direct construction for the 6-dimensional simple R-module V. For readers familiar with the representation theory of the symmetric groups, we can take V to be the 6-dimensional simple $\mathbb{Q}S_5$-module corresponding to the Young diagram

and simply view V as an R-module. This module yields a character χ such that[1]

$$\chi(1) = 6, \quad \chi((12)(34)) = -2, \quad \chi((123)) = 0,$$

$$\chi((12345)) = \chi((13524)) = 1.$$

Since this is exactly the sum of the two irreducible 3-dimensional complex characters (cf. *FC*-p.133), we see that V is the simple R-module we are after.

Ex. 8.22. For any finite group G and any field k, is it true that any irreducible representation of G over k is afforded by a minimal left ideal of kG? (The answer is yes, but don't get very discouraged if you can't prove it.)

Solution. This Exercise is best done by assuming some more facts on group algebras kG not yet developed in *FC*. The most relevant fact needed here is the result that $R = kG$ is a *quasi-Frobenius ring*. For such a ring, we have always the following *Double Annihilator Property*:[2] if I is any left ideal, then

$$\operatorname{ann}_\ell (\operatorname{ann}_r(I)) = I.$$

Assuming this, consider now any simple left R-module V, say $V = R/\mathfrak{m}$, where $\mathfrak{m}$ is a maximal left ideal. By the property quoted above, we see that $\operatorname{ann}_r(\mathfrak{m}) \neq 0$. Fix a nonzero element $a \in R$ such that $\mathfrak{m}a = 0$, and define $\varphi : R \to R$ by $\varphi(r) = ra$. Then φ is a nonzero endomorphism of $_RR$, with $\mathfrak{m} \subseteq \ker(\varphi)$. Since $\mathfrak{m}$ is a maximal left ideal, we have $\mathfrak{m} = \ker(\varphi)$. Therefore

$$\varphi(R) \cong R/\mathfrak{m} = V,$$

so $\varphi(R)$ is the minimal left ideal we sought.

---

[1] See, e.g. F. D. Murnaghan: "The Theory of Group Representations," p. 142, Dover Publications, 1963.

[2] If char $k \nmid |G|$, kG is semisimple so this Double Annihilator Property for kG follows from Exercise 3.12. The harder case occurs when char $k \mid |G|$.

Comment. A ring R is called *left Kasch* if every simple left R-module is isomorphic to a minimal left ideal (i.e. can be embedded in $_RR$). The work done above is the standard proof for the fact that *any quasi-Frobenius ring is left (and right) Kasch*.

Ex. 8.23. If a finite group G has at most three irreducible complex representations, show that $G \cong \{1\}, \mathbb{Z}_2, \mathbb{Z}_3$ or S_3.

Solution. Let r be the number of conjugacy classes in G. By assumption, $r \leq 3$. If $r = 1$, clearly $G = \{1\}$. Now assume $r = 2$. The two conjugacy classes have cardinalities 1 and $|G| - 1$, so we have $(|G| - 1) \mid |G|$. This implies that $|G| = 2$, and hence $G \cong \mathbb{Z}_2$. Finally, assume that $r = 3$. Let $1, a, b$ be the cardinalities of the three conjugacy classes, and let $n = |G|$. Then $a + b = n - 1$, and we may assume that $a \geq (n-1)/2$. Write $n = ac$, where $c \in \mathbb{Z}$. If $c \geq 3$, then

$$n \geq 3a \geq 3(n-1)/2$$

and hence $n \leq 3$. In this case, $G \cong \mathbb{Z}_3$. We may now assume $c = 2$, so

$$b = n - a - 1 = \frac{n}{2} - 1.$$

Since $b \mid n$, we can only have $n = 4$ or $n = 6$. If $n = 4$, G would have 4 conjugacy classes. Therefore, $n = 6$, and since G is not abelian, $G \cong S_3$.

Comment. As pointed out to me by I.M. Isaacs, similar (but somewhat more elaborate) computations can be used to settle the case $r = 4$. In this case, there are exactly four possibilities for G:

$$\mathbb{Z}_4, \quad \mathbb{Z}_2 \oplus \mathbb{Z}_2, \quad A_4, \quad \text{and the dihedral group of order 10.}$$

According to a classical theorem of E. Landau, for any given r, there are only *finitely many* finite groups with exactly r conjugacy classes (or equivalently, r irreducible complex representations). For a proof of Landau's Theorem, see pp. 48–49 of Isaacs' book "Algebra," Brooks/Cole, 1994.

Ex. 8.24. Suppose the character table of a finite group G has the following two rows:

	g_1	g_2	g_3	g_4	g_5	g_6	g_7
μ	1	1	1	ω^2	ω	ω^2	ω
ν	2	-2	0	-1	-1	1	1

$\left(\omega = e^{2\pi i/3}\right)$

Determine the rest of the character table.

Solution. This exercise is easier than it looks. First note that g_1 must be the identity conjugacy class. Let χ_1 be the trivial character, $\chi_2 = \mu$, $\chi_3 = \overline{\mu}$ (complex conjugate of μ), and $\chi_4 = \nu$. We get two more irreducible characters by forming

$$\chi_5 := \chi_2 \nu = (2, -2, 0, -\omega^2, -\omega, \omega^2, \omega),$$
$$\chi_6 := \chi_3 \nu = (2, -2, 0, -\omega, -\omega^2, \omega, \omega^2).$$

It only remains to determine the last irreducible character χ_7 (since there are seven conjugacy classes). We must have $\chi_7 \mu = \chi_7$ (to avoid getting an 8th irreducible character), so $\chi_7 = (a, b, c, 0, 0, 0, 0)$, where $a \in \mathbb{N}$, and b, c are algebraic integers. Using the orthogonality relations for the 1st and 2nd (resp. 1st and 3rd) columns, we get:

$$1^2 + 1^2 + 1^2 - 2^2 - 2^2 - 2^2 + ab = 0 \implies ab = 9,$$
$$1^2 + 1^2 + 1^2 + 0 + 0 + 0 + ac = 0 \implies ac = -3.$$

Therefore, $c = -3/a \in \mathbb{Q}$. Since c is an algebraic integer, we must have $a = 1$ or 3. If $a = 1$, we would be able to use the linear character χ_7 to "twist" ν to get another 2-dimensional irreducible character. Therefore, we must have $a = 3$, $c = -1$, and $b = 3$, which completes the following character table:

	g_1	g_2	g_3	g_4	g_5	g_6	g_7
χ_1	1	1	1	1	1	1	1
$\mu = \chi_2$	1	1	1	ω^2	ω	ω^2	ω
$\bar{\mu} = \chi_3$	1	1	1	ω	ω^2	ω	ω^2
$\nu = \chi_4$	2	-2	0	-1	-1	1	1
$\mu\nu = \chi_5$	2	-2	0	$-\omega^2$	$-\omega$	ω^2	ω
$\bar{\mu}\nu = \chi_6$	2	-2	0	$-\omega$	$-\omega^2$	ω	ω^2
χ_7	3	3	-1	0	0	0	0

To complete the information, we have

$$|G| = 3 \cdot 1^2 + 3 \cdot 2^2 + 3^2 = 24.$$

The seven conjugacy classes are easily seen to be of cardinalities

$$1, 1, 6, 4, 4, 4, 4 \quad \text{(in that order)},$$

by using the orthogonality relations.

Comment. We would be remiss if we did not try to construct an explicit group G with the above character table. As a matter of fact, we can take G to be the binary tetrahedral group

$$G = \{\pm 1, \pm i, \pm j, \pm k, (\pm 1 \pm i \pm j \pm k)/2\},$$

which is the group of units of Hurwitz' ring of integral quaternions (see *FC*-p. 5).[3] We have

$$G/\{\pm 1, \pm i, \pm j, \pm k\} \cong \mathbb{Z}_3,$$

so we have the three linear characters χ_1, χ_2, χ_3. Also, $G/\{\pm 1\} \cong A_4$ (which is why G is called the binary tetrahedral group), so we get the character χ_7 by "lifting" the unique 3-dimensional $\mathbb{C}$-representation of A_4. The real-valued 2-dimensional character ν can be constructed as follows. View the real quaternions $\mathbb{H}$ as $\mathbb{C} \oplus \mathbb{C}j$ (a 2-dimensional $\mathbb{C}$-space) and let G act on $\mathbb{H}$ by

$$g * h = hg^{-1} \quad (g \in G, \ h \in \mathbb{H}).$$

This gives an irreducible 2-dimensional $\mathbb{C}$-representation of G affording ν (cf. Exercise 18). The conjugacy class representatives can be taken to be

$$g_1 = 1, \quad g_2 = -1, \quad g_3 = i, \quad g_4 = -\alpha, \quad g_5 = \alpha^2, \quad g_6 = \alpha, \quad g_7 = -\alpha^2,$$

where

$$\alpha = (1 + i + j + k)/2, \quad \text{and} \quad \alpha^2 = (-1 + i + j + k)/2.$$

The three characters χ_4, χ_5, χ_6 can also be constructed by using induced representations from the quaternion group $\{\pm 1, \pm i, \pm j, \pm k\}$.

In fact, the G above is the only group with the character table worked out in this exercise. I am grateful to I. M. Isaacs who patiently explained to me a nice proof for the uniqueness of G. For readers with a penchant for "proofs by exhaustion," we might mention that the recent book "Representations and Characters of Groups" by James and Liebeck (Cambridge University Press, 1993) contains character tables for *all* groups of order < 32. A look at the character tables of all 15 groups of order 24 will reveal that the table in this exercise occurs only once (incidentally, on p. 404 of *loc. cit.*)!

Ex. 8.25. (Littlewood's Formula) Let $e = \sum_{g \in G} a_g \, g \in kG$ be an idempotent, where k is a field and G is a finite group. Let χ be the character of G afforded by the kG-module $kG \cdot e$. Show that for any $h \in G$,

$$\chi(h) = |C_G(h)| \cdot \sum_{g \in C} a_g,$$

where C denotes the conjugacy class of h^{-1} in G.

Solution. Given $h \in G$, let $\varphi : kG \to kG$ be the linear map defined by $\varphi(\alpha) = h\alpha e$. Consider the decomposition

$$kG = kG \cdot e \oplus kG \cdot (1 - e).$$

---

[3] Another "model" for this group is the classical group $SL_2(\mathbb{F}_3)$.

On $kG \cdot e$, φ is just left multiplication by h, and on $kG \cdot (1 - e)$, φ is the zero map. Therefore, $\chi(h) = \mathrm{tr}(\varphi)$. Now

$$\varphi(\alpha) = h\alpha \sum_{g \in G} a_g \, g = \sum_{g \in G} a_g \, h\alpha g.$$

Thus, if $\varphi_g : kG \to kG$ is defined by $\varphi_g(\alpha) = h\alpha g$, we have

$$\chi(h) = \sum_{g \in G} a_g \, \mathrm{tr}(\varphi_g).$$

We finish by calculating $\mathrm{tr}(\varphi_g)$ $(g \in G)$. Since φ_g takes G to G, $\mathrm{tr}(\varphi_g)$ is just the number of group elements $\alpha \in G$ such that $h\alpha g = \alpha$, i.e. $g = \alpha^{-1}h^{-1}\alpha$. Thus, if g is not conjugate to h^{-1}, $\mathrm{tr}(\varphi_g) = 0$. If g is conjugate to h^{-1}, the number of $\alpha \in G$ such that $g = \alpha^{-1}h^{-1}\alpha$ is $|C_G(h)|$, so $\mathrm{tr}(\varphi_g) = |C_G(h)|$. It follows that

$$\chi(h) = |C_G(h)| \cdot \sum_{g \in C} a_g,$$

where C is the conjugacy class of h^{-1} in G.

Comment. The formula proved above is named after Dudley E. Littlewood, no relation to John E. Littlewood, the longtime collaborator of G. H. Hardy. (A.O. Morris informed me, however, that when D.E. Littlewood was an undergraduate student in Cambridge, his Director of Studies was none other than J.E. Littlewood!) Littlewood's Formula can be found as Theorem I in p. 266 of his classic text "A University Algebra" (2nd edition), reprinted by Dover in 1970.

Ex. 8.26. Let $G = S_3$, and k be any field of characteristic 3.

(a) Show that there are only two irreducible representations for G over k, namely, the trivial representation and the sign representation.
(b) It is known that there are exactly six (finite-dimensional) indecomposable representations for G over k (see Curtis-Reiner: Representation Theory of Finite Groups and Associative Algebras, p.433). Construct these representations.

Solution. (a) Let U be an irreducible kG-module, and let T be the action of (123) on U. Since

$$(T - I)^3 = T^3 - I = 0,$$

T has an eigenvalue 1, so

$$U_0 = \{u \in U : \ Tu = u\} \neq 0.$$

For $u \in U_0$, we have

$$(123)\,((12)u) = (12)(132)u = (12)(123)^2 u = (12)u,$$

so (12) $u \in U_0$ also. This shows that U_0 is a kG-submodule of U, so $U_0 = U$, and we may view U as a $k\overline{G}$-module where

$$\overline{G} = G/\langle(123)\rangle = \left\langle \overline{(12)} \right\rangle.$$

Since (12) has order 2, we have $U = U_+ \oplus U_-$ where

$$U_+ = \{u \in U : (12)u = u\}, \quad \text{and} \quad U_- = \{u \in U : (12)u = -u\}.$$

Therefore, we have either $U = U_+$ (giving the trivial representation), or $U = U_-$ (giving the sign representation).

(b) In the following, let U_+ denote the trivial representation, and U_- the sign representation of kG. *We shall construct four more indecomposable kG-modules: M, M' of dimension 3, and V, V' of dimension 2.* Let $M = ke_1 \oplus ke_2 \oplus ke_3$, on which G acts by permuting the e_i's, and let

$$V = M/k(e_1 + e_2 + e_3).$$

(b1) Let $V_0 = kv$, where $v = \overline{e}_1 - \overline{e}_2 \in V$. We check easily that V_0 is a kG-submodule of V isomorphic to U_-, with $V/V_0 \cong U_+$. Thus, the composition factors for V are $\{U_+, U_-\}$. We claim that V has no submodule $\cong U_+$. In fact, if $0 \neq u = a\overline{e}_1 + b\overline{e}_2$ spans a line $\cong U_+$, then

$$a\overline{e}_1 + b\overline{e}_2 = (12)u = a\overline{e}_2 + b\overline{e}_1 \quad \text{and}$$

$$a\overline{e}_1 + b\overline{e}_2 = (123)u = a\overline{e}_2 + b\overline{e}_3 = -b\overline{e}_1 + (a - b)\overline{e}_2$$

imply that $a = b = 0$, a contradiction. It follows that V is indecomposable, and hence so is $V' : = U_- \otimes V$. Also, since V' contains $U_- \otimes U_- \cong U_+$, we see that $V \not\cong V'$.

(b2) Using the same type of calculations as above, we can verify the following two properties of $M = ke_1 \oplus ke_2 \oplus ke_3$:

(A) *M has no kG-submodule $\cong U_-$.*
(B) *The only kG-submodule of M isomorphic to U_+ is $k(e_1 + e_2 + e_3)$.*

We can now show that M is indecomposable. In fact, if otherwise, (A) implies that $M \cong U_+ \oplus N$ for some N. Since M has composition factors $\{U_+, U_+, U_-\}$, N has composition factors $\{U_+, U_-\}$. By (A) again, N must contain a copy of U_+. But then M contains a copy of $U_+ \oplus U_+$, which contradicts (B). This shows that M is indecomposable, and hence so is $M' : = U_- \otimes M$. Since

$$M \supseteq k(e_1 + e_2 + e_3) \cong U_+,$$

M' contains a copy of $U_- \otimes U_+ \cong U_-$. By (A) again, we have $M' \not\cong M$. This completes the construction of the six indecomposable kG-modules.

Comment. Since by the Krull-Schmidt Theorem FC-(19.22) every finite-dimensional kG-module is uniquely a direct sum of indecomposable modules, the above work together with the result quoted from Curtis-Reiner enable us to construct and classify *all* finite-dimensional kG-modules.

Ex. 8.27. Let G be a cyclic group of prime order $p > 2$. Show that the group of units of finite order in $\mathbb{Q}G$ decomposes into a direct product of G with $\{\pm 1\}$ and another cyclic group of order 2.

Solution. Let ζ be a primitive pth root of unity, and let $G = \langle x \rangle$. The map $x \mapsto \zeta$ extends to a surjective ring homomorphism $\mathbb{Q}G \to \mathbb{Q}(\zeta)$. Therefore, one of the simple components of $\mathbb{Q}G$ is $\mathbb{Q}(\zeta)$ (see Exercise 3.3). On the other hand, $\mathbb{Q}G$ also has a simple component $\mathbb{Q}$. Since

$$\dim_{\mathbb{Q}}(\mathbb{Q} \times \mathbb{Q}(\zeta)) = 1 + p - 1 = p,$$

we must have $\mathbb{Q}G \cong \mathbb{Q} \times \mathbb{Q}(\zeta)$. For any commutative ring R, let $\mathrm{U}_0(R)$ denote the group of units of finite order in R. Then

$$\mathrm{U}_0(\mathbb{Q}G) \cong \mathrm{U}_0(\mathbb{Q}) \times \mathrm{U}_0(\mathbb{Q}(\zeta)) \cong \{\pm 1\} \times \mathrm{U}_0(\mathbb{Q}(\zeta)).$$

We finish by noting that $\mathrm{U}_0(\mathbb{Q}(\zeta))$ is the group of roots of unity in $\mathbb{Q}(\zeta)$, which is just the group $\langle -\zeta \rangle = \langle \pm 1 \rangle \times \langle \zeta \rangle$.

Ex. 8.28. Let G be the group of order 21 generated by two elements a, b with the relations $a^7 = 1$, $b^3 = 1$, and $bab^{-1} = a^2$.

(1) Construct the irreducible complex representations of G, and compute its character table.

(2) Construct the irreducible rational representations of G and determine the Wedderburn decomposition of the group algebra $\mathbb{Q}G$.

(3) How about $\mathbb{R}G$ and the real representations of G?

Solution. (1) In preparation for the computation of the character table of G, we first note that G has *five* conjugacy classes, represented by 1, a, a^3, b, b^2. (This is an easy group-theoretic computation, which we omit.) Thus, we expect to have *five* irreducible complex representations.

Obviously, $[G, G] = \langle a \rangle$, so $G/[G,G] \cong \langle b \rangle$. This shows that there are three 1-dimensional representations $\chi_i : G \to \mathbb{C}^*$, which are trivial on $\langle a \rangle$, with $\chi_1(b) = 1$, $\chi_2(b) = \omega$, and $\chi_3(b) = \omega^2$, where ω is a primitive cubic root of unity.

Next, we construct a 3-dimensional $\mathbb{C}$-representation $D : G \to \mathrm{GL}_3(\mathbb{C})$ by taking

(A) $$D(a) = \begin{pmatrix} \zeta & & \\ & \zeta^2 & \\ & & \zeta^4 \end{pmatrix}, \quad \text{and} \quad D(b) = \begin{pmatrix} 0 & 1 & 0 \\ 0 & 0 & 1 \\ 1 & 0 & 0 \end{pmatrix},$$

where ζ is a primitive 7th root of unity. (It is straightforward to check that the relations between a and b are respected by D.) If D is a reducible representation, it would have to "contain" a 1-dimensional representation. This is easily checked to be not the case. Thus, D is irreducible, and we get another irreducible 3-dimensional $\mathbb{C}$-representation D' by taking

(B)

$$D'(a) = \overline{D(a)} = \begin{pmatrix} \zeta^6 & & \\ & \zeta^5 & \\ & & \zeta^3 \end{pmatrix}, \quad \text{and} \quad D'(b) = \overline{D(b)} = \begin{pmatrix} 0 & 1 & 0 \\ 0 & 0 & 1 \\ 1 & 0 & 0 \end{pmatrix}.$$

Note that we have $D \not\cong D'$, since they have different characters, say χ_4 and χ_5. We have now computed all complex irreducible representations of G, arriving at the following character table:

	1	a	a^3	b	b^2
χ_1	1	1	1	1	1
χ_2	1	1	1	ω	$\overline{\omega}$
χ_3	1	1	1	$\overline{\omega}$	ω
χ_4	3	α	$\overline{\alpha}$	0	0
χ_5	3	$\overline{\alpha}$	α	0	0

$\left(\text{where } \alpha = \zeta + \zeta^2 + \zeta^4\right).$

From the first column of this character table, we see that the Wedderburn decomposition of $\mathbb{C}G$ is:

(C) $\mathbb{C}G \cong \mathbb{C} \times \mathbb{C} \times \mathbb{C} \times M_3(\mathbb{C}) \times M_3(\mathbb{C}).$

(2) We now turn to the case of *rational* representations. Going to the quotient group $G/\langle a \rangle \cong \langle b \rangle$, we have a surjection $\mathbb{Q}G \xrightarrow{\pi} \mathbb{Q}\langle b \rangle \cong \mathbb{Q} \times \mathbb{Q}(\omega)$, where ω is as in (1). This gives two simple $\mathbb{Q}G$-modules: $\mathbb{Q}$ with the trivial G-action, and $\mathbb{Q}(\omega)$ with a acting trivially and b acting by multiplication by ω. The surjection π gives $\mathbb{Q}$ and $\mathbb{Q}(\omega)$ as two simple components of $\mathbb{Q}G$.

Our next step is to construct a 6-dimensional simple $\mathbb{Q}G$-module. We start with the simple $\mathbb{Q}\langle a \rangle$-module $K = \mathbb{Q}(\zeta)$ (where ζ is as before), on which a acts via multiplication by ζ. We can make K into a $\mathbb{Q}G$-module by letting b act via the field automorphism $\sigma \in \text{Gal}(\mathbb{Q}(\zeta)/\mathbb{Q})$ of order 3 sending $\zeta \mapsto \zeta^2$. This gives a well-defined G-action since, for any $k \in K$, $(a^2 b)k = \zeta^2 \sigma(k)$, while

$$(ba)k = b(\zeta k) = \sigma(\zeta k) = \zeta^2 \sigma(k).$$

Thus, K is a simple $\mathbb{Q}G$-module. Let $E = \text{End}(_{\mathbb{Q}G}K)$. Any $f \in E$ is right multiplication by some element $\ell \in K$, and we must have $(bk)f = b(kf)$ (for every $k \in K$); that is,

$$\sigma(k)\ell = \sigma(k\ell) = \sigma(k)\sigma(\ell),$$

which amounts to $\ell \in \mathbb{Q}(\zeta)^\sigma$. Now $F := \mathbb{Q}(\zeta)^\sigma$ is the unique quadratic extension of $\mathbb{Q}$ in K, which is well-known to be $\mathbb{Q}(\sqrt{-7})$. (More explicitly,

$$\alpha = \zeta + \sigma\zeta + \sigma^2\zeta = \zeta + \zeta^2 + \zeta^4 \in F$$

has minimal equation $\alpha^2 + \alpha + 2 = 0$, so $F = \mathbb{Q}(\alpha) = \mathbb{Q}(\sqrt{-7})$.) We have now seen that $E \cong F$. Since $\dim_E K = 3$, a simple component of $\mathbb{Q}G$ is given by

$$\operatorname{End}(K_E) \cong \mathbb{M}_3(E) \cong \mathbb{M}_3(\mathbb{Q}(\sqrt{-7})).$$

This has $\mathbb{Q}$-dimension 18, so we must have

(D) $$\mathbb{Q}G \cong \mathbb{Q} \times \mathbb{Q}(\omega) \times \mathbb{M}_3(\mathbb{Q}(\sqrt{-7})),$$

and $\mathbb{Q}$, $\mathbb{Q}(\omega)$, and K are the three simple $\mathbb{Q}G$-modules.

(3) Tensoring (D) up to $\mathbb{R}$, we get

(E) $$\mathbb{R}G \cong \mathbb{R} \times \mathbb{C} \times \mathbb{M}_3(\mathbb{C}),$$

which means that each of the irreducible $\mathbb{Q}$-representations above remains irreducible over $\mathbb{R}$ (and these give all irreducible $\mathbb{R}$-representations).

Comparing (D) and (E) with (C), we see that the 6-dimensional irreducible rational (resp. real) representation of G tensors up into the direct sum of the two 3-dimensional complex representations of G. (Of course, this is also easy to see by checking directly that $\chi_K = \chi_4 + \chi_5$.)

Comment. While the $\mathbb{C}$-representation D and the 6-dimensional $\mathbb{Q}$-representation of G were constructed above in a matter-of-fact manner, we should point out that these constructions do come from some general facts. To explain this, however, we have to assume some material from FC-§14 on cyclic algebras. Using the notations in the above solution, we note that K/F is a cyclic extension with Galois group $\langle \sigma \rangle$. Thus, we can form the cyclic algebra

$$A = (K/F, \sigma, 1) = K \oplus Kb \oplus Kb^2 \quad (b^3 = 1),$$

where $bk = \sigma(k)b$ for every $k \in K$. Identifying G with the semidirect product $\langle \zeta \rangle \cdot \langle b \rangle \subseteq U(A)$, we see that G acts linearly on the right of the 3-dimensional left K-vector space A. The matrices $D(a)$ and $D(b)$ in (1) are simply those for a and b (with respect to the basis $\{1, b, b^2\}$ on A) under this K-representation. (And, of course, a K-representation may be viewed as a $\mathbb{C}$-representation.) As for the 6-dimensional $\mathbb{Q}$-representation in (2), we note that A above is just the twisted group ring $K * \langle b \rangle$ in Ex. 14.13A, with b acting as σ on K. According to Ex. 14.13B, K can be naturally viewed as an A-module. Again, we can identify G with $\langle \zeta \rangle \cdot \langle b \rangle \subseteq U(A)$, so G acts $\mathbb{Q}$-linearly on K: this gives the 6-dimensional $\mathbb{Q}$-representation constructed in (2).

Ex. 8.29. Let U, V be simple modules of infinite k-dimensions over a group algebra kG (where k is a field), and let $W = U \otimes_k V$, viewed as a kG-module under the diagonal G-action.

(1) Does W have finite length?
(2) Is W semisimple?

Solution. The answers to (1), (2) are both "no", by a single counterexample. Let G be the multiplicative group of any infinite field extension K of k. Let U (resp. V) be a copy of K (as k-vector space) with G-action defined by $g * u = gu$ (resp. $g * v = g^{-1}v$). Clearly, U, V are simple kG-modules. If we view K as a kG-module with the trivial G-action, the k-map

$$\varphi : U \otimes_k V \to K \text{ induced by } u \otimes v \mapsto uv$$

is easily checked to be a kG-epimorphism. Since K has infinite length as a kG-module, so does $W = U \otimes_k V$.

We finish by checking that W is *not* a semisimple kG-module. Indeed, if W is semisimple, φ would be a split kG-epimormorphism, so W would have a submodule with trivial G-action mapping isomorphically onto K. We shall contradict this below by showing that *there are actually no nonzero G-fixed points in W*.

Consider any nonzero element $w = \sum_{i=1}^{n} u_i \otimes v_i \in W$. We may assume all $u_i \neq 0$. Let U_1 be the k-span of $u_1, \ldots, u_n$ in U, and fix a k-space $U_2 \subseteq U$ such that $U = U_1 \oplus U_2$. Since each $u_i^{-1}U_2$ has finite k-codimension in U, so does $\bigcap_{i=1}^{n} u_i^{-1}U_2$. In the infinite-dimensional k-space U, we can thus find a nonzero element $g \in \bigcap_{i=1}^{n} u_i^{-1}U_2$. Then $gu_i \in U_2$ for all i, so we have

$$g * w = \sum_i gu_i \otimes g^{-1}v_i \in U_2 \otimes V.$$

Since $W = (U_1 \otimes V) \oplus (U_2 \otimes V)$ and $0 \neq w \in U_1 \otimes V$, it follows that $g * w \neq w$.

Comment. The argument in the last paragraph shows, in fact, that a nonzero kG-submodule $W_0 \subseteq W$ cannot be contained in $U_1 \otimes V$ for any finite-dimensional k-subspace $U_1 \subseteq U$. In particular, any such W_0 must be infinite-dimensional.

I thank H.W. Lenstra, Jr. for suggesting part (1) of this exercise. Part (2), which I made up, is related to a classical theorem of Chevalley (p. 88 in Vol. 3 of his book "Théorie des Groupes de Lie"), which states that, if U, V are *finite-dimensional* simple kG-modules over a field k *of characteristic 0*, then $W = U \otimes_k V$ is semisimple. (In other words, finite-dimensional semisimple kG-modules are closed under tensor products in case char$(k) = 0$.) The negative answer to (2) in this exercise shows that Chevalley's theorem holds *only* for finite-dimensional representations. The next two exercises show, in turn, that Chevalley's theorem does not hold in characteristic $p > 0$.

Ex. 8.30. For $G = S_3$ and any field k of characteristic 2, view

$$V = ke_1 \oplus ke_2 \oplus ke_3/k(e_1 + e_2 + e_3)$$

as a (simple) kG-module with the permutation action.

(1) Show that $W = V \otimes_k V$ with the diagonal G-action is *not* a semisimple kG-module.

(2) Show that $kG \cong M_2(k) \times (k[t]/(t^2))$, and that $kG/\mathrm{rad}(kG) \cong M_2(k) \times k$.

Solution. (1) With the basis $\{\bar{e}_1, \bar{e}_2\}$ on V, the G-action is described by

$$(12)\bar{e}_1 = \bar{e}_2, \quad (12)\bar{e}_2 = \bar{e}_1; \quad (123)\bar{e}_1 = \bar{e}_2, \quad (123)\bar{e}_2 = \bar{e}_1 + \bar{e}_2.$$

Using these, it is easy to check that the k-span of

$$x = \bar{e}_1 \otimes \bar{e}_1, \quad y = \bar{e}_2 \otimes \bar{e}_2, \quad \text{and} \quad z = \bar{e}_1 \otimes \bar{e}_2 + \bar{e}_2 \otimes \bar{e}_1$$

is a 3-dimensional kG-submodule $A \subseteq W$ (with $B = k \cdot z$ as a trivial kG-submodule). *We claim that A is not a kG-direct summand of W.* Indeed, if $W = A \oplus k \cdot w$ is a kG-decomposition, we must have $w \in W^G$ (G-fixed points in W) since $G/[G,G] \cong \{\pm 1\}$ implies that any 1-dimensional kG-module is trivial. But if

$$w = a(\bar{e}_1 \otimes \bar{e}_1) + b(\bar{e}_2 \otimes \bar{e}_2) + c(\bar{e}_1 \otimes \bar{e}_2) + d(\bar{e}_2 \otimes \bar{e}_1),$$

$(12)w = w$ implies that $c = d$ so $w \in A$, a contradiction. This shows that W is *not* semisimple. (In fact, $(123)w = w$ implies further that $a = 0$, so we have $W^G = k \cdot z = B$. The kG-composition factors of W are the trivial G-modules B, W/A, together with $A/B \cong V$.)

(2) Let $\varphi : kG \to \mathrm{End}_k(V) \cong M_2(k)$ be the k-algebra homomorphism associated with the kG-module V. This map is *onto* since V is absolutely irreducible.[*] Thus, $\dim_k \ker(\varphi) = 2$. Now $e = (1) + (123) + (132)$ is a central idempotent of kG, with

$$kG \cdot e = ke \oplus k\sigma, \quad \text{where} \quad \sigma = \sum_{g \in G} g \in kG.$$

Since $\sigma^2 = 6\sigma = 0$, we have $kG \cdot e \cong k[t]/(t^2)$. By a simple computation, $\varphi(e) = 0$, so $kG \cdot e = \ker(\varphi)$ (both spaces being 2-dimensional). Therefore,

$$kG = kG \cdot (1 - e) \times kG \cdot e \cong M_2(k) \times (k[t]/(t^2)).$$

Computing radicals, we get $\mathrm{rad}(kG) = \mathrm{rad}(kG \cdot e) = k \cdot \sigma$ (consistently with Ex. 6.13(1)), so $kG/\mathrm{rad}(kG) \cong M_2(k) \times k$.

---

[*] It is of interest to note that the $\mathbb{F}_2$-span V_0 of $\{\bar{e}_1, \bar{e}_2\}$ is the Klein 4-group, which is being acted on by $\mathrm{Aut}(V_0) \cong S_3 = G$, and V is simply the scalar extension $k \otimes_{\mathbb{F}_2} V_0$, which is irreducible over *any* $k \supseteq \mathbb{F}_2$.

Ex. 8.31. For any field k of characteristic p, let $G = \mathrm{SL}_2(\mathbb{F}_p)$ act on the polynomial ring $A = k[x, y]$ by linear changes of the variables $\{x, y\}$, and let $V_d \subseteq A$ $(d \geq 0)$ be the kG-submodule of homogeneous polynomials of degree d in A. It is known (and thus you may assume) that $V_0, \ldots, V_{p-1}$ are a complete set of simple modules over kG.[(*)]

(1) Compute the composition factors of V_p, and show that V_p is semisimple over kG if and only if $p = 2$.

(2) If $\{d_1, \ldots, d_n\}$ is any partition of p, show that the tensor product $V_{d_1} \otimes_k \cdots \otimes_k V_{d_n}$ (under the diagonal G-action) is *not* semisimple over kG, unless $p = 2$ and $n = 1$.

Solution. (1) Let $\varepsilon : V_1 \to V_p$ be the k-linear map defined by $\varepsilon(x) = x^p$ and $\varepsilon(y) = y^p$. This is a kG-monomorphism since, at the $\mathbb{F}_p$-level, ε is the Frobenius map $\ell \mapsto \ell^p$. Next, we define a k-linear map $\pi : V_p \to V_{p-2}$ by

$$\pi(f) = \frac{1}{y}\frac{\partial f}{\partial x} = -\frac{1}{x}\frac{\partial f}{\partial y} \qquad (f \in V_p),$$

where the second equality results from Euler's identity

$$x\frac{\partial f}{\partial x} + y\frac{\partial f}{\partial y} = (\deg f) \cdot f = p \cdot f = 0.$$

Note that π maps x^p, y^p to 0, and maps $x^i y^{p-i}$ $(0 < i < p)$ to $ix^{i-1}y^{p-i-1} \in V_{p-2}$, so $\ker(\pi) = \mathrm{im}(\varepsilon)$, and $\mathrm{im}(\pi) = V_{p-2}$. Therefore, we have an exact sequence of kG-modules

$$(*) \qquad\qquad 0 \longrightarrow V_1 \xrightarrow{\varepsilon} V_p \xrightarrow{\pi} V_{p-2} \longrightarrow 0$$

if we can show that $\pi(\sigma \cdot f) = \sigma \cdot \pi(f)$ for any $\sigma = \begin{pmatrix} a & b \\ c & d \end{pmatrix} \in G$. Checking this on the basis elements $f = x^i y^j$ $(i + j = p)$, we have

$$\begin{aligned}
\pi(\sigma \cdot f) &= \pi((ax + cy)^i (bx + dy)^j) \\
&= y^{-1}(ax + cy)^{i-1}(bx + dy)^{j-1}[ia(bx + dy) + jb(ax + cy)] \\
&= y^{-1}(ax + cy)^{i-1}(bx + dy)^{j-1}[(i + j)abx + (iad + jbc)y] \\
&= i(ax + cy)^{i-1}(bx + dy)^{p-i-1}(ad - bc),
\end{aligned}$$

which is just $\sigma \cdot \pi(f)$ since $ad - bc = 1$! From $(*)$, it follows that *the composition factors for V_p are* $\{V_1, V_{p-2}\}$.

If $p = 2$, $x^2 + xy + y^2$ is the only irreducible quadratic form over $\mathbb{F}_2$, so it must be G-invariant. Thus,

$$V_2 = \mathrm{im}(\varepsilon) \oplus k \cdot (x^2 + xy + y^2) \cong V_1 \oplus V_0$$

[(*)] For a proof of this, see J. Alperin's book "Local Representation Theory," pp. 15–16, Cambridge University Press, 1986.

is semisimple. For $p > 2$, we claim that the kG-sequence $(*)$ *does not* split (so V_p is not semisimple). Indeed, if there exists a kG-splitting φ for π, let $f = \varphi(y^{p-2})$ and $g = \varphi(x^{p-2})$. Then, modulo $\mathrm{im}(\varepsilon)$, we have

$$f \equiv xy^{p-1}, \quad \text{and} \quad g \equiv x^{p-1}y/(p-1) \equiv -x^{p-1}y.$$

But for $\sigma = \begin{pmatrix} 0 & 1 \\ 1 & 0 \end{pmatrix} \in G$, we have $\sigma \cdot y^{p-2} = x^{p-2}$, so $\sigma \cdot f = g$. Since $\sigma(\mathrm{im}(\varepsilon)) \subseteq \mathrm{im}(\varepsilon)$, this implies that $yx^{p-1} \equiv -x^{p-1}y$ modulo $\mathrm{im}(\varepsilon)$, which is impossible (for $p > 2$).

(2) Let $d_1 + \cdots + d_n = p$, where each $d_i \geq 1$. Then

$$f_1 \otimes \cdots \otimes f_n \mapsto f_1 \cdots f_n \quad (f_i \in V_{d_i})$$

defines a kG-surjection $V_{i_1} \otimes \cdots \otimes V_{i_n} \to V_p$. If $p > 2$, then $V_{i_1} \otimes \cdots \otimes V_{i_n}$ *cannot* be semisimple over kG, since V_p is not. If $p = 2$, this argument no longer works. However, if $n \neq 1$, the conclusion remains valid; namely, $V_1 \otimes V_1$ is still not semisimple. In fact, for $p = 2$, $G = \mathrm{SL}_2(\mathbb{F}_p)$ is the group S_3, and V_1 is easily seen to be isomorphic to the module V in Ex. 8.30. Thus, the desired conclusion follows from Ex. 8.30!

Comment. This exercise comes from J.-P. Serre's paper "Sur les semi-simplicité des produits tensoriels de répresentations des groupes," Invent. Math. *116* (1994), 513–530. In this paper, Serre proved the following remarkable theorem: *if* $\mathrm{char}(k) = p > 0$, *and* $U_1, \ldots, U_n$ *are finite-dimensional (semi)simple* kG*-modules over any (possibly infinite) group* G *with* $\sum_{i=1}^n (\dim_k U_i - 1) < p$, *then* $U_1 \otimes \cdots \otimes U_n$ *is semisimple.* In this exercise, we have $\dim_k V_i = i + 1$ for all i: Serre invoked part (2) above to show that the dimension bound in his theorem cannot be further improved in any prime characteristic (even for finite groups G). The proof of Serre's theorem depends heavily on the theory of linear algebraic groups; see also a sequel to his paper in J. Algebra *194* (1997), 496–520.

§9. Linear Groups

The theory of linear groups is a beautiful branch of algebra with many truly significant results. The early theorems in the subject, due to Burnside and Schur, planted the seed for the (bounded, and restricted) Burnside Problems, solved only many years later by Novikov, Adjan, and Zelmanov. The delightful Lie-Kolchin Theorem turned out to be a harbinger for the modern theory of linear algebraic groups. The study of classical groups and the so-called groups of Lie type is, according to most algebraists, group theory *par excellence!*

A little bit of ring theory enters the scene as well. If $G \subseteq \mathrm{GL}_n(k)$ is a linear group acting on the vector space $V = k^n$ (where k is a field), G spans a k-subalgebra R of $\mathbb{M}_n(k)$ which is a homomorphic image of the group algebra kG. The space $V = k^n$ may be viewed as an R-module as well as a kG-module. We say that G is *completely reducible* if V is semisimple as a kG-module. Somewhat surprisingly, it turns out that this is equivalent to R being a semisimple ring. In any case, the connection between linear groups and modules over rings is now more apparent.

The exercise set for this section is rather small by design, since our main goal is only to point out the interplay between linear groups and ring theory. The eight problems below deal with certain properties of linear groups, such as irreducibility and complete reducibility. Exercise 3 offers an explicit computation of the subalgebra spanned by a linear group, and Exercises 9.4A–C provide a view of the unipotent linear groups from the perspective of noncommutative rings and their nil ideals.

Exercises for §9

Ex. 9.1. Let $G \subseteq \mathrm{GL}_n(k)$ be a linear group over a field k. Show that G is an f.c. group (i.e. every conjugacy class of G is finite) iff the center $Z(G)$ has finite index in G. Show that every finite group can be realized as a linear group, but not every infinite group can be realized as a linear group.

Solution. The "if" part is clearly true for any group G. Now assume $G \subseteq \mathrm{GL}_n(k)$ is f.c. Choose $g_1, \ldots, g_r \in G$ which span the k-space generated by G in $\mathbb{M}_n(k)$. Note that if $g \in G$ commutes with each g_i, then $g \in Z(G)$. Therefore,

$$Z(G) = \bigcap C_G(g_i).$$

Since each $C_G(g_i)$ has finite index in G, we have $[G : Z(G)] < \infty$.

If G is a finite group, G acts on itself faithfully by left multiplication, so we can embed G as a group of permutation matrices in $\mathrm{GL}_m(\mathbb{Z}) \subseteq \mathrm{GL}_m(\mathbb{Q})$, where $m = |G|$.

We finish by constructing an infinite group G which cannot be realized as a linear group. Take any nonabelian finite group H. Since H is f.c., the direct sum

$$G : = H \oplus H \oplus \cdots.$$

is also f.c. However, since $H \neq Z(H)$,

$$Z(G) = Z(H) \oplus Z(H) \oplus \cdots$$

has infinite index in G. By the first part of the exercise, G cannot be realized as a linear group.

Comment. An alternative construction is as follows. It is known that there are infinite groups G with only a finite number of conjugacy classes. According to Burnside's Second Theorem $(FC\text{-}(9.5))$, such a group cannot be realized as a linear group.

Ex. 9.2. Can every finite group be realized as an irreducible linear group? (A linear group $G \subseteq \mathrm{GL}(V)$ is said to be irreducible if G acts irreducibly on V.)

Solution. Consider an irreducible linear group $G \subseteq \mathrm{GL}(V)$, where V is a finite-dimensional vector space over a field k. *We claim that $Z(G)$ is cyclic.* This will show, for instance, that the Klein 4-group $\mathbb{Z}_2 \oplus \mathbb{Z}_2$ cannot be realized as an irreducible linear group. To prove the claim, consider $D = \mathrm{End}(_{kG}V)$, which is, by Schur's Lemma, a (finite dimensional) division algebra over k. Since every $g \in Z(G)$ acts as a kG-automorphism of V, we can think of $Z(G)$ as embeded in D^*. The k-algebra F generated by $Z(G)$ in D is a finite-dimensional k-domain, so F is a field. By a well-known theorem in field theory, $Z(G) \subseteq F^*$ must be a cyclic group.

Ex. 9.3. Let k be any field of characteristic 3, $G = S_3$ and let V be the kG-module
$$ke_1 \oplus ke_2 \oplus ke_3/k(e_1 + e_2 + e_3),$$
on which G acts by permuting the e_i's. Show that this realizes G as a linear group in $\mathrm{GL}(V)$. Is G a completely reducible linear group? What is its unipotent radical? Determine $\mathrm{Span}_k(G)$ and its Jacobson radical.

Solution. Since (123) does not act trivially on V, the representation homomorphism $\varphi : G \to \mathrm{GL}(V)$ must be *injective*. Therefore, φ realizes G as a linear group in $\mathrm{GL}(V)$. It is easy to see that $V_o = k(\bar{e}_1 - \bar{e}_2)$ is a kG-module affording the sign representation (of G), with V/V_o affording the trivial representation: cf. Ex. 8.26. In this exercise, it was also shown that $_{kG}V$ is indecomposable. Therefore, $G \subseteq \mathrm{GL}(V)$ is not completely reducible.

The unipotent radical of G is the unique maximal normal unipotent subgroup. By the discussion in the proof of $FC\text{-}(9.22)$, this unipotent radical consists of $g \in G$ which act trivially on the composition factors of $_{kG}V$. In our case, this subgroup is clearly $\langle(123)\rangle$.

Next we try to determine $S := \mathrm{Span}_k(G)$. Using $\{\bar{e}_1 - \bar{e}_2, \bar{e}_1\}$ as a basis for V, we have

$$\varphi(123) = \begin{pmatrix} 1 & -1 \\ 0 & 1 \end{pmatrix} \quad \text{and} \quad \varphi(12) = \begin{pmatrix} -1 & -1 \\ 0 & 1 \end{pmatrix},$$

so $S \subseteq T$, the k-subalgebra of all upper triangular matrices in $\mathbb{M}_2(k)$. Since S is noncommutative, we must have $S = T$. It follows that

$$\mathrm{rad}\, S = \mathrm{rad}\, T = k \cdot \begin{pmatrix} 0 & 1 \\ 0 & 0 \end{pmatrix} = k \cdot (1 - \varphi(123)).$$

Ex. 9.4A. Let J be a nil ideal in an algebra R over a field k of characteristic 0, and let G be the group $1 + J \subseteq U(R)$.

(1) Show that the map $\exp : J \to G$ defined by the Taylor series of the exponential function is a one-one correspondence between J and G, with inverse given by the log-function.

(2) Show that two elements x, $x' \in J$ commute iff $\exp(x)$ and $\exp(x')$ commute in G, in which case we have

(A) $$\exp(x) \cdot \exp(x') = \exp(x + x') \in G.$$

Similarly, show that two elements y, $y' \in G$ commute iff $\log(y)$ and $\log(y')$ commute in J, in which case we have

(B) $$\log(yy') = \log(y) + \log(y') \in J.$$

(3) For any $y \in G$ and $\alpha \in k$, define

(C) $$y^{\alpha} = \exp(\alpha \log(y)) \in G.$$

Show that $\log(y^{\alpha}) = \alpha \log(y)$, $(y^{\alpha})^{\beta} = y^{\alpha\beta}$, and $y^{\alpha}y^{\beta} = y^{\alpha+\beta}$ for any α, $\beta \in k$.

(4) Show that G is a *uniquely divisible* group; that is, for any positive integer r, each element of G has a unique rth root in G. (In particular, the group G is torsionfree.)

(5) For any $y \in G$, let $y^k := \{y^{\alpha} : \alpha \in k\}$. If $y \neq 1$, show that y^k is a subgroup of G isomorphic to the additive group k.

(6) Show that two subgroups y^k and z^k of G are either the same or have trivial intersection, and G is the union of all such subgroups.

(7) If J is commutative under multiplication, show that $G \cong J$ as (abelian) groups.

Solution. (1) We define $\exp : J \to G$ and $\log : G \to J$ by

(D)
$$\exp(x) = 1 + \frac{x}{1!} + \frac{x^2}{2!} + \cdots \in G, \text{ and } \log(1 + x) = x - \frac{x^2}{2} + \frac{x^3}{3} - \cdots \in J,$$

for any $x \in J$. Since x is *nilpotent*, these formulas make sense, and "exp" and "log" are mutually inverse as usual.

(2) If $xx' = x'x$ in J, certainly $\exp(x)$ commutes with $\exp(x')$ in G, and (A) holds here as it holds already as a power series identity over $\mathbb{Q}$ in the commuting variables x and x'. Conversely, if $\exp(x) = 1 + s$ commutes with $\exp(x') = 1 + s'$ (where s, $s' \in J$), then $ss' = s's$. This implies that $x = \log(1 + s)$ commutes with $x' = \log(1 + s')$, since they are, respectively,

polynomials in s and s'. The second conclusion in (2) can be proved similarly, or derived from the first by using inverse functions.

(3) To begin with, $\log(y^\alpha) = \log[\exp(\alpha \log y)] = \alpha \log y$. Using this, we get

$$\log(y^\alpha)^\beta = \beta \log(y^\alpha) = \beta\alpha \log(y) = \log(y^{\alpha\beta}),$$

so the injectivity of "log" implies that $(y^\alpha)^\beta = y^{\alpha\beta} \in G$. Finally, since $\alpha \log y$ and $\beta \log y$ do commute in J, (A) gives

$$y^\alpha y^\beta = \exp(\alpha \log y) \cdot \exp(\beta \log y)$$
$$= \exp((\alpha + \beta) \log y) = y^{\alpha+\beta}.$$

(4) Note that, for $\alpha \in \mathbb{Z} \subseteq k$, y^α has its usual meaning. (This we can believe! It is a formal fact that the integral power functions y^n can be identified with $\exp(n \log y)$.) Thus, for any $y \in G$ and any positive integer r, $y^{1/r}$ (defined as in (C)) is an rth root of y in G. Moreover, if $z \in G$ is such that $z^r = y$, taking the $(1/r)$-th powers yields $z = y^{1/r}$. This proves (4).

(5) By (3), y^k is a subgroup of G (for every $y \in G$), and "exp" and "log" define mutually inverse group isomorphisms between $k \cdot \log y$ and y^k. If $y \neq 1$, the former group is isomorphic to k.

(6) This is clear since $k \cdot \log y$ and $k \cdot \log z$ are either the same or have zero intersection in J, and J is the union of all of these.

(7) If J is commutative under multiplication, the law (A) implies that $\exp : J \to G$ is a *group isomorphism* (with inverse isomorphism given by "log").

Comment. The exponential expressions y^α $(y \in G, \alpha \in k)$ could also have been defined by writing $y = 1 + x$ $(x \in J)$ and using Newton's binomial expansion

$$(1+x)^\alpha = 1 + \alpha x + \frac{\alpha(\alpha-1)}{2!} x^2 + \frac{\alpha(\alpha-1)(\alpha-2)}{3!} x^3 + \cdots \in G,$$

which makes sense (again) since x is nilpotent.

In case $J^{n+1} = 0$ for some n, we have shown in Ex. 5.11 that $G = 1 + J$ is a nilpotent group of class $\leq n$. This need not be the case here, but the statement (4) proved above is related to following fact: *in any torsionfree nilpotent group* G, $z^r = w^r \Longrightarrow z = w$ *for any integer* $r \geq 1$. This can be proved by induction on the nilpotency class of G, using the fact that z and $[G, G]$ generate a nilpotent group of *smaller* class in G.

For more relevant information on this exercise, see the *Comment* on Ex. 9.4B.

Ex. 9.4B. Keep the notations in Ex. 9.4A, but assume, instead, that $\mathrm{char}(k) = p > 0$. Let $G_1 = \{y \in G : y^p = 1\}$. Show that y^k can still be defined as long as $y \in G_1$, and that the subgroups $y^k \subseteq G$ (for $y \in G_1$) have the same properties as before (in Ex. 9.4A), with union equal to G_1.

Solution. Note that G_1 is only a *subset* (not necessarily a subgroup) of G. Letting $J_1 := \{x \in J : x^p = 0\}$, we have $G_1 = 1 + J_1$. We can define $\exp : J_1 \to G_1$ and $\log : G_1 \to J_1$ by the same formulas as in (D) in Ex. 9.4A, except that here, we drop all terms x^n with $n \geq p$. (Note that the denominators of the remaining terms are all p-free and are thus invertible in k.) This gives a one-one correspondence $J_1 \longleftrightarrow G_1$. For $y \in G_1$ and $\alpha \in k$, we can define $y^\alpha = \exp(\alpha \log y) \in G_1$ just as before, noting that J_1 is closed under multiplication by k. The exponential (and logarithmic) laws in Ex. 9.4A(3) still hold for $y \in G_1$ (by the same proofs), and again we have

$$k \cong k \cdot \log(y) \cong y^k \subseteq G_1 \quad \text{if } y \neq 1.$$

Here, J_1 is the union of the "lines" $k \cdot \log y$ $(1 \neq y \in G_1)$, two of which are either the same or intersect at $\{0\}$. Thus, G_1 is the union of the groups y^k $(1 \neq y \in G_1)$, two of which are either the same or intersect at $\{1\}$.

Comment. Much of the information contained in this and the last exercise is folklore in groups and rings, but our present exposition is inspired by two papers of I.M. Isaacs: "Equally partitioned groups", Pacific J. Math. *49* (1973), 109–116, and "Characters of groups associated with finite algebras", J. Algebra *177* (1995), 708–730. In the latter paper, J is the Jacobson radical of a finite-dimensional k-algebra. Here, we work more generally with nil ideals in any k-algebra.

In Isaacs' terminology, Ex. 9.4A(6) says that the group G is "equally partitioned" by the distinct subgroups $\{y^k\}$. That was in characteristic 0. In characteristic p, this exercise shows that G_1 is also "equally partitioned" by the distinct subgroups $\{y^k\}$, except that G_1 is no longer a group. Isaacs' main result in the first paper cited above was that, if a *finite* group G can be equally partitioned into proper subgroups, then G must have prime exponent. Of course, if a group G has prime exponent p, then G can be equally partitioned into its distinct subgroups of order p. The next exercise will offer examples of nonabelian (finite or infinite) groups of exponent p admitting equal partitions into subgroups of size $> p$.

Ex. 9.4C. Let G be a maximal unipotent subgroup of $\mathrm{GL}_n(k)$ over a field k.

(1) If $\mathrm{char}(k) = 0$, show that G is a torsionfree, nilpotent, and uniquely divisible group, and that it is the union of a family of subgroups $H_i \cong k$ such that $H_i \cap H_j = \{1\}$ for all $i \neq j$.

(2) If $\mathrm{char}(k) = p > 0$, show that G is a p-group. (In particular, G is uniquely r-divisible for any positive integer r that is prime to p.) If, moreover, $n \leq p$, show that G has exponent p, and that it is the union of a family of subgroups $H_i \cong k$ such that $H_i \cap H_j = \{1\}$ for all $i \neq j$.

Solution. By the Lie-Kolchin Theorem (see the version stated in *FC*-(9.19)), G is conjugate to $\mathrm{UT}_n(k)$, the unitriangular group of upper triangular matrices with 1's on the diagonal. To simplify notations, let us assume that $G = UT_n(k)$. Then $G = 1 + J$ for $J = rad(R)$, where R is the ring of $n \times n$ upper triangular matrices over k. We have here $J^n = 0$, so G is nilpotent of class $\leq n - 1$ by Ex. 5.11(2). (In fact, the class is exactly $n - 1$.) The rest of (1) follows from Ex. 9.4A.

If $char(k) = p > 0$, Ex. 5.11(3) shows that G is a p-group. If, moreover, $n \leq p$, then for any $x \in J$, we have $x^n = 0$, and hence

$$(1 + x)^p = 1 + x^p = 1 + x^n x^{p-n} = 1.$$

Thus, G has exponent p, and the rest of (2) follows from Ex. 9.4B.

Comment. In connection with this exercise, it is certainly worth recalling that, in the case $k = \mathbb{F}_q$ where $q = p^s$, $G = UT_n(k)$ *is a p-Sylow subgroup of* $GL_n(k)$: see *FC*-§9 (around (9.20)). In the second paper of Isaacs cited in the *Comment* on Ex. 9.4B, he studied the irreducible complex characters of finite groups of the form $G = 1 + J$, where J is the Jacobson radical of a finite-dimensional $\mathbb{F}_q$-algebra. The main result of that paper is that *all irreducible characters of G have q-power degrees.* In particular, this nice conclusion holds for the p-Sylow subgroups G of the group $GL_n(\mathbb{F}_q)$.

Ex. 9.5. Let k be an algebraically closed field, and $G \subseteq GL_n(k)$ be a completely reducible linear group. Show that G is abelian iff G is conjugate to a subgroup of the group of diagonal matrices in $GL_n(k)$.

Solution. The "if" part is obvious, so it suffices to prove the "only if" part. Assume G is abelian, and decompose $V = k^n$ into $V_1 \oplus \cdots \oplus V_r$ where the V_i's are simple kG-modules. Consider an element $g \in G$, and let λ_i be an eigenvalue of g on V_i. Then

$$\{v \in V_i : \quad gv = \lambda_i v\}$$

is a nonzero kG-submodule of V_i, and hence equal to V_i. This means each $g \in G$ acts as a scalar multiplication on V_i, so $\dim_k V_i = 1$ for all i. Writing $V_i = kv_i$ and using the basis $\{v_1, \ldots, v_n\}$ on V, we have then represented G as a group of diagonal matrices over k.

Ex. 9.6. Let k be a field of characteristic zero. Let $G \subseteq GL_n(k)$ be a linear group, and H be a subgroup of finite index in G. Show that G is completely reducible iff H is.

Solution. We think of $V = k^n$ as a kG-module. If $_{kH}V$ is semisimple, Maschke's Theorem, in the generalized form of Exercise 6.1, implies that $_{kG}V$ is semisimple. This gives the "if" part. Conversely, assume $_{kG}V$ is semisimple. Since $[G : H] < \infty$, the intersection A of the finite number of conjugates of H is a normal subgroup of finite index in G. By Clifford's Theorem (*FC*-(8.5)), $_{kA}V$ is semisimple. Since

$$[H : A] \leq [G : A] < \infty,$$

the first part of the exercise implies that $_{kH}V$ is also semisimple, so H is completely reducible, as desired.

Chapter 4
Prime and Primitive Rings

§10. The Prime Radical; Prime and Semiprime Rings

The notion of prime ideals in commutative rings extends naturally to non-commutative rings. One takes the "same" definition, but replaces the use of elements by ideals of the ring. The Zorn's Lemma construction of prime ideals disjoint from a multiplicative set in the commutative setting finds a natural generalization, if we just replace the multiplicative set with an "m-system": cf. FC-(10.5). (A nonempty set $S \subseteq R$ is called an m-system if, for any $a, b \in S$, $arb \in S$ for some $r \in R$.)

For any ideal $\mathfrak{A} \subseteq R$, one can define $\sqrt{\mathfrak{A}}$ to be the set of all $s \in R$ such that every m-system containing s meets $\mathfrak{A}$. Then $\sqrt{\mathfrak{A}}$ is exactly the intersection of the prime ideals of R containing $\mathfrak{A}$. Those $\mathfrak{A}$ for which $\mathfrak{A} = \sqrt{\mathfrak{A}}$ are called *semiprime ideals*; they are just the intersections of prime ideals. A ring R is called *prime* (resp. *semiprime*) if (0) is a prime (resp. semiprime) ideal. For a proper understanding of how these notions are derived from their commutative counterparts, the chart on FC-p.153 is highly recommended.

The exercises in this section involve three more radicals of a ring R (beyond the Jacobson radical). The *lower nilradical* (or Baer-McCoy radical), denoted by Nil_*R, is simply $\sqrt{(0)}$. This is the intersection of all the prime ideals of R, and, for this reason, it is also known as the *prime radical*. The *upper nilradical*, Nil^*R, is the unique largest nil ideal of R. Finally, the *Levitzki radical*, L-rad R, is the unique largest locally nilpotent ideal of R. (An ideal, or more generally, a subset $I \subseteq R$ is *locally nilpotent* if, for any finite subset $\{a_1, \ldots, a_n\} \subseteq I$, there exists an integer N such that any product of N elements from $\{a_1, \ldots, a_n\}$ is zero.)

In general, the four radicals are related as follows:

$$\text{Nil}_* R \subseteq L\text{-rad } R \subseteq \text{Nil}^* R \subseteq \text{rad } R.$$

Any of these inclusions may be strict. However, for some rings R, we may have one or more equalities. For instance, a classical theorem of Levitzki $(FC\text{-}(10.30))$ says that, if R is a 1-sided noetherian ring, then

$$\text{Nil}_* R = L\text{-rad } R = \text{Nil}^* R,$$

and this is the largest nilpotent (1-sided or 2-sided) ideal in R.

The exercises in this section deal with properties of prime and semiprime rings, and of the various radicals. Again, there are many folklore results: every prime contains a minimal prime (Exercise 14), and every ring with ACC on ideals has only finitely many minimal primes (Exercise 15). We do not feel justified in giving an axiomatic treatment of the theory of radicals in general. As a compromise, Exercise 20 offers a glimpse into what properties an "abstract" radical might be expected to enjoy.

Needless to say, no discussion of nilradicals can be complete without at least a mention of the *Köthe Conjecture*. This famous conjecture says that *if a ring R has no nonzero nil ideals, then it has also no nonzero nil right ideals.* An equivalent form is the equally provocative statement that the sum of two nil right ideals in R is always nil. No one knows if either statement is true or false. Exercises 24 and 25 offer various other equivalent forms of the Conjecture, found by Krempa and Amitsur. Matrix enthusiasts will perhaps give a shrewd nod to the following equivalent form: *If I is a nil ideal in R, then $\mathbb{M}_n(I)$ is a nil ideal in $\mathbb{M}_n(R)$.* According to Exercise 25, if you could prove this, you would have proved the long-standing Köthe Conjecture!

Exercises for §10

Ex. 10.0. Show that a nonzero central element of a prime ring R is not a zero-divisor in R. In particular, the center $Z(R)$ is an integral domain, and char R is either 0 or a prime number.

Solution. Let $0 \neq a \in Z(R)$, and say $ab = 0$. Then $aRb = Rab = 0$, so $b = 0$ since R is a prime ring. This says that a is not a zero-divisor in R, and the other conclusions of the exercise follow immediately.

Comment. Any commutative domain A is the center of a suitable *noncommutative* prime ring R. In fact, we can take R to be $\mathbb{M}_n(A)$ for any $n \geq 2$. (As observed after $FC\text{-}(10.20)$, $\mathbb{M}_n(A)$ is a prime ring for any domain A.)

Ex. 10.1. For any semiprime ring R, show that $Z(R)$ is reduced, and that char R is either 0 or a square-free integer.

Solution. Let $a \in Z(R)$ be such that $a^2 = 0$. Then $aRa = Ra^2 = 0$, so $a = 0$. This shows that $Z(R)$ is a reduced (commutative) ring, and the second conclusion of the exercise follows immediately.

Ex. 10.2. Let $\mathfrak{p} \subset R$ be a prime ideal, $\mathfrak{A}$ be a left ideal and $\mathfrak{B}$ be a right ideal. Does $\mathfrak{A}\mathfrak{B} \subseteq \mathfrak{p}$ imply that $\mathfrak{A} \subseteq \mathfrak{p}$ or $\mathfrak{B} \subseteq \mathfrak{p}$?

Solution. The answer is "no." For instance, let R be any prime ring with an idempotent $e \neq 0, 1$. (We can take $R = \mathbb{M}_n(\mathbb{Z})$ with $n \geq 2$.) Then, $\mathfrak{p} = 0$ is a prime ideal in R. However, for $\mathfrak{A} = Re \neq 0$ and $\mathfrak{B} = (1 - e)R \neq 0$, we have $\mathfrak{A}\mathfrak{B} = Re(1 - e)R = 0$.

Comment. The case when $\mathfrak{A}$ is a left ideal and $\mathfrak{B}$ is a right ideal is "the only bad case." The following statement in the mixed case turns out to be valid (along with FC-(10.2)): $\mathfrak{p} \subsetneq R$ *is prime iff, for any right ideal* $\mathfrak{A}$ *and any left ideal* $\mathfrak{B}$, $\mathfrak{A}\mathfrak{B} \subseteq \mathfrak{p}$ *implies that either* $\mathfrak{A} \subseteq \mathfrak{p}$ *or* $\mathfrak{B} \subseteq \mathfrak{p}$. The "if" part follows directly from FC-(10.2). For the "only if" part, assume $\mathfrak{A}\mathfrak{B} \subseteq \mathfrak{p}$, where $\mathfrak{A}$ is a *right* ideal and $\mathfrak{B}$ is a *left* ideal. Then $R\mathfrak{A}$ and $\mathfrak{B}R$ are ideals, with $R\mathfrak{A} \cdot \mathfrak{B}R \subseteq R\mathfrak{p}R \subseteq \mathfrak{p}$. Therefore, we have either $R\mathfrak{A} \subseteq \mathfrak{p}$ or $\mathfrak{B}R \subseteq \mathfrak{p}$. This implies that either $\mathfrak{A} \subseteq \mathfrak{p}$ or $\mathfrak{B} \subseteq \mathfrak{p}$.

Ex. 10.3. Show that a ring R is a domain iff R is prime and reduced.

Solution. First assume R is a domain. Then $a^n = 0 \Longrightarrow a = 0$, so R is reduced. Also,

$$aRb = 0 \Longrightarrow ab = 0 \Longrightarrow a = 0 \text{ or } b = 0,$$

so R is prime. Conversely, assume R is prime and reduced. Let $a, b \in R$ be such that $ab = 0$. Then, for any $r \in R$,

$$(bra)^2 = br(ab)ra = 0,$$

so $bra = 0$. This means that $bRa = 0$, so $b = 0$ or $a = 0$, since R is prime.

Ex. 10.4. Show that in a right artinian ring R, every prime ideal $\mathfrak{p}$ is maximal. (Equivalently, R is prime iff it is simple.)

Solution. $R/\mathfrak{p}$ is semiprime and right artinian, so it is semisimple by FC-(10.24). Since $R/\mathfrak{p}$ is in fact prime, it can have only one simple component. Therefore, $R/\mathfrak{p}$ is simple, so $\mathfrak{p}$ is a maximal ideal.

Comment. In commutative ring theory, a much sharper statement is possible. In fact, for R commutative, it is well-known that R *is artinian iff R is noetherian and every prime ideal of R is maximal:* see FC-(23.12).

Ex. 10.4*. For any given division ring k, list all the prime and semiprime ideals in the ring R of 3×3 upper triangular matrices over k.

Solution. We first make a general observation on the semiprime ideals in a right artinian ring R. Say $\mathfrak{A} \subseteq R$ is semiprime. Then rad $R \subseteq \mathfrak{A}$ since rad R is nilpotent. On the other hand, if $\mathfrak{A}$ is any ideal containing rad R, then $\mathfrak{A}/\text{rad } R$ must be the sum of some simple components of the semisimple ring $R/\text{rad } R$. But then $R/\mathfrak{A}$ is semisimple and hence $\mathfrak{A}$ is automatically semiprime. On the other hand, for such $\mathfrak{A}$ to be prime, we need $\mathfrak{A}/\text{rad } R$ to be the sum of all but one simple components of $R/\text{rad } R$ (cf. the last exercise).

To apply this to the ring R in the exercise, recall that rad R is the ideal of R consisting of matrices with zero diagonal, and

$$R/\text{rad } R \cong k \times k \times k$$

(cf. *FC*-p. 60). In particular, the prime (or maximal) ideals of R are:

$$\begin{pmatrix} 0 & k & k \\ 0 & k & k \\ 0 & 0 & k \end{pmatrix}, \quad \begin{pmatrix} k & k & k \\ 0 & 0 & k \\ 0 & 0 & k \end{pmatrix}, \quad \text{and} \quad \begin{pmatrix} k & k & k \\ 0 & k & k \\ 0 & 0 & 0 \end{pmatrix}.$$

There are only five more semiprime ideals in R, namely:

$$\begin{pmatrix} 0 & k & k \\ 0 & 0 & k \\ 0 & 0 & 0 \end{pmatrix}, \quad \begin{pmatrix} k & k & k \\ 0 & 0 & k \\ 0 & 0 & 0 \end{pmatrix}, \quad \begin{pmatrix} 0 & k & k \\ 0 & k & k \\ 0 & 0 & 0 \end{pmatrix}, \quad \begin{pmatrix} 0 & k & k \\ 0 & 0 & k \\ 0 & 0 & k \end{pmatrix}, \quad \text{and } R.$$

Comment. The method above can be applied to some nonartinian rings as well. For instance, if R is the ring of 3×3 upper triangular matrices over $\mathbb{Z}$, then rad R is the ideal of R consisting of matrices with zero diagonal, and

$$R/\text{rad } R \cong \mathbb{Z} \times \mathbb{Z} \times \mathbb{Z}.$$

The prime ideals of R are

$$\begin{pmatrix} p\mathbb{Z} & \mathbb{Z} & \mathbb{Z} \\ 0 & \mathbb{Z} & \mathbb{Z} \\ 0 & 0 & \mathbb{Z} \end{pmatrix}, \quad \begin{pmatrix} \mathbb{Z} & \mathbb{Z} & \mathbb{Z} \\ 0 & p\mathbb{Z} & \mathbb{Z} \\ 0 & 0 & \mathbb{Z} \end{pmatrix}, \quad \begin{pmatrix} \mathbb{Z} & \mathbb{Z} & \mathbb{Z} \\ 0 & \mathbb{Z} & \mathbb{Z} \\ 0 & 0 & p\mathbb{Z} \end{pmatrix}$$

where p is zero or a prime number, and the semiprime ideals of R are

$$\begin{pmatrix} n_1\mathbb{Z} & \mathbb{Z} & \mathbb{Z} \\ 0 & n_2\mathbb{Z} & \mathbb{Z} \\ 0 & 0 & n_3\mathbb{Z} \end{pmatrix}$$

where each n_i is either zero or a square-free integer.

Ex. 10.5. Show that the following conditions on a ring R are equivalent:

(1) All ideals $\neq R$ are prime.
(2) (a) The ideals of R are linearly ordered by inclusion, and
    (b) All ideals $I \subseteq R$ are idempotent (i.e. $I^2 = I$).

If R is commutative, show that these conditions hold iff R is either (0) or a field.

Solution. (1) $\Longrightarrow$ (2). To show (2a), let I, J be two ideals $\neq R$. By (1), $I \cap J$ is prime, so $IJ \subseteq I \cap J$ implies that either $I \subseteq I \cap J$ or $J \subseteq I \cap J$. Thus, we have either $I \subseteq J$ or $J \subseteq I$. To show (2b), we may assume that $I \neq R$. By (1), I^2 is a prime ideal. Since $I \cdot I \subseteq I^2$, we must have $I \subseteq I^2$ and hence $I = I^2$.

(2) $\Longrightarrow$ (1). Let $\mathfrak{p}$ be any ideal $\neq R$, and let $I, J \supseteq \mathfrak{p}$ be two ideals such that $IJ \subseteq \mathfrak{p}$. We wish to show that $I = \mathfrak{p}$ or $J = \mathfrak{p}$. By (2a), we may assume that $I \subseteq J$. By (2b), $I = I^2 \subseteq IJ \subseteq \mathfrak{p}$, so we have $I = \mathfrak{p}$.

Finally, let $R \neq 0$ be a commutative ring in which all proper (principal) ideals are prime. Then R is a domain (since (0) is prime). If R is *not* a field, some $a \neq 0$ is a non-unit. Then $a^2 R$ is prime, so we must have $a = a^2 b$ for some $b \in R$. But then $ab = 1$, a contradiction.

Comment. (A) The last part of this exercise is well-known to commutative algebraists. In fact, it is the very first exercise in Kaplansky's book "Commutative Rings"!

(B) *In general,* (2a) *alone does not imply* (1). For instance, in a discrete valuation ring R with maximal ideal $\mathfrak{p} \neq 0$, (2a) holds since the only ideals in R are

$$R \supseteq \mathfrak{p} \supseteq \mathfrak{p}^2 \supseteq \cdots \supseteq (0).$$

However, (1) does not hold since $\mathfrak{p}^2, \mathfrak{p}^3, \ldots$ are not prime ideals.

(C) *In general,* (2b) *also fails to imply* (1). For instance, let

$$R = B_1 \oplus \cdots \oplus B_r$$

be a semisimple ring with simple components $B_1, \ldots, B_r$ ($r \geq 2$). It is easy to see that (2b) holds (since any ideal is a sum of the B_i's). However, (0) is not a prime ideal, since $B_i B_j = 0$ whenever $i \neq j$. Of course, (2a) also fails to hold, since there are no inclusion relations among the B_i's.

(D) In the noncommutative case, a trivial class of rings satisfying (1) and (2) in the above exercise is given by the simple rings. A more interesting class of noncommutative examples is given in the next exercise.

Ex. 10.6. Let $R = \text{End}(V_k)$ where V is a vector space over a division ring k. Show that R satisfies the properties (1), (2) of the exercise above. In particular, every nonzero homomorphic image of R is a prime ring.

Solution. If $\dim_k V < \infty$, R is a simple ring. Therefore, it suffices to treat the case when V is infinite-dimensional. By Exercise 3.16, the ideals of R are linearly ordered by inclusion. To show that they are all idempotent, consider any ideal $I \neq 0, R$. By Exercise 3.16, there exists an infinite cardinal $\beta < \dim_k V$ such that

$$I = \{f \in R : \dim_k f(V) < \beta\}.$$

For any $f \in I$, let $f' \in R$ be such that f' is the identity on the $f(V)$, and zero on a direct complement of $f(V)$. Clearly, $f' \in I$, and $f = f'f$. Therefore, $f \in I^2$, and we have proved that $I = I^2$. (It is also not difficult to check directly that each such I, as well as the zero ideal, is prime.)

Ex. 10.7. For any integer $n > 0$, show that

(1) $R = \begin{pmatrix} \mathbb{Z} & n\mathbb{Z} \\ \mathbb{Z} & \mathbb{Z} \end{pmatrix}$ is a prime ring, but $R' = \begin{pmatrix} \mathbb{Z} & n\mathbb{Z} \\ 0 & \mathbb{Z} \end{pmatrix}$ is not, and

(2) R is not isomorphic to the prime ring $P = M_2(\mathbb{Z})$ if $n > 1$.

Solution. (1) In R', there is a nilpotent ideal $\begin{pmatrix} 0 & n\mathbb{Z} \\ 0 & 0 \end{pmatrix} \neq 0$, so R' is not semiprime, let alone prime. To see that R is prime, view it as a subring of $P = M_2(\mathbb{Z})$. Note that $nP \subseteq R$. If $a, b \in R$ are such that $aRb = 0$, then $naPb \subseteq aRb = 0$, and hence $aPb = 0$. Since P is a prime ring (see FC-(10.20)), we conclude that $a = 0$ or $b = 0$.

(2) Assume $n > 1$. To see that $R \not\cong P$, note that, by FC-(3.1), the ideals of P are of the form

$$M_2(k\mathbb{Z}) = kM_2(\mathbb{Z}) = kP,$$

where $k \in \mathbb{Z}$. Now R has an ideal $M_2(n\mathbb{Z})$ (which is, in fact, an ideal of the larger ring P). Since this ideal of R is obviously not of the form kR for any integer k, it follows that $R \not\cong P$.

Comment. (A) Obviously, the argument in the second part of the solution permits the following generalization. *Let R be a subring of a prime ring P such that, for any $s \in P$, there exists $\alpha \in Z(P)\setminus\{0\}$ with $\alpha s \in R$. Then R is also a prime ring.* (Note that, by Exer. 10.0, α is not a zero-divisor in P.) In particular, R can be any subring containing αP for some $\alpha \in Z(P)\setminus\{0\}$.

(B) Many examples of prime rings can be constructed by using (A). For instance, if A is a prime ring and $\beta, \gamma \in Z(A)\setminus\{0\}$, then $R = \begin{pmatrix} A & \beta A \\ \gamma A & A \end{pmatrix}$ is also a prime ring, for we can take $P = M_2(A)$ and $\alpha = \beta\gamma$ (viewed as a central element in P). Taking $A = \begin{pmatrix} \mathbb{Z} & n\mathbb{Z} \\ \mathbb{Z} & \mathbb{Z} \end{pmatrix}$ and $\beta = n \neq 0$, $\gamma = m \neq 0$ for instance, we get a prime ring

$$\begin{pmatrix} \mathbb{Z} & n\mathbb{Z} & n\mathbb{Z} & n^2\mathbb{Z} \\ \mathbb{Z} & \mathbb{Z} & n\mathbb{Z} & n\mathbb{Z} \\ m\mathbb{Z} & mn\mathbb{Z} & \mathbb{Z} & n\mathbb{Z} \\ m\mathbb{Z} & m\mathbb{Z} & \mathbb{Z} & \mathbb{Z} \end{pmatrix}.$$

Ex. 10.7*. Let R be a subring of a right noetherian ring Q with a set $S \subseteq R \cap U(Q)$ such that every element $q \in Q$ has the form rs^{-1} for some $r \in R$ and $s \in S$. Show that:

(1) If B is an ideal of R, then BQ is an ideal of Q.

(2) If Q is prime (resp. semiprime), then so is R.

(3) The converse of (2) is true even without assuming Q to be right noetherian.

Solution. (1) For any $q \in Q$, we need to show that $qBQ \subseteq BQ$. Write $q = rs^{-1}$ as above. The ascending sequence of right ideals $s^{-1}BQ \subseteq s^{-2}BQ \subseteq \cdots$ in the right noetherian ring Q shows that $s^{-n}BQ = s^{-(n+1)}BQ$ for some $n \geq 1$, and thus $BQ = s^{-1}BQ$. From this we see that

$$qBQ = r(s^{-1}BQ) = rBQ \subseteq BQ.$$

(2) Let $a, b \in R$ be such that $aRb = 0$. Applying (1) to the ideal $B = RbR$ in R, we have

$$aQb \subseteq aQ \cdot (RbR)Q \subseteq a(RbR)Q = 0,$$

so $a = 0$ or $b = 0$. The semiprime case is similar, by setting $a = b$.

(3) Assume R is prime, and $(as^{-1})Q(bt^{-1}) = 0$ where $a, b \in R$ and $s, t \in S$. Then $0 = (as^{-1})(sR)b = aRb$, so $a = 0$ or $b = 0$. The semiprime case is similar, by setting $a = b$.

Comment. This exercise leads to many examples of prime (resp. semiprime) rings, by taking suitable subrings R of, say, simple artinian (resp. semisimple) rings Q. For instance, if $Q = \mathbb{M}_2(\mathbb{Q})$, we get once more the prime ring examples $\begin{pmatrix} \mathbb{Z} & n\mathbb{Z} \\ m\mathbb{Z} & \mathbb{Z} \end{pmatrix}$ (with $nm \neq 0$) obtained in the previous exercise. The idea of Ex. 10.7* comes from the general theory of "rings of quotients". For an exposition of this theory, see §§10–11 in the author's *Lectures on Modules and Rings*, Springer GTM, Vol. 189, 1999.

Ex. 10.8. (a) Show that a ring R is semiprime iff, for any two ideals $\mathfrak{A}, \mathfrak{B}$ in R, $\mathfrak{A}\mathfrak{B} = 0$ implies that $\mathfrak{A} \cap \mathfrak{B} = 0$.

(b) Let $\mathfrak{A}, \mathfrak{B}$ be left (resp. right) ideals in a semiprime ring R. Show that $\mathfrak{A}\mathfrak{B} = 0$ iff $\mathfrak{B}\mathfrak{A} = 0$. If $\mathfrak{A}$ is an ideal, show that $\mathrm{ann}_r(\mathfrak{A}) = \mathrm{ann}_l(\mathfrak{A})$.

Solution. (a) First assume R is semiprime, and let $\mathfrak{A}\mathfrak{B} = 0$. For $\mathfrak{C} := \mathfrak{A} \cap \mathfrak{B}$, we have $\mathfrak{C}^2 = \mathfrak{C}\mathfrak{C} \subseteq \mathfrak{A}\mathfrak{B} = 0$, so $\mathfrak{C} = 0$. Conversely, assume the implication property about $\mathfrak{A}, \mathfrak{B}$ in (a), and consider any ideal $\mathfrak{C}$ with $\mathfrak{C}^2 = 0$. Then $\mathfrak{C}\mathfrak{C} = 0$ implies $0 = \mathfrak{C} \cap \mathfrak{C} = \mathfrak{C}$, so R is semiprime.

(b) It suffices to prove the "only if" part, so assume $\mathfrak{A}\mathfrak{B} = 0$, where $\mathfrak{A}, \mathfrak{B}$ are (say) left ideals. We have

$$(\mathfrak{B}\mathfrak{A})^2 = \mathfrak{B}\mathfrak{A}\mathfrak{B}\mathfrak{A} = 0,$$

so $\mathfrak{B}\mathfrak{A} = 0$ since R is semiprime. For the last statement in (b), we shall prove a little bit more: *if $\mathfrak{A}$ is a left ideal, then* $\mathrm{ann}_l(\mathfrak{A}) \subseteq \mathrm{ann}_r(\mathfrak{A})$. (This clearly gives the desired statement, by symmetry.) Indeed, let $\mathfrak{B} = \mathrm{ann}_l(\mathfrak{A})$, which is also a left ideal. By definition, $\mathfrak{B}\mathfrak{A} = 0$, so the first part of (b) gives $\mathfrak{A}\mathfrak{B} = 0$, that is, $\mathfrak{B} \subseteq \mathrm{ann}_r(\mathfrak{A})$.

Comment. The inclusion $\mathrm{ann}_l(\mathfrak{A}) \subseteq \mathrm{ann}_r(\mathfrak{A})$ above may be strict for a left ideal $\mathfrak{A}$. For instance, if $\mathfrak{A} = Re$ for an idempotent e in a semiprime ring R, we have $\mathrm{ann}_r(\mathfrak{A}) = (1-e)R$, which may not lie in $\mathrm{ann}_l(\mathfrak{A})$ since $(1-e)Re$ may not be zero.

Ex. 10.8*. Show that, with respect to inclusion, the set S of semiprime ideals in any ring R forms a lattice having a smallest element and a largest element. Give an example to show, however, that the sum of two semiprime ideals need not be semiprime.

Solution. For $A, B \in S$, we have $A \cap B \in S$. Thus, $\inf\{A, B\}$ is given simply by $A \cap B$. For $\sup\{A, B\}$, we take $\sqrt{A+B} \in S$, where the "radical" of an ideal is defined as in *FC*-(10.6). A semiprime ideal C contains both A and B iff $C \supseteq A+B$, iff $C \supseteq \sqrt{A+B}$. Thus, $\sqrt{A+B}$ is indeed the supremum of A and B in S. This shows that S is a lattice. Clearly, S has a largest element, R, and a smallest element, $\mathrm{Nil}_* R$.

In the above construction, we cannot replace $\sqrt{A+B}$ by $A+B$, since $A + B$ may not be semiprime. For an explicit example of this, consider $R = \mathbb{Z}[x]$, in which $A = (x)$ and $B = (x-4)$ are (semi)prime ideals (since $R/A \cong \mathbb{Z} \cong R/B$). Here,

$$A + B = (x,\ x-4) = (x, 4)$$

is *not* semiprime (since $R/(A+B) \cong \mathbb{Z}_4$), and we have

$$\sup\{A, B\} = \sqrt{A+B} = (2, x).$$

Alternatively, we could have also taken $A = (2)$ and $B = (x^2 - 2)$, for which $A + B = (2, x^2)$ is not semiprime. Here $\sup\{A, B\}$ is again $(2, x)$.

In spite of these examples, there are many rings in which we do have $\sup\{A, B\} = A + B$ for semiprime ideals A and B. These include, for instance, von Neumann regular rings, and left (right) artinian rings, as the reader can easily verify. The ring $\mathbb{Z}$ is another example: here, $A + B$ is semiprime as long as *one of* A, B is semiprime!

Comment. The S in this exercise is actually a *complete* lattice, in the sense that "sup" and "inf" exist for arbitrary subsets in S. If $\{A_i : i \in I\} \subseteq S$, the infimum is given as before by the semiprime ideal $\bigcap_i A_i \in S$, and the supremum is given by the semiprime ideal $\sqrt{\sum_i A_i}$.

Ex. 10.9. Let $I \subseteq R$ be a right ideal containing no nonzero nilpotent right ideals of R. (For instance, I may be any right ideal in a semiprime ring.) Show that the following are equivalent: (1) I_R is an artinian module; (2) I_R is a finitely generated semisimple module. In this case, show that (3) $I = eR$ for an idempotent $e \in I$.

Solution. (2) $\implies$ (1) is clear. Conversely, assume (1). If $I \neq 0$, it contains a minimal right ideal A_1. Since A_1 cannot be nilpotent, Brauer's Lemma FC-(10.22) implies that $A_1 = e_1 R$ for some idempotent e_1. Let

$$B_1 = I \cap (1 - e_1)R,$$

so $I = A_1 \oplus B_1$. If $B_1 \neq 0$, the same argument shows that B_1 contains a minimal right ideal $A_2 = e_2 R$ where $e_2^2 = e_2$. Since $e_2 \in (1 - e_1)R$, we have $e_1 e_2 = 0$. By a straightforward calculation,

$$e_2' := e_1 + e_2(1 - e_1)$$

is an idempotent, and $A_1 \oplus A_2 = e_2' R$. Now let $B_2 = I \cap (1 - e_2')R$. Since

$$1 - e_2' = (1 - e_1) + (1 - e_1)e_2(1 - e_1) \in (1 - e_1)R,$$

we have $B_2 \subseteq B_1$. If $B_2 \neq 0$, we can take a minimal right ideal $A_3 = e_3 R \subseteq B_2$, and continue this construction. Since $B_1 \supsetneq B_2 \supsetneq \cdots$, this construction process must stop in a finite number of steps. If, say, $B_n = 0$, then

$$I_R = A_1 \oplus \cdots \oplus A_n$$

is semisimple, and equal to $e_n' R$ for the idempotent e_n', as desired.

Comment. The Exercise above may be thought of as a generalization of the fact that any semiprime right artinian ring is semisimple. (Cf. Exercise 22.3C below.)

Ex. 10.10A. Let $N_1(R)$ be the sum of all nilpotent ideals in a ring R.

(1) Show that $N_1(R)$ is a nil subideal of $\mathrm{Nil}_* R$ which contains all nilpotent one-sided ideals of R.
(2) If $N_1(R)$ is nilpotent, show that $N_1(R) = \mathrm{Nil}_* R$.
(3) Show that the hypothesis and conclusion in (2) both apply if the ideals in R satisfy DCC.

Solution. Let $N = N_1(R)$.

(1) Note that any nilpotent ideal is in $\mathrm{Nil}_* R$, so $N \subseteq \mathrm{Nil}_* R$, and N is nil. If $\mathfrak{A}$ is a nilpotent (say) left ideal, then $\mathfrak{A}R$ is a nilpotent ideal, so $\mathfrak{A} \subseteq \mathfrak{A}R \subseteq N$.

(2) Say $N^n = 0$. Then R/N has no nonzero nilpotent ideals. (For, if I/N is a nilpotent ideal of R/N, then $I^m \subseteq N$ for some m, and hence $I^{mn} \subseteq N^n = 0$. This implies that $I \subseteq N$.) Therefore, N is semiprime, so $N = \mathrm{Nil}_* R$.

(3) Assume now ideals in R satisfy DCC. Then $N^n = N^{n+1}$ for some n. We finish by showing that $M := N^n$ is zero. To see this, assume instead $M \neq 0$. Then there exist ideals $A \subseteq M$ with $MAM \neq 0$ (for instance $A =$

M). Among such ideals A, choose a B that is minimal. Then $MBM \neq 0$, so $MbM \neq 0$ for some $b \in B$. Since

$$MbM \subseteq B \quad \text{and} \quad M(MbM)M = MbM \neq 0,$$

we must have $MbM = B$. In particular, there exists an equation

$$b = \sum_{i=1}^{r} m_i \, bm_i', \quad \text{where} \quad m_i, m_i' \in M.$$

Now consider the ideal $J \subseteq N$ generated by

$$\{m_i, m_i' : \quad 1 \leq i \leq r\}.$$

Since N is a sum of nilpotent ideals, J lies in a *finite* sum of nilpotent ideals, so $J^k = 0$ for some k. Since $b \in JbJ$, we now have by repeated substitution $b \in J^k b J^k = 0$, a contradiction.

Comment. Of course, if R is a *commutative* ring, then $N_1(R) = \text{Nil}_* R = \text{Nil}^* R$ (although this may not be nilpotent). If R is a Banach algebra, the same equation is known to hold, and the common nilradical is nilpotent: see Theorem 4.4.11 in T. Palmer's recent book "Banach Algebras and the Theory of *-Algebras," Vol. 1, Cambridge University Press, 1994.

Ex. 10.10B. Keeping the notations of Exercise 10A, give an example of a (necessarily noncommutative) ring R in which $N_1(R) \subsetneq \text{Nil}_* R$.

Solution. We follow here a construction in p. 233 of Jacobson's book "Structure of Rings." Fix a commutative ring S with a nilradical J that is not nilpotent, and consider $R_0 = \begin{pmatrix} J & S \\ J & J \end{pmatrix} \subseteq M_2(S)$. This R_0 is a "subring" of $M_2(S)$, except for the fact that it may not possess an identity element.

Step 1. R_0 is a "nil rng" (i.e. *every element of R_0 is nilpotent*). Indeed, let $\alpha = \begin{pmatrix} x & s \\ y & z \end{pmatrix} \in R_0$, where $x, y, z \in J$ and $s \in S$. Let

$$I = xS + yS + zS,$$

which is a nilpotent ideal. We shall show by induction that

$$(1) \ \alpha^{2m} \in \begin{pmatrix} I^m & I^m \\ I^{m+1} & I^m \end{pmatrix}, \quad \text{and} \quad (2) \ \alpha^{2m+1} \in \begin{pmatrix} I^{m+1} & I^m \\ I^{m+1} & I^{m+1} \end{pmatrix}.$$

(Clearly, these imply that $\alpha^k = 0$ for sufficiently large k.) The induction is started by noting that $\alpha \in \begin{pmatrix} I & S \\ I & I \end{pmatrix}$. If we have (2) for some m, then

$$\alpha^{2m+2} \in \begin{pmatrix} x & s \\ y & z \end{pmatrix} \begin{pmatrix} I^{m+1} & I^m \\ I^{m+1} & I^{m+1} \end{pmatrix} \subseteq \begin{pmatrix} I^{m+1} & I^{m+1} \\ I^{m+2} & I^{m+1} \end{pmatrix}.$$

On the other hand, if (1) holds for some m, then

$$\alpha^{2m+1} \in \begin{pmatrix} x & s \\ y & z \end{pmatrix} \begin{pmatrix} I^m & I^m \\ I^{m+1} & I^m \end{pmatrix} \subseteq \begin{pmatrix} I^{m+1} & I^m \\ I^{m+1} & I^{m+1} \end{pmatrix}.$$

This completes the induction.

Step 2. We adjoin an identity to R_0 to form a ring $R := R_0 \oplus \mathbb{Z}$. Step 1 clearly implies that $\mathrm{Nil}^* R = R_0$. Therefore, $\mathrm{Nil}_* R \subseteq R_0$. Let

$$N = N_1(R) \subseteq \mathrm{Nil}_* R$$

be the sum of all nilpotent ideals of R. *We claim that the matrix* $\beta = \begin{pmatrix} 0 & 1 \\ 0 & 0 \end{pmatrix}$ *does not belong to* N. Indeed, assume $\beta \in N$. Then, β is contained in a finite sum of nilpotent ideals of R, so $(R\beta R)^r = 0$ for some r. But $R\beta R$ contains

$$\begin{pmatrix} 0 & 1 \\ 0 & 0 \end{pmatrix} \begin{pmatrix} 0 & 0 \\ J & 0 \end{pmatrix} = \begin{pmatrix} J & 0 \\ 0 & 0 \end{pmatrix}.$$

Hence $J^r = 0$, in contradiction to the choice of J.

Step 3. We finish by showing that $\beta \in \mathrm{Nil}_* R$ (which implies that $N \subsetneq \mathrm{Nil}_* R$). For this, it suffices to show, by Exercise 17 below, that for any sequence $\beta_1, \beta_2, \ldots \in R$ such that $\beta_1 = \beta$ and $\beta_{n+1} \in \beta_n R \beta_n$ for every n, we have $\beta_n = 0$ for sufficiently large n. Since $\beta_1^2 = 0$, an explicit calculation shows that $\beta_2 = \begin{pmatrix} 0 & y \\ 0 & 0 \end{pmatrix}$ for some $y \in J$. Similarly, the fact that $\beta_2^2 = 0$ implies $\beta_3 = \begin{pmatrix} 0 & y^2 y' \\ 0 & 0 \end{pmatrix}$ for some $y' \in J$, etc. Since y is nilpotent, we see that $\beta_n = 0$ for sufficiently large n.

Comment. On p. 233 of Jacobson's book referenced above, it is shown that $R_0^2 \subseteq N$. Therefore, one has in fact $\mathrm{Nil}_* R = R_0 = \mathrm{Nil}^* R$.

Ex. 10.11. (Levitzki) For any ring R and any ordinal α, define $N_\alpha(R)$ as follows. For $\alpha = 1$, $N_1(R)$ is defined as in Exercise 10A. If α is the successor of an ordinal β, define

$$N_\alpha(R) = \{r \in R : r + N_\beta(R) \in N_1(R/N_\beta(R))\}.$$

If α is a limit ordinal, define

$$N_\alpha(R) = \bigcup_{\beta < \alpha} N_\beta(R).$$

Show that $\mathrm{Nil}_* R = N_\alpha(R)$ for any ordinal α with Card $\alpha >$ Card R.

Solution. The $N_\alpha(R)$'s form an ascending chain of ideals in Nil_*R. Write $P(R) = N_\alpha(R)$ where α is an ordinal with $\text{Card } \alpha > \text{Card } R$. Then, for any ordinal α' with $\text{Card } \alpha' > \text{Card } R$, we have $P(R) = N_{\alpha'}(R)$. Since $P(R) \subseteq \text{Nil}_*R$, it suffices to show that $\text{Nil}_*R \subseteq P(R)$. Now $R/P(R)$ has no nonzero nilpotent ideals, so it is a semiprime ring. This means that $P(R)$ is a semiprime ideal. Hence $\text{Nil}_*R \subseteq P(R)$ since Nil_*R is the smallest semiprime ideal of R.

Comment. For the example R constructed in Exercise 10B, we have $N_1(R) \subsetneq \text{Nil}_*R$, and in the *Comment* on that exercise, we pointed out that $N_2(R) = \text{Nil}_*R$. More generally, for *any* ordinal α, Amitsur has constructed a ring $R = R_\alpha$ for which $N_\alpha(R) \subsetneq N_{\alpha+1}(R)$. This means that, for a general ring R, we might need an arbitrarily large ordinal α for $N_\alpha(R)$ to reach Nil_*R. See Amitsur's article "Nil radicals. Historical notes and some new results" in *Rings, Modules and Radicals*, Coll. Math. Soc. János Bolyai 6, Keszthely (Hungary), 47–65, North Holland, 1971. For a sketch of Amitsur's construction, see p. 318 of Rowen's "Ring Theory, I," Academic Press, 1988.

Ex. 10.12. Let I be a left ideal in a ring R such that, for some integer $n \geq 2$, $a^n = 0$ for all $a \in I$. Show that $a^{n-1}Ra^{n-1} = 0$ for all $a \in I$.

Solution. (A. A. Klein) Given any $r \in R$, let $s = ra^{n-1} \in I$. Then $sa = 0$, and a quick induction on m shows that

$$(s+a)^m = s^m + as^{m-1} + a^2s^{m-2} + \cdots + a^m.$$

Taking $m = n$ and using $s^n = a^n = (s+a)^n = 0$, we have

$$0 = as^{n-1} + \cdots + a^{n-2}s^2 + a^{n-1}s = (at+1)a^{n-1}s$$

for some $t \in R$. (Note that $s^2 = ra^{n-1}s$, etc.) Since $(at)^{n+1} = a(ta)^nt = 0$, we have $at + 1 \in U(R)$, so

$$0 = a^{n-1}s = a^{n-1}ra^{n-1},$$

as desired.

Ex. 10.13. (Levitzki, Herstein) Let $I \neq 0$ be a left ideal in a ring R such that, for some integer n, $a^n = 0$ for all $a \in I$.

(1) Show that I contains a nonzero nilpotent left ideal, and R has a nonzero nilpotent ideal.
(2) Show that $I \subseteq \text{Nil}_*R$.

Solution. We may assume n is chosen minimal. Since $I \neq 0$, $n \geq 2$. Fix an element $a \in I$ with $a^{n-1} \neq 0$. By Exercise 12, $a^{n-1}Ra^{n-1} = 0$, so $(Ra^{n-1}R)^2 = 0$. Therefore $Ra^{n-1}R$ is a nonzero nilpotent ideal, and I contains the nonzero nilpotent left ideal Ra^{n-1}. For (2), let $J = \text{Nil}_*R$. If the image $\bar{I}$ of I in R/J is nonzero, (1) would give a nonzero nilpotent ideal in R/J, which is impossible. Therefore, we must have $\bar{I} = 0$, that is, $I \subseteq \text{Nil}_*R$.

Comment. For any left ideal $I \subseteq R$, the above (for $n = 2$) shows that every $a \in I$ has square zero iff I is a union of left ideals of square zero. However, I itself need not be nilpotent, even if R is commutative. For instance, let R be the commutative $\mathbb{F}_2$-algebra generated by $x_1, x_2, \ldots$ with the relations

$$x_1^2 = x_2^2 = \cdots = 0.$$

Then every element in the ideal $I = \sum Rx_i$ has square zero, but each power I^m contains a nonzero element $x_1 x_2 \cdots x_m$, so I is not nilpotent.

Ex. 10.14. (Krull-McCoy) Show that any prime ideal $\mathfrak{p}$ in a ring R contains a minimal prime ideal. Using this, show that the lower nilradical Nil_*R is the intersection of all the minimal prime ideals of R.

Solution. The second conclusion follows directly from the first, since Nil_*R is the intersection of all the prime ideals of R (cf. FC-(10.13)). To prove the first conclusion, we apply Zorn's Lemma to the family of prime ideals $\subseteq \mathfrak{p}$. It suffices to check that, for any chain of prime ideals $\{\mathfrak{p}_i : i \in I\}$ in $\mathfrak{p}$, the intersection $\mathfrak{p}' = \bigcap \mathfrak{p}_i$ is prime. Let $a, b \notin \mathfrak{p}'$. Then $a \notin \mathfrak{p}_i$ and $b \notin \mathfrak{p}_j$ for some $i, j \in I$. If, say, $\mathfrak{p}_i \subseteq \mathfrak{p}_j$, then both a, b are outside $\mathfrak{p}_i$, so we have $arb \notin \mathfrak{p}_i$ for some $r \in R$. But then $arb \notin \mathfrak{p}'$, and we have checked that $\mathfrak{p}'$ is prime.

Ex. 10.15. Show that if the ideals in R satisfy *ACC* (e.g. when R is left noetherian), then R has only finitely many minimal prime ideals.

Solution. Assuming $R \neq 0$, we first prove that any ideal in R contains a finite product of prime ideals. Suppose, instead, that the family $\mathfrak{F}$ of ideals which do not contain a finite product of prime ideals is nonempty. Let I be a maximal member of $\mathfrak{F}$. Certainly $I \neq R$, and I is not prime. Therefore, there exist ideals $A, B \supsetneq I$ such that $AB \subseteq I$. But each of A, B contains a finite product of primes, and hence so does I, a contradiction.

Applying the above conclusion to the zero ideal, we see that there exist prime ideals $\mathfrak{p}_1, \ldots, \mathfrak{p}_n$ such that $\mathfrak{p}_1 \cdots \mathfrak{p}_n = 0$. We claim that any minimal prime $\mathfrak{p}$ is among the $\mathfrak{p}_i$'s. Indeed, from $\mathfrak{p}_1 \cdots \mathfrak{p}_n \subseteq \mathfrak{p}$, we must have $\mathfrak{p}_i \subseteq \mathfrak{p}$ for some i. Hence $\mathfrak{p} = \mathfrak{p}_i$. Therefore, R has at most n minimal primes.

Ex. 10.16. (McCoy) For any ideal $\mathfrak{A}$ in a ring R, show that $\sqrt{\mathfrak{A}}$ consists of $s \in R$ such that every n-system containing s meets $\mathfrak{A}$. [An n-system is a subset $N \subseteq R$ such that, for any $s \in N$, $srs \in N$ for some $r \in R$.]

Solution. In FC-(10.6), $\sqrt{\mathfrak{A}}$ is defined to be the set of $s \in R$ such that every m-system containing s meets $\mathfrak{A}$. The desired conclusion, therefore, follows from the following two facts: (1) every m-system is an n-system, and (2) if N is an n-system and $s \in N$, then there exists an m-system M such that $s \in M \subseteq N$ (cf. the proof of FC-(10.10)).

Ex. 10.17. (Levitzki) An element a of a ring R is called *strongly nilpotent* if every sequence $a_1, a_2, a_3, \ldots$ such that $a_1 = a$ and $a_{n+1} \in a_n R a_n$ $(\forall n)$ is eventually zero. Show that $\mathrm{Nil}_* R$ is precisely the set of all strongly nilpotent elements of R.

Solution. First assume $a \notin \mathrm{Nil}_* R$. By FC-(10.6) and (10.13), there exists an m-system M containing a and not containing 0. Take $a_1 = a$, and inductively

$$a_{n+1} \in (a_n R a_n) \cap M.$$

Then we get a sequence $a_1, a_2, a_3, \ldots$ of the desired type which is never 0. Therefore, a is not strongly nilpotent. Conversely, if a is not strongly nilpotent, there exists a set $M = \{a_i : i \geq 1\}$ of nonzero elements such that $a_1 = a$ and $a_{n+1} \in a_n R a_n (\forall n)$. As observed in the proof of FC-(10.10), M is an m-system. Since $0 \notin M$, it follows from FC-(10.6) that $a \notin \mathrm{Nil}_* R$.

Ex. 10.18A. (1) Let $R \subseteq S$ be rings. Show that $R \cap \mathrm{Nil}_*(S) \subseteq \mathrm{Nil}_*(R)$.

(2) If $R \subseteq Z(S)$, show that $R \cap \mathrm{Nil}_*(S) = \mathrm{Nil}_*(R)$.
(3) Let R, K be algebras over a commutative ring k such that R is k-projective and $K \supseteq k$. Show that $R \cap \mathrm{Nil}_*(R \otimes_k K) = \mathrm{Nil}_*(R)$.

Solution. (1) Let $r \in R \cap \mathrm{Nil}_*(S)$. Consider any m-system M of the ring R such that $r \in M$ (see FC-(10.3)). By definition, M is also an m-system of S, so $0 \in M$ since $r \in \mathrm{Nil}_*(S)$. It follows that $r \in \mathrm{Nil}_*(R)$. (We could have also proved this part by Exercise 17.)

(2) Under the assumption $R \subseteq Z(S)$, any prime ideal $\mathfrak{p}$ of S contracts to a prime $\mathfrak{p}_0 := \mathfrak{p} \cap R$ of R. Indeed, if $a, b \in R$ are such that $ab \in \mathfrak{p}_0$, then

$$aSb = abS \subseteq \mathfrak{p}_0 S \subseteq \mathfrak{p},$$

so we have, say, $a \in \mathfrak{p} \cap R = \mathfrak{p}_0$. Therefore, any $r \in \mathrm{Nil}_* R$ lies in all primes $\mathfrak{p}$ of S, and so $r \in \mathrm{Nil}_* S$. This (together with (1)) gives the equation in (2).

(3) Let $S = R \otimes_k K$. Since R is k-projective, tensoring the inclusion $k \to K$ by R gives an inclusion

$$R \otimes_k k \to R \otimes_k K.$$

We identify $R \otimes_k k$ with R, and view it as a subring of $S = R \otimes_k K$. We finish again by showing that any prime ideal $\mathfrak{p}$ of S contracts to a prime ideal $\mathfrak{p}_0 = \mathfrak{p} \cap R$ of R. Indeed, let $aRb \subseteq \mathfrak{p}_0$, where $a, b \in R$. Then

$$aSb = a(R \otimes_k K)b \subseteq \sum aRb \otimes_k K \subseteq \sum \mathfrak{p}_0(1 \otimes_k K) \subseteq \mathfrak{p},$$

so we must have, say, $a \in \mathfrak{p} \cap R = \mathfrak{p}_0$.

Comment. The prime contraction property used in (2) and (3) does not hold in general for a ring extension $R \subseteq S$. For instance, let k be a field, and let

$$S = \mathbb{M}_2(k), \quad \text{and} \quad R = \begin{pmatrix} k & k \\ 0 & k \end{pmatrix}.$$

Then (0) is a prime ideal of S, but is not a prime ideal of R. Here, $\text{Nil}_*(S) = 0$ and $\text{Nil}_*(R) = \begin{pmatrix} 0 & k \\ 0 & 0 \end{pmatrix}$, so the inclusion

$$R \cap \text{Nil}_*(S) \subseteq \text{Nil}_*(R)$$

in (1) is now a strict inclusion.

Ex. 10.18B. Let R be a k-algebra where k is a field. Let K/k be a separable algebraic field extension.

(1) Show that R is semiprime iff $R^K = R \otimes_k K$ is semiprime.
(2) Show that $\text{Nil}_*(R^K) = (\text{Nil}_*(R))^K$.

Solution. (1) The "if" part is clear (for *any* field extension), since, for any nilpotent ideal $I \subseteq R$, I^K is a nilpotent ideal in R^K. For the converse, assume that R is semiprime. Repeating the entire argument of *FC*-(5.16) with the lower nilradical replacing the Jacobson radical, we can deduce that R^K is also semiprime. (Note that the proof of *FC*-(5.16) uses only the change-of-rings properties of the radical, so it applies equally well here, in view of Exercise 18A above.)

(2) Since $\text{Nil}_*(R) \subseteq \text{Nil}_*(R^K)$, we have $(\text{Nil}_*(R))^K \subseteq \text{Nil}_*(R^K)$. On the other hand,

$$R^K / (\text{Nil}_* R)^K \cong (R/\text{Nil}_* R)^K$$

is semiprime by (1), so $(\text{Nil}_*(R))^K \supseteq \text{Nil}_*(R^K)$. Therefore, the desired equality follows.

Comment. Again, the separability condition on K/k is essential for both conclusions above: see the standard counterexample in *FC*-p. 77 for the inseparable case.

Ex. 10.19. For a ring R, consider the following conditions:

(1) Every ideal of R is semiprime.
(2) Every ideal I of R is idempotent (i.e. $I^2 = I$).
(3) R is von Neumann regular.

Show that $(3) \Rightarrow (2) \Leftrightarrow (1)$, and that $(1) \Rightarrow (3)$ if R is commutative.

Solution. (3) $\Rightarrow$ (2). Let I be any ideal of R. For any $a \in I$, we have $a \in aRa \subseteq I^2$, so clearly $I^2 = I$.

(2) $\Rightarrow$ (1). Let J be any ideal of R. Suppose I is an ideal such that $I^2 \subseteq J$. Then $I = I^2 \subseteq J$, so J is semiprime.

(1) $\Rightarrow$ (2). If (2) does not hold, there exists an ideal I such that $I^2 \subsetneq I$. Then R/I has a nonzero ideal I/I^2 of square zero. This means I^2 is not a semiprime ideal, so (1) does not hold.

(1) $\Rightarrow$ (3) (assuming now R is commutative). Let $a \in R$. Then a^2R is a semiprime ideal; that is, R/a^2R is reduced. This implies that $a \in a^2R$, so we have (3).

Comment. The fact that we left (1) $\Rightarrow$ (3) open for general rings suggested that it may not hold for noncommutative R. To see that this is indeed the case, we appeal to a later exercise. In Exercise 12.2 below, a noncommutative domain A is constructed which has exactly three ideals, (0), A, and some L. Clearly, (0), L must be both prime and idempotent, so (1), (2) above are both satisfied. However, since A is a domain (but not a division ring), it cannot be von Neumann regular.

It is of interest to compare this exercise with Exercise 5. In the latter exercise, the condition (1) here is strengthened to every ideal $I \subsetneq R$ being prime, and (2) is strengthened by adding the condition that the ideals of R are linearly ordered by inclusion.

Ex. 10.20. Let Rad R denote one of the two nilradicals, or the Jacobson radical, or the Levitzki radical of R. Show that Rad R is a semiprime ideal. For any ideal $I \subseteq$ Rad R, show that Rad $(R/I) = ($Rad $R)/I$. Moreover, for any ideal $J \subseteq R$ such that Rad $(R/J) = 0$, show that $J \supseteq$ Rad R.

Solution. The case of the Jacobson radical is already covered in the text of *FC*, and in earlier exercises. The case of the lower nilradical follows easily from the interpretation of Nil_*R as the smallest semiprime ideal of R. Now consider the upper nilradical Nil^*R. If $N \supseteq \mathrm{Nil}^*R$ is an ideal with $N^2 \subseteq \mathrm{Nil}^*R$, then N is clearly nil, and so $N = \mathrm{Nil}^*R$. This checks that Nil^*R is semiprime, and we can check the other two properties without difficulty. Finally, we deal with the Levitzki radical L-rad R. Again, consider an ideal $N \supseteq L$-rad R with $N^2 \subseteq L$-rad R. We claim that N is locally nilpotent. In fact, if $a_1, \ldots, a_n \in N$, form the products

$$a_{ij} := a_i a_j \in L\text{-rad } R.$$

There exists an integer m such that any product of m elements from the set $\{a_{ij}\}$ is zero. Therefore, any product of $2m$ elements from the set $\{a_i\}$ is zero. This shows that $N \subseteq L$-rad R, so L-rad R is semiprime. If I is an ideal $\subseteq L$-rad R, a similar argument shows that

$$L\text{-rad } (R/I) = (L\text{-rad } R)/I.$$

For the last part, consider any ideal $J \subseteq R$ such that L-rad $(R/J) = 0$. The image of L-rad R in R/J is still locally nilpotent, so it must be zero, as L-rad $(R/J) = 0$. Thus, we must have L-rad $R \subseteq J$.

Ex. 10.21. Let $R[T]$ be a polynomial ring over R, and let $N = R \cap$ rad $R[T]$. Show that N is a semiprime ideal of R and that L-rad $R \subseteq N \subseteq \mathrm{Nil}^* R$, where L-rad R denotes the Levitzki radical of R.

Solution. That $N \subseteq \mathrm{Nil}^* R$ is part of Amitsur's Theorem (FC-(5.10)). To see that N is semiprime, it suffices to show that the ring $(R/N)[T]$ is semiprime (see FC-(10.18)). By Amitsur's Theorem:

$$(R/N)[T] = \frac{R[T]}{N[T]} = \frac{R[T]}{\mathrm{rad}\ R[T]}.$$

This ring is semiprimitive, and therefore semiprime, as desired. Finally, since L-rad R is locally nilpotent, $(L$-rad $R)[T]$ is nil, so it lies in rad $R[T]$. Contracting back to R, we see that L-rad $R \subseteq N$.

Ex. 10.22. For any ring R, show that $\mathrm{Nil}_* \mathbb{M}_n(R) = \mathbb{M}_n(\mathrm{Nil}_* R)$.

Solution. Let $N = \mathrm{Nil}_* R$. Then R/N is semiprime, so $\mathbb{M}_n(R/N)$ is also semiprime by FC-(10.20). But then $\mathbb{M}_n(R)/\mathbb{M}_n(N)$ is semiprime, so $\mathrm{Nil}_* \mathbb{M}_n(R) \subseteq \mathbb{M}_n(N)$. Write the ideal $\mathrm{Nil}_* \mathbb{M}_n(R)$ of $\mathbb{M}_n(R)$ in the form $\mathbb{M}_n(I)$, where I is an ideal in R (see FC-(3.1)). Then

$$\mathbb{M}_n(R/I) \cong \mathbb{M}_n(R)/\mathrm{Nil}_* \mathbb{M}_n(R)$$

is semiprime, and so is R/I by FC-(10.20). This implies that $N \subseteq I$, so we have

$$\mathbb{M}_n(N) \subseteq \mathbb{M}_n(I) = \mathrm{Nil}_* \mathbb{M}_n(R),$$

and equality holds.

Comment. Although it was easy to show that

$$\mathrm{Nil}_* \mathbb{M}_n(R) = \mathbb{M}_n(\mathrm{Nil}_* R) \quad \text{(for all } n\text{)},$$

it is not known if the analogous equation

$$\mathrm{Nil}^*(\mathbb{M}_n(R)) = \mathbb{M}_n(\mathrm{Nil}^* R)$$

holds for the *upper* nilradicals. In fact, it turns out that the truth of the above equation (for all n and for all rings R) is *equivalent* to the famous Köthe Conjecture (cf. FC-(10.28)). For more details on this, see Exercises 23 and 25 below.

Ex. 10.23. Let I be a nil left ideal of a ring R.

(1) Show that the set of matrices in $\mathbb{M}_n(R)$ whose kth column consists of elements of I and whose other columns are zero is a nil left ideal of $\mathbb{M}_n(R)$.

(2) If $T_n(R)$ is the ring of $n \times n$ upper triangular matrices over R, show that $T_n(I)$ is a nil left ideal in $T_n(R)$.

(3) If R is a commutative ring, show that $\mathbb{M}_n(I)$ is a nil ideal in $\mathbb{M}_n(R)$. Using this, show that $\mathrm{Nil}^*(\mathbb{M}_n(R)) = \mathbb{M}_n(\mathrm{Nil}^*(R))$.

Solution. (1) It is clear that S is a left ideal in $\mathbb{M}_n(R)$. Let $A \in S$, and let the kth column of A be $(a_1, \ldots, a_n)^t$. By explicit multiplication, we see that A^{m+1} has kth column $(a_1 a_k^m, \ldots, a_n a_k^m)^t$, and other columns zero. Choosing m such that $a_k^m = 0$, we have $A^{m+1} = 0$, so S is nil.

(2) Again, it is clear that $T_n(I)$ is a left ideal in $T_n(R)$. Let $B \in T_n(I)$, say with diagonal entries $b_1, \ldots, b_n$. Then B^m has diagonal entries $b_1^m, \ldots, b_n^m$. Choosing m such that $b_i^m = 0$ for all i, we obtain an upper triangular matrix B^m with a zero diagonal. Such a matrix is always nilpotent; hence so is the given matrix $B \in T_n(I)$.

(3) We assume here that R is a commutative ring. It is easy to see that $J := \mathbb{M}_n(I)$ is an ideal of $\mathbb{M}_n(R)$. To show that J is nil, consider a power α^N where $\alpha = (a_{ij}) \in J$. Each entry $(\alpha^N)_{ij}$ is a sum of N-fold products of elements from the set $\{a_{ij}\}$. Fix an integer r such that $a_{k\ell}^r = 0$ for all k, ℓ. If N is large enough, *some* $a_{k\ell}$ will occur at least r times in any given summand of $(\alpha^N)_{ij}$. Since R is commutative, such a summand has a factor $a_{k\ell}^r$, and is therefore 0. This shows that $\alpha^N = 0$, as desired.

   Applying the above to $I = \mathrm{Nil}^*(R)$, we see that $\mathbb{M}_n(\mathrm{Nil}^* R) \subseteq \mathrm{Nil}^*(\mathbb{M}_n(R))$. On the other hand, if we write the ideal $\mathrm{Nil}^*(\mathbb{M}_n(R))$ in the form $\mathbb{M}_n(I')$ where I' is a suitable ideal in R (see *FC*-(3.1)), I' must be nil, so $I' \subseteq I$. Therefore, $\mathrm{Nil}^*(\mathbb{M}_n(R)) \subseteq \mathbb{M}_n(\mathrm{Nil}^* R)$ *for any ring R*, and equality holds for any *commutative* ring R.

Ex. 10.23*. (Analogue of Ex. 4.12C for the nilradicals.)

(1) For a triangular ring $T = \begin{pmatrix} R & M \\ 0 & S \end{pmatrix}$ (where M is an (R, S)-bimodule), show that $\mathrm{Nil}_*(T) = \begin{pmatrix} \mathrm{Nil}_*(R) & M \\ 0 & \mathrm{Nil}_*(S) \end{pmatrix}$. Apply this to compute the lower nilradical of the ring $T_n(k)$ of upper triangular matrices over any ring k.

(2) Do the same for the upper nilradical.

Solution. (1) We proceed as in the solution to Ex. 4.12C. For the ideal $N = \begin{pmatrix} N_1 & M \\ 0 & N_2 \end{pmatrix} \subseteq T$, where $N_1 = \mathrm{Nil}_*(R)$ and $N_2 = \mathrm{Nil}_*(S)$, $T/N \cong (R/N_1) \times (S/N_2)$ is semiprime, so $N \supseteq \mathrm{Nil}_*(T)$. This will be an equality if we can show that $N \subseteq \mathfrak{p}$ for any prime ideal $\mathfrak{p}$ of T. Now $\mathfrak{p}$ must contain the nilpotent ideal $\begin{pmatrix} 0 & M \\ 0 & 0 \end{pmatrix}$, so $\mathfrak{p}$ "corresponds" to a prime ideal in the ring $R \times S$. Thus, we have either $\mathfrak{p} = \begin{pmatrix} \mathfrak{p}_1 & M \\ 0 & S \end{pmatrix}$, or $\mathfrak{p} = \begin{pmatrix} R & M \\ 0 & \mathfrak{p}_2 \end{pmatrix}$, where $\mathfrak{p}_1$ (resp. $\mathfrak{p}_2$) is a prime ideal in R (resp. S). But then $N_1 \subseteq \mathfrak{p}_1$, and $N_2 \subseteq \mathfrak{p}_2$, so in either case, $N \subseteq \mathfrak{p}$, as desired.

Using the above computation and induction, we deduce as in Ex. 4.12C that $\mathrm{Nil}_*(T_n(k))$ consists of matrices in $T_n(k)$ with diagonal entries in $\mathrm{Nil}_*(k)$.

(2) Consider again the ideal $N = \begin{pmatrix} N_1 & M \\ 0 & N_2 \end{pmatrix}$, where now $N_1 = \mathrm{Nil}^*(R)$, $N_2 = \mathrm{Nil}^*(S)$. Since $T/N \cong (R/N_1) \times (S/N_2)$ has no nonzero nil ideals (by a simple application of Ex. 1.8), we have $\mathrm{Nil}^*(T) \subseteq N$. On the other hand, the idea used in the solution of Ex. 10.23(2) shows that N is a nil ideal in T, so we have $N \subseteq \mathrm{Nil}^*(T)$ also. Using induction, we can show as before that $\mathrm{Nil}^*(T_n(k))$ consists of matrices in $T_n(k)$ with diagonal entries in $\mathrm{Nil}^*(k)$.

Ex. 10.24. (Krempa-Amitsur) Let I be an ideal of a ring R such that, for all n, $\mathbb{M}_n(I)$ is a nil ideal in $\mathbb{M}_n(R)$. Show that $I[t] \subseteq \mathrm{rad}\, R[t]$.

Solution. It suffices to show that, for any $a_i \in I$,

$$1 + a_1 t + \cdots + a_n t^n \quad \text{is invertible in } R[t].$$

Let $\sum_i b_i t^i$ be the "formal" inverse of $1 + a_1 t + \cdots + a_n t^n$ in the power series ring $R[[t]]$. Then we have:

$$b_0 = 1, \quad b_1 = -a_1, \quad b_2 = -(b_1 a_1 + a_2), \quad b_3 = -(b_2 a_1 + b_1 a_2 + a_3), \text{ etc.}$$

Let A be the companion matrix

$$\begin{pmatrix} 0 & & & -a_n \\ 1 & 0 & & -a_{n-1} \\ & \ddots & \ddots & \vdots \\ & & 1 & -a_1 \end{pmatrix}.$$

By direct matrix multiplication, we see that, for $e = (0, \ldots, 0, 1)$:

$$eA = (0, \ldots, 1, -a_1) = (0, \ldots, 0, b_0, b_1)$$
$$eA^2 = (0, \ldots, 0, b_0, b_1, -b_0 a_2 - b_1 a_1) = (0, \ldots, 0, b_0, b_1, b_2),$$
$$\cdots \cdots$$
$$eA^n = (b_1, \ldots, b_{n-1}, b_n)$$
$$eA^{n+1} = (b_2, \ldots, b_n, b_{n+1}), \text{ etc.}$$

Also, we can check that the entries of A^n "involve" only the a_i's and not the element 1, so $A^n \in \mathbb{M}_n(I)$. By the hypothesis, A^n is nilpotent, so the equations for eA^k above show that $b_k = 0$ for sufficiently large k. Therefore, the formal inverse for $1 + \sum_i a_i t^i$ lies in $R[t]$, as desired.

Ex. 10.25. Using Ex. 23 and 24, show that the following are equivalent:

(1) Köthe's Conjecture. ("The sum of two nil left ideals in any ring is nil": see FC-(10.28).)
(1)' The sum of two nil 1-sided ideals in any ring is nil.
(2) If I is a nil ideal in any ring R, then $\mathbb{M}_n(I)$ is nil for any n.
(2)' If I is a nil ideal in any ring, then $\mathbb{M}_2(I)$ is nil.
(3) Nil*$(\mathbb{M}_n(R)) = \mathbb{M}_n(\text{Nil}^*(R))$ for any ring R and any n.
(4) If I is a nil ideal in any ring R, then $I[t] \subseteq \text{rad } R[t]$.
(5) rad $(R[t]) = (\text{Nil}^*R)[t]$ for any ring R.

Solution. First, (1) $\Longleftrightarrow$ (1)' is clear, since, for any element b in a nil right ideal, Rb is a nil left ideal. Second, we have (2) $\Longleftrightarrow$ (2)'. For, if (2)' holds, then for any nil ideal I, $\mathbb{M}_{2^t}(I) \cong \mathbb{M}_2(\mathbb{M}_{2^{t-1}}(I))$ is also nil, by induction on t. Since any $n \in \mathbb{N}$ is bounded by 2^t for some t, we have (2). In the following, we may therefore focus on the conditions (1)–(5) only.

(1) $\Longrightarrow$ (2). Note that $\mathbb{M}_n(I) = \sum_k J_k$ where J_k is the left ideal in $\mathbb{M}_n(R)$ consisting of matrices with kth column entries from I and all other entries zero. By Exercise 23, each J_k is nil, so by (1) (and induction), $\mathbb{M}_n(I)$ is nil.

(2) $\Longleftrightarrow$ (3). This follows from the argument in the last paragraph of the solution to Ex. 10.23(3).

(2) $\Longrightarrow$ (4) follows from Exercise 24.

(4) $\Longrightarrow$ (5). The inclusion "$\subseteq$" in (5) follows from Amitsur's Theorem (FC-(5.10)). The reverse inclusion "$\supseteq$" follows from (4).

(5) $\Longrightarrow$ (1). We shall prove (1) in the alternative form: If R is a ring such that Nil*$R = 0$, then every nil left ideal $I \subseteq R$ is zero (see FC-(10.28)). *It suffices to show that, for any $f(t) \in I[t]$, $1 + f(t)$ has a left inverse in $R[t]$*, for then $I[t] \subseteq \text{rad } R[t]$, and rad $R[t] = 0$ by (5). Consider the ring $R' = I \oplus \mathbb{Z}$, obtained by formally adjoining an identity $1' = (0,1)$ to the "rng" I. It is easy to see that Nil*$R' = I$, so by (5), rad $R'[t] = I[t]$. Given $f(t) \in I[t]$, $1' + f(t)$ has then a left inverse, say $1' + g(t)$, in $R'[t]$. From

$$1' = (1' + g(t))(1' + f(t)),$$

we have

$$-g(t) = f(t) + g(t)f(t) \in I[t],$$

since $I[t]$ is an ideal in $R'[t]$. Now the equation

$$1 = (1 + g(t))(1 + f(t))$$

implies that $1 + f(t)$ has a left inverse $1 + g(t) \in R[t]$, as desired.

Comment. Note that, if (5) is true, it would give a much sharper form of Amitsur's Theorem (*FC*-(5.10)) on the Jacobson radical of a polynomial extension. But, of course, (5) may very well be false. Köthe's Conjecture has remained open as of this day.

Amitsur had once conjectured that, if I is any nil ring (a ring without identity in which every element is nil), then the polynomial ring $I[t]$ is also nil. This *would* imply the truth of (4) (for then $1 + I[t] \subseteq \mathrm{U}(R[t])$). However, A. Smoktunowicz has found a counterexample to Amitsur's conjecture; see her paper in J. Algebra *233* (2000), 427–436. This is an important counterexample, but it did not alter the open status of (4), or of Köthe's Conjecture.

§11. Structure of Primitive Rings; the Density Theorem

A central notion in noncommutative ring theory is that of a left primitive ring. Following Jacobson, we call a ring $R \neq 0$ *left primitive* if R has a faithful simple left module. Similarly, we call R *semiprimitive* if R has a faithful semisimple (left) module. An easy check shows that R is semiprimitive iff R is *J*-semisimple.

If $M = {}_R V_k$ is an (R, k)-bimodule (where R, k are given rings), we say that R acts *densely* on V_k if we can "approximate" any $f \in \mathrm{End}(V_k)$ on a finite number of elements $v_1, \ldots, v_n \in V$ by an element of R (that is, there exists $r \in R$ such that $rv_i = f(v_i)$ for all i). The Jacobson-Chevalley Density Theorem expresses a fundamental fact in abstract algebra: *If ${}_R V$ is a semisimple module over any ring R, and $k = \mathrm{End}({}_R V)$, then R acts densely on V_k.* The power of this result lies in its simplicity: there is only one assumption needed, that ${}_R V$ be semisimple. And needless to say, the result is not true for nonsemisimple modules.

If k is a division ring above, then V_k is a k-vector space. In this case, R acts densely on V_k iff, given any vectors

$$v_1, \ldots, v_n, u_1, \ldots, u_n \in V$$

where the u_i's are k-linearly independent, there exists $r \in R$ such that $ru_i = v_i$ for all i. The Density Theorem in this context can be restated as follows: *A nonzero ring R is left primitive iff it is isomorphic to a dense ring of linear transformations on a vector space V_k over some division ring k.* In this case, there are two possibilities. If R is left artinian, then $R \cong \mathbb{M}_n(k)$ for some n; if R is *not* left artinian, then, for any $n > 0$, *some* subring $R_n \subseteq R$ admits a ring homomorphism onto $\mathbb{M}_n(k)$.

In the commutative category, primitive rings are just fields. In the noncommutative category, however, left primitive rings abound, and are considerably more general than division rings. For instance, if k is a field, then

$$k \langle x, y \rangle / (xy - 1) \quad \cdot$$

is primitive, and so is the free algebra $k \langle x, y \rangle$ (cf. *FC*-pp. 184–185). If char $k = 0$, the Weyl algebra

$$k \langle x, y \rangle / (xy - yx - 1)$$

is also primitive. If K is a division ring containing an element not algebraic over its center, then the polynomial ring $K[t]$ is primitive (cf. *FC*-(11.14)). The important role played by left primitive rings in the structure theory of noncommutative rings was brought to the fore by Jacobson in a series of fundamental papers in the 1940's. This work of Jacobson culminated in his A.M.S. Colloquium Publications volume "Structure of Rings," which has remained a standard reference for classical ring theory as of this day.

The exercise set for this section collects sundry facts and examples about primitive rings. For instance, left primitivity is not inherited by subrings, homomorphic images, or polynomial rings, but is inherited by matrix rings, and by passage from R to eRe for any idempotent e. We also see that a free algebra $k \langle x, y \rangle$ has more than one faithful simple module (Exercise 10). The Density Theorem is seen to have somewhat surprising applications to simple rings (Exercise 6) and to differential operators (Exercise 12).

Among the left primitive rings, those with nonzero socle deserve special attention (cf. *FC*-(11.11)). These rings turn out to be dense rings of linear transformations which contain projection operators of finite (nonzero) rank. Their minimal (left or right) ideals are generated by projections onto 1-dimensional subspaces. These facts are developed in Exercises 16, 17 and 18 below.

Exercises for §11

Ex. 11.1. Show that a homomorphic image of a left primitive ring need not be left primitive.

Solution. For any field k, let

$$R = k \langle x, y \rangle \quad \text{and} \quad S = k \times k.$$

We have a surjective k-algebra homomorphism $\varphi : R \to S$ defined by

$$\varphi(x) = (1, 0) \quad \text{and} \quad \varphi(y) = (0, 1).$$

By *FC*-(11.23), R is left primitive. However, S is not. (Recall that commutative primitive rings are fields.) Many other examples can be given. For instance, let k be a field with a non-surjective endomorphism σ, and let $R = k[x; \sigma]$ (with the Hilbert twist $xb = \sigma(b)x$). By *FC*-(11.13), R is a left primitive ring. However, its homomorphic image $S = R/Rx^2$ is not left primitive. (The nonzero ideal $Rx/Rx^2 \subseteq S$ has square zero, so S is not even semiprime!)

Ex. 11.2. Show that a ring R can be embedded into a left primitive ring iff either char R is a prime number $p > 0$, or $(R, +)$ is a torsion-free abelian group.

Solution. First assume $R \subseteq S$, where S is a left primitive ring. Then S is a prime ring, so by Exercise 10.0, char $R =$ char S is either a prime number p, or char $R = 0$. In the latter case, for any integer $n \geq 1$, $n \cdot 1$ is not a 0-divisor in S by Exercise 10.0. Clearly, this implies that $(R, +)$ is torsion-free.

(Alternatively, in the spirit of the solution of Ex. 3.0, let $_S M$ be a faithful simple S-module. Then $\mathrm{End}_S(M)$ is a division ring, with prime field $F \cong \mathbb{Q}$ or $\mathbb{F}_p$ (for some prime p). Since S acts faithfully on M as a ring of F-linear transformations, the desired conclusions follow for S, and thus also for R.)

Conversely, assume char R is a prime p, or that $(R, +)$ is torsion-free. In either case, R can be embedded into a k-algebra A over some field k. (In the former case, choose $k = \mathbb{F}_p \subseteq R$ and let $A = R$. In the latter case, choose $k = \mathbb{Q}$ and embed R into $A = R \otimes_{\mathbb{Z}} \mathbb{Q}$.) Now the "left regular representation"

$$\varphi : A \longrightarrow \mathrm{End}_k(A)$$

defined by $\varphi(a)(b) = ab$ for $a, b \in A$ is an embedding of A (and hence of R) into the left (and incidentally also right) primitive k-algebra $\mathrm{End}_k A$.

Comment. There is an analogue of the above result for the embeddability of a ring R into a semiprimitive (i.e. J-semisimple) ring: The necessary and sufficient condition is that, for any prime number $p > 0$, R has no element of additive order p^2. This is a result of O. Goldman: see his paper in Bull. Amer. Math. Soc. *52* (1946), 1028–1032.

Ex. 11.3. Let R be a left primitive ring. Show that for any nonzero idempotent $e \in R$, the ring $A = eRe$ is also left primitive.

Solution. Let V be a faithful simple left R-module. It suffices to show that $U = eV$ is a faithful simple left A-module. First

$$A \cdot U = eRe \cdot eV = eReV \subseteq eV = U,$$

so U is indeed an A-module. Let $a = ere \in A$, where $r \in R$. Then $ae = ere^2 = a$. If $aU = 0$, then $0 = aeV = aV$ implies that $a = 0$, so $_A U$ is faithful. To check that $_A U$ is simple, let us show that for $0 \neq u \in U$ and $u' \in U$, we have $u' \in Au$. Note that $u, u' \in eV$ implies $u = eu$, $u' = eu'$. We have $u' = ru$ for some $r \in R$, so

$$u' = eu' = eru = (ere)u \in Au,$$

as desired.

Ex. 11.4. Which of the following implications are true?

(a) R left primitive $\Longleftrightarrow$ $\mathbb{M}_n(R)$ left primitive.
(b) R left primitive $\Longleftrightarrow$ $R[t]$ left primitive.

Solution. (a) *Both implications here are true.* In fact, assume $S :=\mathbb{M}_n(R)$ is left primitive (for some n). For the idempotent $e = E_{11}$ (matrix unit), eSe is also left primitive by Exercise 3. Since

$$eSe = RE_{11} \cong R \quad \text{(as rings)},$$

we conclude that R is left primitive. Conversely, assume R is left primitive, and fix a faithful simple left R-module V. Let U be the group of column n-tuples $(v_1, \ldots, v_n)^T$, where $v_i \in V$. We make U into a left S-module by using "matrix multiplication" as the action. We are done if we can show that U is a faithful simple S-module. If $(r_{ij})U = 0$, we have $r_{ij}V = 0$, so $r_{ij} = 0$, for all i, j. To show that U is simple, it suffices to show that, given any

$$u = (v_1, \ldots, v_n)^T \neq 0 \quad \text{and} \quad u' = (v'_1, \ldots, v'_n)^T,$$

we have $u' \in Su$. Say $v_i \neq 0$. Then there exist $r_1, \ldots, r_n \in R$ such that $v'_j = r_j v_i$ for all j. For the matrix $\sigma \in S$ with ith column $(r_1, \ldots, r_n)^T$ and other columns zero, we have clearly $\sigma u = u'$. Alternatively, we can use the characterization of left primitive rings given in *FC*-(11.28) ("R is left primitive iff there exists a left ideal $\mathfrak{A} \neq R$ which is comaximal with any nonzero ideal $\mathfrak{B}$.") Fix such a left ideal $\mathfrak{A}$ in R. Any nonzero ideal in S has the form $\mathbb{M}_n(\mathfrak{B})$ for some ideal $\mathfrak{B} \neq 0$ in R (according to *FC*-(3.1)). Since $\mathfrak{A} + \mathfrak{B} = R$, the left ideal $\mathbb{M}_n(\mathfrak{A}) \subsetneqq S$ is clearly comaximal with $\mathbb{M}_n(\mathfrak{B})$. Therefore, *FC*-(11.28) implies that S is left primitive.

(b) The *forward* implication is certainly not true in general. For instance, if R is a field, then R is primitive; however, $R[t]$ is commutative but not a field, so it is not primitive. The *backward* implication is also not true, but to see this requires more work. The following construction of a counterexample is based on a crucial lemma due to George Bergman, included here with his kind permission.

Lemma. *Let $A = A_0 \oplus A_1 \oplus \cdots$ be a nonzero graded ring with a faithful module $_AV$ such that $A_n v = V$ for any $n \geq 0$ and any $v \neq 0$ in V. Then the ring[1] $R := A_0 \times A_1 \times \cdots$ is not (left or right) primitive, but the polynomial ring $T = R[t]$ is left primitive.*

---

[1] Multiplication on R is defined by

$$(a_0, a_1, \ldots)(b_0, b_1, \ldots) = (a_0 b_0, \ a_0 b_1 + a_1 b_0, \ a_0 b_2 + a_1 b_1 + a_2 b_0, \ldots).$$

In some sense, R is a "completion" of the ring A.

Proof. As in FC-(1.5), we can check that any element $(1, a_1, a_2, \dots)$ is a unit in R, so

$$A_1 \times A_2 \times \cdots \subseteq \operatorname{rad} R.$$

Since each $A_n \neq 0$, rad $R \neq 0$, so R is not even semiprimitive (let alone primitive). To see that $R[t]$ is left primitive, let M be the additive group of sequences $(v_i)_{i \in \mathbb{Z}}$ where $v_i \in V$ are eventually 0 on the left. We make M into a left R-module by letting $a_n \in A_n$ act by its given action on the "coordinates" v_i followed by a shift n steps to the right. Thus,

$$(a_0, a_1, a_2, \dots) \cdot (\dots, 0, v_i, v_{i+1}, v_{i+2}, \dots)$$
$$= (\dots,\ 0,\ a_0 v_i,\ a_0 v_{i+1} + a_1 v_i,\ a_0 v_{i+2} + a_1 v_{i+1} + a_2 v_i,\ \dots).$$

From the hypothesis on $_A V$, we see that, if $v_i \neq 0$:

$$(1) \qquad R \cdot (\dots, 0, v_i, v_{i+1}, \dots) = \{(u_j)_{j \in \mathbb{Z}} :\quad u_j = 0 \quad \text{for } j < i\}.$$

Next, we make M into a (left) module over $T = R[t]$ by letting t "shift coordinates" one step to the left (which clearly commutes with the R-action). From (1), we see easily that

$$(2) \qquad T \cdot (\dots, 0, v_i, v_{i+1}, \dots) = M \quad \text{whenever} \quad v_i \neq 0,$$

so M is a *simple* T-module. We finish by showing that $_T M$ is *faithful* (for then T is a left primitive ring). To simplify the notation, let us show that, for $\alpha, \beta \in R$:

$$(3) \qquad (\alpha + \beta t) \cdot M = 0 \Longrightarrow \alpha = \beta = 0 \in R.$$

(The general case when $\alpha + \beta t$ is replaced by

$$\alpha + \beta t + \cdots + \gamma t^m \in T$$

can be handled similarly.) Writing

$$\alpha = (a_0, a_1, \dots) \quad \text{and} \quad \beta = (b_0, b_1, \dots)$$

(where $a_n, b_n \in A_n$), (3) gives

$$0 = (\alpha + \beta t) \cdot (\dots, 0, v_i, v_{i+1}, \dots)$$
$$= (\dots,\ 0,\ b_0 v_i,\ b_0 v_{i+1} + (a_0 + b_1) v_i,\ b_0 v_{i+2} + (a_0 + b_1) v_{i+1}$$
$$+ (a_1 + b_2) v_i,\ \dots).$$

Since this holds for all $v_i, v_{i+1}, \dots \in V$ and $_A V$ is faithful, we must have

$$b_0 = 0, \quad a_0 + b_1 = 0, \quad a_1 + b_2 = 0, \text{ etc.}$$

Therefore, $a_0 = b_1 = 0$, $a_1 = b_2 = 0$, etc., so $\alpha = \beta = 0 \in R$, as desired.
QED

To produce an explicit example, we must construct the data (A, V) as in the Lemma. This is fairly easy if we look in the right direction. Let k be a field with an automorphism σ of infinite order, and let $A = k[x; \sigma]$ be the twisted polynomial ring over k constructed from σ. Then

$$A = A_0 \oplus A_1 \oplus \cdots$$

is a graded ring with $A_n = kx^n$ $(n \geq 0)$. Taking $V = k$, with the A-action

$$(4) \qquad (ux^n) \cdot v = u\sigma^n(v) \quad (u, v \in k, \quad n \geq 0),$$

we see from FC-(11.13) that $_A V$ is a *faithful* module. Furthermore, for any $v \neq 0$ in V and any $n \geq 0$, (4) shows that

$$A_n \cdot v = k\sigma^n(v) = k = V,$$

so the hypotheses in the lemma are satisfied. The ring

$$R = A_0 \times A_1 \times \cdots$$

constructed in the lemma is just the twisted power series ring $k[[x; \sigma]]$. Also, the M in the proof of the lemma may be identified with the Laurent series ring $k((x; \sigma))$. The left R-action of R on $M = k((x; \sigma))$ is now just the multiplication in M, since, under this multiplication

$$(ux^n)(vx^m) = u\sigma^n(v)x^{n+m},$$

which is exactly the ux^n-action followed by a shift of n steps to the right. The t-action on $_T M$ is just *right* multiplication by x^{-1} on M (which commutes with the left R-action in view of the associative law). With the above data, R is the (local) domain with rad $R = Rx \neq 0$, but $T = R[t]$ is a left primitive ring with a simple faithful module $M = k((x; \sigma))$ with the R-action and t-action described above.

Ex. 11.5. Let R be a ring which acts faithfully and irreducibly on a left module V. Let $v \in V$ and $\mathfrak{A}$ be a nonzero right ideal in R. Show that $\mathfrak{A} \cdot v = 0 \Longrightarrow v = 0$.

Solution. If $v \neq 0$, then we must have $Rv = V$. Since

$$\mathfrak{A} \cdot (Rv) = (\mathfrak{A}R) \cdot v = \mathfrak{A} \cdot v = 0,$$

the faithfulness of V implies that $\mathfrak{A} = 0$, a contradiction.

Ex. 11.5*. For any left ideal I in a ring R, define the *core* of I to be the sum of all ideals in I. Thus, core (I) is the (unique) largest ideal of R contained in I.

(1) Show that core $(I) = \text{ann}(V)$ where V is the left R-module R/I. (In particular, V is faithful iff core $(I) = 0$.)
(2) Show that $_R(R/I)$ is faithful only if $I \cap Z(R) = 0$, where $Z(R)$ is the center of R.

Solution. First, $\text{ann}(V) \subseteq I$, and $\text{ann}(V)$ is an ideal, so $\text{ann}(V) \subseteq \text{core}(I)$. Secondly,

$$\text{core}(I) \cdot R \subseteq \text{core}(I) \subseteq I \Longrightarrow \text{core}(I) \cdot V = 0,$$

so $\text{core}(I) \subseteq \text{ann}(V)$. This establishes (1). If R/I is indeed faithful, then, in view of (1) and $I \cap Z(R) \subseteq \text{core}(I)$, we have $I \cap Z(R) = 0$, establishing (2).

Ex. 11.6. (Artin-Whaples) Let R be a simple ring with center k (which is a field by Exercise 3.4). Let $x_1, \ldots, x_n \in R$ be linearly independent over k. Show that, for any $y_1, \ldots, y_n \in R$, there exist $a_1, \ldots, a_m$ and $b_1, \ldots, b_m$ in R such that $y_i = \sum_{j=1}^{m} a_j x_i b_j$ for every i.

Solution. Let $A := R \otimes_k R^{\text{op}}$ act on R by the rule

$$(a \otimes b^{\text{op}})x = axb.$$

Since the A-submodules of $_A R$ are just the ideals of R, $_A R$ is a simple module. An endomorphism φ of this module is an additive map $R \to R$ commuting with both left and right multiplication, so φ is just multiplication by an element of k. We have now $\text{End}(_A R) = k$, so by the Density Theorem (FC-(11.16)), A acts densely on R_k. Fix $f \in \text{End}(R_k)$ such that $f(x_i) = y_i$ for all i. By density, there exists

$$r = \sum_{j=1}^{m} a_j \otimes b_j^{\text{op}} \in A$$

such that $rx_i = f(x_i)$ for all i. Thus,

$$y_i = f(x_i) = \sum_{j=1}^{m} a_j x_i b_j$$

for $1 \leq i \leq n$.

Ex. 11.6* Let M be a (not necessarily finitely generated) semisimple left module over a semiprimary ring R, and let $k = \text{End}_R(M)$. Show that

(1) M_k is a finitely generated k-module; and
(2) the natural ring homomorphism $R \to \text{End}(M_k)$ is surjective.

Solution. By the Density Theorem, R acts densely on M_k. Therefore, as soon as we prove (1), (2) follows from FC-(11.17). To prove (1), note that, since $(\text{rad } R) \cdot M = 0$, we may replace R by $R/\text{rad}(R)$ to assume that R is a semisimple ring. Let $S_1, \ldots, S_r$ be a representative set of simple left R-modules, and let M_i be the isotypic component of M with respect to S_i. Then $k = k_1 \times \cdots \times k_r$ where $k_i = \text{End}_R(M_i)$, so it suffices to show that each M_i is finitely generated over k_i.

Let D_i be the division ring $\text{End}_R S_i$. By the Artin-Wedderburn Theorem, we know that $(S_i)_{D_i}$ has a finite basis, say $\{v_1, \ldots, v_n\}$. (Here, we *regard i as fixed*). Let $M_i = S_i^{(J)}$ (direct sum of J copies of S_i). Fix an index, say, $1 \in J$. Let v'_j $(1 \le j \le n)$ be the element in M_i with component v_j in the "first" copy of S_i and zero components in the other copies. Clearly, $\sum_{j=1}^n v'_j k_i$ contains the "first" copy of S_i in the direct sum $S_i^{(J)}$. Using "translations" (which are elements of k_i), we see that $\sum_{j=1}^n v'_j k_i$ contains *any* copy of S_i in the direct sum $S_i^{(J)}$. Thus, $M_i = \sum_{j=1}^n v'_j k_i$, as desired.

Comment. Note that the key property needed from R is that $R/\mathrm{rad}\, R$ be semisimple. Rings with this property are called *semilocal* rings; see FC-§20.

Ex. 11.7. Let $E = \mathrm{End}(V_k)$ where V is a right vector space over the division ring k. Let R be a subring of E and $\mathfrak{A}$ be a nonzero ideal in R. Show that R is dense in E iff $\mathfrak{A}$ is dense in E.

Solution. It suffices to prove the "only if," for which we may assume that $\mathfrak{A} = (a)$, the ideal generated by a nonzero element $a \in R$. Fix a vector $w \in V$ such that $aw \ne 0$. Given $v_1, \ldots, v_n \in V$, and k-linearly independent vectors $u_1, \ldots, u_n \in V$, we want to find $b \in \mathfrak{A}$ such that $bu_j = v_j$ for all j. Since $R \cdot aw = V$, $v_j = r_j aw$ for suitable $r_j \in R$. By density, there also exist $s_1, \ldots, s_n \in R$ such that $s_i u_j = \delta_{ij} w$. For

$$b := \sum_i r_i a s_i \in \mathfrak{A},$$

we have, for all j:

$$b u_j = \sum_i r_i a s_i u_j = \sum_i r_i a \delta_{ij} w = r_j a w = v_j.$$

Ex. 11.7*. Let $E = \mathrm{End}(V_k)$ be as in Exercise 7, and let $R \subseteq E$ be a dense subring.

(1) For any $a \in R$ with finite rank, show that $a = ara$ for some $r \in R$.
(2) Deduce that the set

$$S = \{a \in R : \mathrm{rank}(a) < \infty\}$$

is a von Neumann regular ring (possibly without identity).

Solution. (1) The proof here is a slight modification of the usual proof for E to be von Neumann regular (cf. FC-(4.27)). Let $K = \ker(a)$, and fix a k-subspace $U \subseteq V$ such that $V = K \oplus U$. Then

$$n := \dim_k U = \mathrm{rank}(a) < \infty.$$

Let $\{u_1, \ldots, u_n\}$ be a k-basis for U. Then $a(u_1), \ldots, a(u_n)$ remain k-linearly independent, so by density, there exists $r \in R$ such that $ra(u_i) = u_i$ for all

i. We have now $ara = a$ since both sides vanish on K, and take u_i to $a(u_i)$ on the basis elements $\{u_i\}$ of U.

(2) It is easy to see that S is an ideal of R, so we may view S as a ring, possibly without identity. To prove (2), we may assume that $S \neq 0$. By Exercise 7, S also acts densely on V. Noting that the proof of (1) above did not make use of the identity element, we conclude from (1) that S is a von Neumann regular ring.

Comment. The ideal S above turns out to be the (left and right) socle of the ring R: see Exercise 18 below.

Ex. 11.8. Let $V = \bigoplus_{i=1}^{\infty} e_i k$ where k is a field. For any n, let S_n be the set of endomorphisms $\lambda \in E = \text{End}(V_k)$ such that λ stabilizes $\sum_{i=1}^{n} e_i k$ and $\lambda(e_i) = 0$ for $i \geq n+1$. Show that

$$S = \bigcup_{n=1}^{\infty} S_n \subseteq E$$

is a dense set of linear transformations. For any i, j, let $E_{ij} \in E$ be the linear transformation which sends e_j to e_i and all $e_{j'}$ ($j' \neq j$) to zero. Show that any k-subalgebra R of E containing all the E_{ij}'s is dense in E and hence left primitive.

Solution. Let $v_1, \dots, v_m, u_1, \dots, u_m \in V$, where the u_i's are k-linearly independent. Pick a large integer n such that

$$u_j, v_j \in \sum_{i=1}^{n} e_i k \quad \text{for all} \quad j.$$

Take

$$\lambda \in \text{End}_k \left(\sum_{i=1}^{n} e_i k \right)$$

such that $\lambda(u_j) = v_j$ for all j, and extend λ to an endomorphism of V by defining $\lambda(e_i) = 0$ for $i \geq n+1$. Then $\lambda \in S_n \subseteq S$, and we have proved the denseness of S. The last statement in the Exercise is clear, since the k-algebra R contains S_n for all n.

Ex. 11.9. (Jacobson) Keep the notations above and define $f, g \in E$ by $g(e_i) = e_{i+1}$, $f(e_i) = e_{i-1}$ (with the convention that $e_0 = 0$). Let R be the k-subalgebra of E generated by f and g.

(1) Use Exercise 8 to show that R acts densely on V_k.
(2) Show that R is isomorphic to $S := k \langle x, y \rangle / (xy - 1)$, with a k-isomorphism matching f with $\bar{x}$ and g with $\bar{y}$.

Solution. (1) For $i, j \geq 1$, we claim that $g^{i-1}f^{j-1} - g^i f^j$ equals the E_{ij} defined in Exercise 8. In fact,

$$(g^{i-1}f^{j-1} - g^i f^j)(e_j) = g^{i-1}(e_1) - g^i(0) = e_i,$$
$$(g^{i-1}f^{j-1} - g^i f^j)(e_\ell) = g^{i-1}(f^{j-1}e_\ell) - g^i(f^j e_\ell) = 0 \quad \text{for } \ell < j,$$
$$(g^{i-1}f^{j-1} - g^i f^j)(e_\ell) = g^{i-1}(e_{\ell-j+1}) - g^i(e_{\ell-j})$$
$$= e_{\ell-j+i} - e_{i+\ell-j} = 0 \quad \text{for } \ell > j$$

Therefore, $E_{ij} \in R$, and Exercise 8 implies that R is dense in E. (A somewhat different proof for the density of R was given in FC-p. 184.)

(2) To simplify the notations, let us continue to write x, y for the elements $\bar{x}, \bar{y} \in S$. Since $fg = 1 \in R$, we have a well-defined k-algebra homomorphism $\varepsilon : S \to R$ with $\varepsilon(x) = f$ and $\varepsilon(y) = g$. We need to show that ε is injective. In view of $xy = 1$, any element in S has the form

$$z = \alpha(y) + \beta(y)x + \gamma(y)x^2 + \cdots,$$

where $\alpha, \beta, \gamma, \ldots$ are polynomials in y. Suppose $\varepsilon(z) = 0$. Applying $\varepsilon(z)$ to e_1, we get $\alpha(g)e_1 = 0$. Clearly this implies that $\alpha \equiv 0$. Applying $\varepsilon(z)$ to e_2, we get

$$0 = \beta(g)f(e_2) = \beta(g)e_1,$$

so $\beta = 0$. Repeating this argument, we see that $z = 0 \in S$.

Ex. 11.10. For a field k, construct two left modules V, V' over the free algebra $R = k\langle x, y \rangle$ as follows. Let $V = V' = \sum_{i=1}^{\infty} e_i k$. Let R act on V by:

$$xe_i = e_{i-1}, \quad ye_i = e_{i^2+1},$$

and let R act on V' by

$$xe_i = e_{i-1}, \quad ye_i = e_{i^2+2}$$

(with the convention that $e_0 = 0$). Show that V, V' are nonisomorphic faithful simple left R-modules.

Solution. The fact that R acts faithfully and irreducibly on V was proved in FC-(11.23). The same argument also applies to V'. To see that V, V' are not isomorphic as R-modules, note that

$$x^2 y(e_1) = x^2 e_2 = 0$$

in V. We are done if we can show that $x^2 y$ does not annihilate any nonzero vector in V'. But

$$x^2 y(e_i) = x^2 e_{i^2+2} = e_{i^2}$$

in V' for any $i \geq 1$. Therefore, if

$$0 \neq v' = e_j a_j + e_{j+1} a_{j+1} + \cdots \in V'$$

with $a_j \neq 0$, then

$$x^2 y(v') = e_{j^2} a_j + e_{(j+1)^2} a_{j+1} + \cdots \neq 0,$$

as desired.

Ex. 11.11. Let A be a subring of a field K. Show that the subring S of $K \langle x_1, \ldots, x_n \rangle$ $(n \geq 2)$ consisting of polynomials with constant terms in A is a left primitive ring with center A.

Solution. We know that $R = K \langle x_1, \ldots x_n \rangle$ is a left-primitive ring, by FC-(11.26). Let V be a faithful simple left R-module, and $k = \mathrm{End}(_R V)$. Then R acts densely on V_k. By Exercise 7, the ideal $(x_1, \ldots, x_n)$ of R generated by the x_i's also acts densely on V_k. The ring S in question is just $A + (x_1, \ldots, x_n)$. Since S acts densely on V_k, FC-(11.20) implies that S is a left primitive ring. Finally, it is easy to see that

$$Z(S) = Z(R) \cap S = K \cap S = A.$$

Ex. 11.12. Let k be a field of characteristic zero. Represent the Weyl algebra $R = k \langle x, y \rangle / (xy - yx - 1)$ as a dense ring of linear transformations on an infinite-dimensional k-vector space V, and restate the Density Theorem in this context as a theorem on differential operators.

Solution. Let $_R V = k[y]$ with $y \in R$ acting as left multiplication by y and $x \in R$ acting as $D = d/dy$ (formal differentiation). Choosing the k-basis

$$\{e_n = y^n : n \geq 0\}$$

on V, we have

$$y e_n = e_{n+1} \quad \text{and} \quad x e_n = n e_{n-1} \quad (\text{with } e_{-1} = 0).$$

Using these action rules, we can check directly that the R-module $_R V$ is simple, with $\mathrm{End}\,(_R V) = k$. (We suppress the calculations here, since more general conclusions have already been obtained in Exercise 3.18.) Thanks to char $k = 0$, R is a simple domain by FC-(3.17), so $_R V$ is automatically faithful. It follows that R is k-isomorphic to the (dense) ring of linear transformations on V generated by the actions of y and x. Stated in the language of differential operators, the density of R means the following: Given any polynomials

$$v_1, \ldots, v_n \in k[y],$$

and k-linearly independent polynomials

$$u_1, \ldots, u_n \in k[y],$$

there exists a differential operator

$$\sum a_{ij} y^i D^j \quad (a_{ij} \in k)$$

which takes u_ℓ to v_ℓ for all ℓ $(1 \leq \ell \leq n)$.

Ex. 11.13. Let R be a left primitive ring such that $a(ab - ba) = (ab - ba)a$ for all $a, b \in R$. Show that R is a division ring.

Solution. Let $_RV$ be a faithful simple R-module with $k = \text{End}(_RV)$. It suffices to show that $\dim_k V = 1$ (for then $R \cong k$, which is a division ring). Assume instead there exist k-linearly independent vectors $u, v \in V$. By the Density Theorem, there exist $a, b \in R$ such that $au = u, av = 0$ and $bu = 0, bv = u$. But then

$$a(ab - ba)(v) = a^2u = u, \quad (ab - ba)a(v) = 0,$$

a contradiction.

Comment. Similar arguments were used in FC-§12 to prove certain commutativity theorems in ring theory. In fact, the choice of a, b above corresponds precisely to the choice of the matrix units $a = E_{11}, b = E_{12}$ in the proof of FC-(12.11). As observed in that proof, $ab - ba = b$, and hence

$$(ab - ba)a = 0 \neq a(ab - ba) = b.$$

Ex. 11.14. Let R be a left primitive ring such that $1 + r^2$ is a unit for any $r \in R$. Show that R is a division ring.

Solution. We repeat the argument in the last exercise. If the independent vectors $u, v \in V$ exist, find $r \in R$ such that $r(u) = -v$ and $r(v) = u$. Then

$$(1 + r^2)(u) = u - u = 0,$$

contradicting the assumption that $1 + r^2 \in U(R)$.

Comment. Again, in terms of matrices, the idea is that

$$I + \begin{pmatrix} 0 & 1 \\ -1 & 0 \end{pmatrix}^2 = I - I = 0.$$

In the next four exercises (15 through 18), let R be a left primitive ring, $_RV$ be a faithful simple R-module, and k be the division ring $\text{End}(_RV)$. By the Density Theorem, R acts densely on V_k.

Ex. 11.15. For any k-subspace $W \subseteq V$, let

$$\text{ann}(W) = \{r \in R : rW = 0\},$$

and, for any left ideal $\mathfrak{A} \subseteq R$, let

$$\text{ann}(\mathfrak{A}) = \{v \in V : \mathfrak{A}v = 0\}.$$

Suppose $n = \dim_k W < \infty$. *Without assuming the Density Theorem*, show by induction on n that $\text{ann}(\text{ann}(W)) = W$. From this equation, deduce that R acts densely on V_k. If, in addition, R is left artinian, show that $\dim_k V < \infty$, $R = \text{End}(V_k)$, and $\text{ann}(\text{ann}(\mathfrak{A})) = \mathfrak{A}$ for any left ideal $\mathfrak{A} \subseteq R$. In this case, $W \mapsto \text{ann}(W)$ gives an inclusion-reversing one-one correspondence between the subspaces of V_k and the left ideals of the (simple artinian) ring R.

Solution. If $\dim_k W = 0$, we have $\mathrm{ann}(\mathrm{ann}(W)) = \mathrm{ann}(R) = 0$. To handle the inductive step, fix a k-basis $w_1, \ldots, w_n$ for W, with $n \geq 1$. Let $v \in V$ be such that $\mathrm{ann}(W) \cdot v = 0$. Our job is to show that $v \in W$. Consider the left ideal

$$\mathfrak{A} = \mathrm{ann}\left(\sum_{i=2}^{n} w_i k\right).$$

By the inductive hypothesis, $\mathfrak{A} \cdot w_1 \neq 0$, so we must have $\mathfrak{A} \cdot w_1 = V$. Define a map $\alpha : V \to V$ by the rule $(aw_1)\alpha = av$, for any $a \in \mathfrak{A}$. This map is well-defined, for, if $aw_1 = 0$, then

$$a \in \mathfrak{A} \cap \mathrm{ann}(w_1 k) = \mathrm{ann}(W),$$

and so $av = 0$. Clearly, then, $\alpha \in \mathrm{End}(_R V) = k$. Since $a(v - w_1 \alpha) = 0$ for all $a \in \mathfrak{A}$, we have (by the inductive hypotheses)

$$v - w_1 \alpha \in \sum_{i=2}^{n} w_i k,$$

and therefore $v \in W$, as desired.

To prove that R acts densely on V, let $w_1, \ldots, w_n$ be as above, and $u_1, \ldots, u_n \in V$. Recalling the equation $\mathfrak{A} \cdot w_1 = V$ above, we see that there exists $r_1 \in R$ that takes w_1 to u_1, and $w_2, \ldots, w_n$ to zero. Similarly, there exists $r_i \in R$ that takes w_i to u_i and the other w_j's to zero. The sum

$$r = r_1 + \cdots + r_n \in R$$

therefore takes w_i to u_i for all i.

Now assume that R is left artinian. Then we must have $\dim_k V < \infty$, for otherwise an ascending chain

$$W_1 \subsetneq W_2 \subsetneq \cdots$$

of finite-dimensional subspaces would lead to a descending chain of left ideals in R (in view of $\mathrm{ann}(\mathrm{ann}(W_i)) = W_i$). The denseness of R then implies that $R = \mathrm{End}(V_k)$. Let $\mathfrak{A}$ be any left ideal in R. We finish by showing that $\mathrm{ann}(\mathrm{ann}\,\mathfrak{A}) \subseteq \mathfrak{A}$. Since R is a simple artinian ring, $\mathfrak{A} = Re$ for some $e = e^2 \in R$, and so $V = \ker(e) \oplus eV$. Let

$$r \in \mathrm{ann}(\mathrm{ann}\,\mathfrak{A}) = \mathrm{ann}(\ker(e)).$$

Then re and r both vanish on $\ker(e)$, and have the same effect on eV. Therefore,

$$r = re \in Re = \mathfrak{A},$$

as desired.

Comment. The above solution is adapted from an argument given in Artin's article "The influence of J.H.M. Wedderburn on the development of modern algebra," Bull. Amer. Math. Soc. *56* (1950), 65–72. In a footnote to this paper, Artin acknowledged that his treatment of the Density Theorem "(follows) a presentation by Mr. J.T. Tate."

Ex. 11.16. For any $r \in R$ of rank m (i.e. $\dim_k rV = m$), show that there exist $r_1, \ldots, r_m \in Rr$ of rank 1 such that $r = r_1 + \cdots + r_m$.

Solution. Let $\{v_1, \ldots, v_m\}$ be a k-basis for rV. By the density of R, there exist $s_1, \ldots, s_m \in R$ such that $s_i(v_j) = \delta_{ij}v_j$ (where δ_{ij} are the Kronecker deltas). For any $v \in V$, write

$$r(v) = \sum_{j=1}^{m} v_j \alpha_j$$

where $\alpha_j \in k$. Then

$$r(v) = \sum_{i,j} \delta_{ij} v_j \alpha_j = \sum_{i,j} (s_i(v_j)) \alpha_j = \sum_i s_i \left(\sum_j v_j \alpha_j \right) = \sum_i s_i r(v).$$

Setting $r_i = s_i r \in Rr$, we have $r = \sum_i r_i$, with

$$r_i V = s_i \left(\sum_j v_j k \right) = v_i k,$$

so $\mathrm{rank}(r_i) = 1$ for all i.

Ex. 11.17. (1) Show that $\mathfrak{A} \subseteq R$ is a minimal left ideal of R iff $\mathfrak{A} = Re$ where $e \in R$ has rank 1, and that (2) $\mathfrak{B} \subseteq R$ is a minimal right ideal of R iff $\mathfrak{B} = eR$ where $e \in R$ has rank 1.

Solution. (1) Let $\mathfrak{A} = Re$, where $eV = vk, v \neq 0$. To check that $\mathfrak{A}$ is a minimal left ideal, we show that, for any $re \neq 0$ ($r \in R$), we have $e = sre$ for some $s \in R$. Since $0 \neq reV = (rv)k$, there exists $s \in R$ such that $s(rv) = v$. Now

$$(1 - sr)eV = (1 - sr)vk = 0,$$

so $(1 - sr)e = 0$, as desired. Conversely, let $\mathfrak{A}$ be any minimal left ideal in R. Since R is semiprime, Brauer's Lemma (*FC*-(10.22)) implies that $\mathfrak{A} = Re$ for some $e = e^2 \in R$. We finish by proving that e has rank 1. Assume, for the moment, that eV contains two k-linearly independent elements v, v'. Then $ev = v$, $ev' = v'$. By density, R has an element s such that $sv' = 0$ and $sv = v$. Consider the left ideal

$$\mathfrak{A}_0 = \{r \in \mathfrak{A} : rv' = 0\} \subseteq \mathfrak{A}.$$

Since $(se)v' = sv' = 0$ and $(se)v = sv = v$, we have $0 \neq se \in \mathfrak{A}_0$, so $\mathfrak{A}_0 = \mathfrak{A}$. But then $e \in \mathfrak{A}_0$, so $0 = ev' = v'$, a contradiction.

(2) Suppose $e \in R$ has rank 1. By (1), Re is a minimal left ideal. Since R is semiprime, FC-(11.9) implies that eR is a minimal right ideal. Conversely, let $\mathfrak{B} \subseteq R$ be a minimal right ideal. As before, $\mathfrak{B} = eR$ for some $e = e^2 \in R$. By FC-(11.9) again, Re is a minimal left ideal. From the proof of part (1), we infer that e has rank 1.

Ex. 11.18. Show that the following statements are equivalent:

(1) $\operatorname{soc}(R) \neq 0$,
(2) R contains a projection of V onto a line,
(3) there exists a nonzero $r \in R$ of finite rank, and
(4) for any finite-dimensional k-subspace $W \subseteq V$, R contains a projection of V onto W.

Finally, show that $\operatorname{soc}(R) = \{r \in R : \operatorname{rank}(r) < \infty\}$.

Solution. (1) $\Longrightarrow$ (2). (Recall that, in any semiprime ring R, $\operatorname{soc}(_R R) = \operatorname{soc}(R_R)$: cf. FC-p.175.) Let $\mathfrak{A}$ be a minimal left ideal in R. By the solution of Exercise 17(1), $\mathfrak{A} = Re$ where e is a projection of V onto a line.

(4) $\Longrightarrow$ (2) $\Longrightarrow$ (3) are trivial.

(3) $\Longrightarrow$ (1). By Exercise 16, (3) implies that R has an element e of rank 1. By Exercise 17(1), $Re \subseteq \operatorname{soc}(_R R)$.

(2) $\Longrightarrow$ (4). Let $e \in R$ be a (fixed) projection of R onto a line uk ($u \in V$). We shall prove (4) by induction on $n = \dim_k W$. First assume $n = 1$; say $W = wk$. Choose $a, b \in R$ such that $a(u) = w$ and $b(w) = u$. We have

$$aeb(w) = ae(u) = a(u) = w, \quad \text{and}$$
$$aeb(V) = ae(bV) \subseteq a(uk) = wk = W.$$

Therefore, aeb is a projection of V onto W. To handle the general case, let $\{w_1, \dots, w_n\}$ be a basis of W. By the inductive hypothesis, there exists an $f \in R$ that is a projection of V onto $\sum_{i=2}^{n} w_i k$. Write

$$f(w_1) = \sum_{i=2}^{n} w_i \alpha_i \quad (\text{where } \alpha_i \in k),$$

and consider the vector

$$w = w_1 - \sum_{i=2}^{n} w_i \alpha_i \in W.$$

Note that $\{w, w_2, \ldots, w_n\}$ is a basis of W, and that

$$f(w) = f(w_1) - \sum_{i=2}^{n} f(w_i)\alpha_i = f(w_1) - \sum_{i=2}^{n} w_i\alpha_i = 0.$$

By the case we have already handled, there exists $g \in R$ which is a projection of V onto wk. *We claim that*

$$h := f + g - gf \in R$$

is a projection of V onto W. In fact, for $2 \leq i \leq n$:

$$h(w_i) = f(w_i) + g(w_i) - gf(w_i) = w_i + g(w_i) - g(w_i) = w_i, \quad \text{and}$$
$$h(w) = f(w) + g(w) - gf(w) = g(w) = w.$$

Finally,

$$h(V) \subseteq f(V) + g(V) = \sum_{i=2}^{n} w_i k + wk = W.$$

This completes the proof of our claim.

We finish by proving the equation for the socle. If $r \in R\backslash\{0\}$ has finite rank m, then, by Exercise 16,

$$r = r_1 + \cdots + r_m$$

where each $r_i \in R$ has rank 1. By Exercise 17, we have $r_i \in \mathrm{soc}(R)$ for all i, so $r \in \mathrm{soc}(R)$. This shows that

$$\{r \in R : \mathrm{rank}(r) < \infty\} \subseteq \mathrm{soc}(R).$$

For the reverse inclusion, it suffices to show that any minimal left ideal $\mathfrak{A}$ of R consists of linear transformations of finite rank. By Exercise 17, $\mathfrak{A} = Re$ where $e \in R$ has rank 1. Therefore, any $a \in \mathfrak{A}$ has the form re for some $r \in R$. In particular, $\mathrm{rank}(a) \leq \mathrm{rank}(e) = 1$.

Comment. The proof of (2) $\Longrightarrow$ (4) actually shows quite a bit more: *If $e \in R$ has finite rank $m > 0$, then, for any finite-dimensional k-subspace $W \subseteq V$, the ideal (e) generated by e in R contains a projection of V onto W.* In fact, first decompose e into $e_1 + \cdots + e_m$ where each $e_i \in Re$ has rank 1 (cf. Exercise 16). Since $(e_i) \subseteq (e)$, we are reduced to the case $\mathrm{rank}(e) = 1$. In view of the proof of Exercise 17(1), we may further assume that $e = e^2$. In this case, the proof of (2) $\Longrightarrow$ (4) gives what we want, since $aeb \in (e)$, and we may choose $f, g \in (e)$ as well, so that the final choice

$$h = f + g - gf$$

is in (e), as desired.

Left primitive rings R with $\operatorname{soc}(R) \neq 0$ have a *unique* faithful simple left module, and are always right primitive, according to FC-(11.11). These rings were very well-studied in the classical literature. The main result in this direction is that such rings always arise from "a pair of dual vector spaces": see, e.g. Jacobson's "Structure of Rings," p. 75, Amer. Math. Soc. Colloq. Publ., Vol. *37*, 1956.

Ex. 11.19. Show that a ring R is left primitive iff R is prime and R has a faithful left module M of finite length.

Solution. We need only prove the "if" part. Let

$$0 = M_0 \subseteq M_1 \subseteq \cdots \subseteq M_n = M$$

be a composition series for M. Then $I_j = \operatorname{ann}(M_j/M_{j-1})$ are ideals in R with

$$I_1 \cdots I_n M_n \subseteq I_1 \cdots I_{n-1} M_{n-1} \subseteq \cdots \subseteq I_1 M_1 \subseteq M_0 = 0.$$

Since $M = M_n$ is faithful, we have $I_1 \ldots I_n = 0$ in R, and hence $I_j = 0$ for some j. Now M_j/M_{j-1} is a simple faithful left R-module, so R is left primitive.

Ex. 11.20. For any division ring k with center C, show that the following are equivalent:

(1) $k[x]$ is left primitive,
(2) $k[x]$ is right primitive,
(3) there exists $0 \neq f \in k[x]$ such that $k[x]f \cap C[x] = 0$.

Solution. Note first that

$$0 \neq gf \in C[x] \Longrightarrow gf = fg.$$

(Multiply by f from the left and cancel off f from the right.) This implies that (3) is a left-right symmetric condition. Therefore, it suffices to prove (1) $\Longleftrightarrow$ (3). Let $R = k[x]$ and note that $Z(R) = C[x]$.

(1) $\Longrightarrow$ (3). Since R is a principal left ideal domain, a faithful simple left R-module V can be represented as R/Rf, where $f \in R$ is necessarily nonzero. By Exercise 5*, the faithfulness of V implies that

$$0 = Rf \cap Z(R) = Rf \cap C[x].$$

(The polynomial f here is irreducible, but this fact is not needed.)

(3) $\Longrightarrow$ (1). Let $V = R/Rf$, where f is as in (3). Since $_k V$ is finite-dimensional, $_R V$ has a composition series. By Exercise 1.16(3), the hypothesis that $Rf \cap C[x] = 0$ implies that $\operatorname{core}(Rf) = 0$, so by Exercise 5*, $_R V$ is faithful. Since R is a domain and hence a prime ring, Exercise 19 shows that R is left primitive.

Comment. The conditions (1), (2), (3) above are satisfied by some division rings k, and not by others. For instance, if k is not algebraic over C, then the conditions hold, by FC-(11.14). But if $\dim_C k < \infty$, then the conditions do not hold, by Exercise 16.9 below. Both of these results are special cases of a theorem of Amitsur and Small (Israel J. Math. *31* (1978), 353–358), which gives a criterion for a polynomial ring

$$R_n = k[t_1, \ldots, t_n]$$

to be left primitive. According to this theorem, R_n *is left primitive iff some matrix ring* $M_r(k)$ *contains a field of transcendence degree n over C.* The proof of this theorem makes crucial use of a noncommutative version of Hilbert's Nullstellensatz, to the effect that any simple left R_n-module is finite-dimensional over k.

Ex. 11.21. Let $_RV$ be a nonsemisimple module, and $k = \operatorname{End}(_RV)$. Show that R may not act densely on V_k. (In other words, the Jacobson-Chevalley Density Theorem FC-(11.6) may not hold for nonsemisimple modules.)

Solution. To construct a counterexample, let k be a field, and let $R = \begin{pmatrix} k & k \\ 0 & k \end{pmatrix}$ act on $V = \begin{pmatrix} k \\ k \end{pmatrix} = k^2$ by left matrix multiplication. Then $_RV$ is not semisimple and $\operatorname{End}(_RV) = k$ (cf. FC-p. 110). But no element of R can take $\begin{pmatrix} 1 \\ 0 \end{pmatrix}$ to $\begin{pmatrix} 0 \\ 1 \end{pmatrix}$, so R is not even 1-transitive. For an even simpler counterexample, take $R = \mathbb{Z}$, $V = \mathbb{Q}$, with $k = \operatorname{End}(_\mathbb{Z}\mathbb{Q}) = \mathbb{Q}$. Here, no element of R can take 1 to 1/2 in V.

§12. Subdirect Products and Commutativity Theorems

Subdirect product representations of rings offer a nice way to deal with their structure theory. A *subdirect product representation* of R by other rings R_i is an injective ring homomorphism $\varepsilon: R \to \prod_i R_i$ for which each "projection map" $R \to R_i$ is onto. Some basic facts are, for instance:

(1) $R \neq 0$ is semiprimitive iff R is a subdirect product of left primitive rings,
(2) $R \neq 0$ is semiprime iff R is a subdirect product of prime rings, and
(3) $R \neq 0$ is reduced iff R is a subdirect product of domains.
(4) Any $R \neq 0$ is a subdirect product of subdirectly irreducible rings ("Birkhoff's Theorem").

Here, a *subdirectly irreducible* ring is a nonzero ring R for which there is no "nontrivial way" of expressing R as a subdirect product (see FC-(12.1)).

These are rings R which have an ideal $L \neq 0$ that is contained in all other nonzero ideals. We call L the "little ideal" of R; in the literature, L is also known as the "heart" of R.

The role of subdirect product representations in the general structure theory of rings is explained in detail in FC-p. 196. One of the best known applications of this is to the study of commutativity theorems in ring theory. Although "nowadays pretty outré stuff" according to L. Small[2], these commutativity theorems have remained an excellent initiation to the subject for generations of fledgling ring theorists. Small himself recalled his delight in learning about these classical theorems; there will certainly be others to come. Speaking from my own teaching experience, even the slowest and the most timid student in algebra can be counted upon to take out pencil and paper and to dabble at the proof of an intriguing commutativity theorem!

Without going overboard, this section offers a few elementary commutativity exercises (see (4) and (7)–(10)). The proofs for the cases $n = 2, 3, 4, 5, 6$ of Jacobson's "$a^n = a$" theorem (FC-(12.10)) are the best I have known so far. The *Comment* offered after Exercise 8A for rings without identity is possibly a bit surprising.

Exercises 5 and 6A, 6B, 6C deal with a subclass of von Neumann regular rings which are known as *strongly regular* (or *abelian regular*) rings. These rings were discovered and named by Arens and Kaplansky, in their study of topological representations of algebras (c. 1948). In many ways, this study is akin to the study of subdirect product representations of rings. Strongly regular rings also arise in the context of rings of operators. They turn out to be always *unit-regular* in the sense of Exercise 4.14B: see Exercise 6C below.

Exercises for §12

Ex. 12.0. (1) Characterize rings R that are subdirect products of fields.
(2) Characterize rings S that can be embedded into direct products of fields.
(3) Characterize rings T that can be embedded into a direct product of $\mathbb{F}_2$'s and $\mathbb{F}_3$'s.

Solution. (1) Since fields are primitive, any subdirect product of fields is (commutative and) semiprimitive by FC-(12.5). Conversely, let R be a commutative semiprimitive ring. By FC-(12.5) again, R is a subdirect product of left primitive rings $\{R_i\}$. Since the R_i's are commutative, they are fields by FC-(11.8), so R is a subdirect product of fields.

(2) If $S \subseteq \prod_i S_i$ where the S_i's are fields, clearly S is commutative and reduced. Conversely, if S is any commutative reduced ring, then $(0) = \bigcap_i \mathfrak{p}_i$

---

[2] See his review of FC in the Amer. Math. Monthly *100* (1993), 698–699.

where $\{\mathfrak{p}_i\}$ are the prime ideals of S. It follows that S can be embedded into $\prod_i S_i$ where S_i is the quotient field of $S/\mathfrak{p}_i$.

(3) If $T \subseteq \prod_i T_i$ where each T_i is either $\mathbb{F}_2$ or $\mathbb{F}_3$, clearly, every element $t \in T$ satisfies $t^2 = \pm t$. Conversely, let T be any ring with the property that $t^2 = \pm t$ for every $t \in T$. Then $t^2 = 0$ implies $t = 0$, so T is a reduced ring. By FC-(12.7), we have a subdirect product representation

$$(\varphi_i): \ T \longrightarrow \prod_i T_i$$

where the T_i's are domains. Since $\varphi_i(T) = T_i$, we have $t_i^2 = \pm t_i$ for every $t_i \in T_i$, so $t_i \in \{0, \pm 1\}$. This implies that $T_i \cong \mathbb{F}_2$ or $T_i \cong \mathbb{F}_3$, so T is a subdirect product of $\mathbb{F}_2$'s and $\mathbb{F}_3$'s. (In particular, T must be commutative.)

Ex. 12.1. Let R be a subdirectly irreducible ring. Show that if R is semiprimitive (resp. semiprime, reduced), then R is left primitive (resp. prime, a domain). In particular, show that R is left primitive iff R is right primitive.

Solution. First assume R is semiprimitive. By FC-(12.5), there exists a subdirect product representation

$$\varepsilon = (\varepsilon_i): \ R \longrightarrow \prod_i R_i$$

where the R_i's are left primitive rings. Since R is subdirectly irreducible, some $\varepsilon_i : R \to R_i$ is an isomorphism, so R is left primitive. The two other cases are handled in the same way, using FC-(12.5) again, and FC-(12.7). The last statement in the Exercise now follows from left-right symmetry.

Alternative Solution. We offer here a direct proof of the Exercise without "resolving" R into a subdirect product. Let $L \neq 0$ be the "little ideal" of R (see FC-p. 192).

(1) Assume R is semiprimitive, but not left primitive. Then, for any maximal left ideal $\mathfrak{m} \subset R$, $R/\mathfrak{m}$ cannot be faithful, so $\mathrm{ann}(R/\mathfrak{m})$ must contain L. This implies that $L \subseteq \mathfrak{m}$. But then

$$L \subseteq \bigcap \mathfrak{m} = \mathrm{rad}\,R = 0,$$

a contradiction.

(2) Assume R is semiprime, but not prime. Then there exist nonzero ideals I, J with $IJ = 0$. But then $L \subseteq I$, $L \subseteq J$, and we have $L^2 \subseteq IJ = 0$, a contradiction to the fact that R is semiprime.

(3) Assume R is reduced. Then R is semiprime, and hence prime by (2). From Exercise 10.3, it follows that R is a domain.

Ex. 12.2. (1) Show that a commutative domain cannot have exactly three ideals.

(2) Show that there exist (noncommutative) domains A which have exactly three ideals.

(3) Give an example of a nonsimple subdirectly irreducible domain.

Solution. Suppose A is a domain with exactly three ideals, (0), A, and L. Then A is not a simple ring, but A is clearly subdirectly irreducible, with L as its little ideal. Such a domain A cannot be commutative, since a commutative subdirectly irreducible domain is a field (cf. FC-(12.4)). Therefore, it suffices to construct a domain A as in (2). The construction of A below follows an example in 1.3.10 of the book "Noncommutative Noetherian Rings" by McConnell and Robson (AMS, 2001).

   Let k_0 be any field of characteristic 0, and R be the (first) Weyl algebra over k_0, with generators x, y and relation $xy - yx = 1$. Letting δ be the differentiation operator d/dy on $k : = k_0[y]$, we think of R as the skew polynomial ring $k[x; \delta]$ (cf. FC-(1.9)). Exploiting the fact that R is a simple k_0-domain (FC-(3.17)), we propose to prove:

Lemma. $L : = Rx = \{a_m x^m + \cdots + a_1 x : \ a_i \in k\}$ *is a simple domain without identity.*

   Once this lemma is proved, it follows easily that

(4) $A : = L \oplus k_0 = \{a_m x^m + \cdots + a_1 x + a_0 : \ a_0 \in k_0, \ a_1, \ldots, a_m \in k\}$

is a k_0-domain with exactly three ideals, (0), A, and L. (Use the fact that any ideal in A intersects L at (0) or L.)

Proof. Let J be any nonzero ideal in the ring (without identity) L. Since R is a simple ring (*loc. cit.*), we have $RxJR = R$. Right multiplying by x, we get

(5) $$L = Rx = RxJRx = LJL \subseteq J,$$

and therefore $J = L$. This shows that L is a simple domain (without identity), as desired.

Comment. (A) The ring A constructed in (4) is in fact the "idealizer" of the left ideal $L = Rx$ in R, that is:

$$A = I_R(L) : = \{h \in R : \ Lh \subseteq L\}.$$

The inclusion "$\subseteq$" is clear since $k_0 \subseteq Z(R)$. For "$\supseteq$", note that if

$$h = a_m x^m + \cdots + a_0 \in I_R(L) \quad (\text{where } a_i \in k),$$

then $a_0 \in I_R(L)$ too, and $xa_0 = a_0 x + \delta(a_0)$ implies that $\delta(a_0) = 0$. Hence $a_0 \in k_0$ and $h \in Rx + k_0 = A$.

(B) Since A is a domain with only three ideals (0), A, and L, we must have $L = L^n$ for all $n \geq 1$. In fact, applying (5) to $J = L$, we have $L = L^3$. From this, it follows that $L = L^n$ for all $n \geq 1$. Note that A is a good example of a noncommutative domain satisfying the properties (1) and (2) of Exercise 10.5.

(C) Note that the ring A above is a primitive domain. To see this, note that, since $x + 1$ has no inverse in R, $A(x + 1)$ is contained in a maximal left ideal M of A. We check easily that A/M is a simple faithful left A-module (since its annihilator cannot be L), so A is a left primitive domain. By considering $(x + 1)A$, we see similarly that A is a right primitive domain. (Of course these conclusions are also clear from Exercise 1, since A is subdirectly irreducible, and $x \notin \mathrm{rad}\, A$ implies that $\mathrm{rad}\, A = 0$.)

(D) It is known that R is a (2-sided) noetherian ring. As it turns out, the ring A is also noetherian. For a proof of this, see 1.3.10 in the book of McConnell and Robson referred to above.

Ex. 12.3A. Let k be a field of characteristic p and G be a finite p-group, where p is a prime. Show that the ring $A = kG$ is subdirectly irreducible, with its little ideal given by $k\sigma$ where $\sigma = \sum_{g \in G} g \in A$.

Solution. Every nonzero left ideal of A contains a minimal left ideal. Therefore, we are done if we can prove the following stronger fact: *the only minimal left ideal of A is $k\sigma$*. Let I be any minimal left ideal. By FC-(8.4), $_A I$ is the 1-dimensional trivial G-module. Let $I = k\alpha$, where $\alpha \neq 0$. Since $g\alpha = \alpha$ for all $g \in G$, $\alpha = a\sigma$ for some $a \in k^*$. Therefore, $I = k \cdot a\sigma = k\sigma$.

Comment. In the terminology of §19, $A = kG$ is a local ring with its unique maximal (left) ideal given by the augmentation ideal $\mathfrak{m} = \ker \varepsilon$, where $\varepsilon : kG \to k$ is defined by

$$\varepsilon \left(\sum \alpha_g g \right) = \sum \alpha_g.$$

Since $\alpha\sigma = \sigma\alpha = \varepsilon(\alpha)\sigma$, we have

$$\mathrm{ann}_\ell(\sigma) = \mathrm{ann}_r(\sigma) = \mathfrak{m},$$

so $\mathfrak{m}$ is exactly the set of zero-divisors in A (together with zero). By FC-(8.8), we have in fact $\mathfrak{m}^{|G|} = 0$.

Ex. 12.3B. The following exercise appeared in an algebra text: "The zero-divisors in a subdirectly irreducible ring (together with zero) form an ideal." Give a counterexample! Then suggest a remedy.

Solution. Take R to be $\mathbb{M}_n(k)$, where k is a field and $n \geq 2$. Then R is a simple ring, and it is subdirectly irreducible (with little ideal R). There are various (left, right, 2-sided) zero-divisors in R, but surely they don't form an ideal (or even a 1-sided ideal). To rectify the statement, assume that R

is *commutative*. Let I be the set of zero-divisors in R, together with 0. To see that I is an ideal, it suffices to show that I is closed under addition. Let $a_1, a_2 \in I\backslash\{0\}$. Then $\mathrm{ann}(a_i) \neq 0$, and so $\mathrm{ann}(a_i)$ contains the little ideal, say L, of R. Therefore, $La_i = 0$ for $i = 1, 2$, and we have $L(a_1 + a_2) = 0$. This clearly implies that $a_1 + a_2 \in I$.

Comment. In FC-(12.4), the commutative subdirectly irreducible rings R that are reduced are completely determined: they turn out to be just fields. If R is *not* assumed to be reduced, its structure is more interesting. In the next exercise, we offer more information on the ideal I of 0-divisors in a nonreduced commutative subdirectly irreducible ring R.

Ex. 12.3C. (McCoy) Let R be a commutative subdirectly irreducible ring which is not a field. Let L be the little ideal of R. Prove the following statements:

(1) $L = aR$ for some a with $a^2 = 0$.
(2) $\mathfrak{m} : = \mathrm{ann}(a)$ is a maximal ideal of R.
(3) $\mathfrak{m} = I$ (the set of all 0-divisors of R together with 0).
(4) $\mathrm{ann}(\mathfrak{m}) = aR$, and $a \in \mathrm{Nil}(R) \subseteq \mathfrak{m}$.
(5) If R is noetherian, then $\mathrm{Nil}(R) = \mathfrak{m}$, and R is an artinian local ring.

Solution. (1) Fix any nonzero element $a \in L$. Clearly, $L = aR$. By Brauer's Lemma $(FC$-(10.22)$)$, we have either $L^2 = 0$ or $L = eR$ for some idempotent $e \in R$. In the latter case, we must have $e \neq 1$ (for otherwise $L = R$, and R is a field). But then $L \cap (1 - e)R = 0$ gives a contradiction. Therefore, we must have $L^2 = 0$.

(2) Since $a \neq 0$, $\mathfrak{m} \neq R$. Let $b \notin \mathfrak{m}$. Then $ab \neq 0$ implies $a \in abR$. Writing $a = abc$ for some $c \in R$, we have $a(1 - bc) = 0$ so $1 - bc \in \mathfrak{m}$. This proves (2).

(3) It suffices to show that $I \subseteq \mathfrak{m}$. Let $r \in I$. From $\mathrm{ann}(r) \neq 0$, we see that $aR \subseteq \mathrm{ann}(r)$. This means that $ar = 0$, so $r \in \mathfrak{m}$.

(4) We need only show that $\mathrm{ann}(\mathfrak{m}) \subseteq aR$. Say $x\mathfrak{m} = 0$, where $x \neq 0$. Write $a = xy$ for some $y \in R$. Then $y \notin \mathfrak{m}$ (since $x\mathfrak{m} = 0$) so by (2), $yz - 1 \in \mathfrak{m}$ for some $z \in R$. Multiplying by x, we see that

$$x = xyz = az \in aR.$$

The inclusion $\mathrm{Nil}(R) \subseteq \mathfrak{m}$ is trivial since $\mathfrak{m}$ is a prime ideal.

(5) Let $b \in \mathfrak{m}$. Using the ACC on ideals, we see that $\mathrm{ann}(b^n) = \mathrm{ann}(b^{n+1})$ for some $n \geq 1$. If $b^n \neq 0$, we would have $a = xb^n$ for some x. Then $0 = ab = xb^{n+1}$ implies $0 = xb^n = a$, a contradiction. Therefore, $b^n = 0$, and we have $b \in \mathrm{Nil}(R)$. Since now $\mathrm{Nil}(R) = \mathfrak{m}$, this is the *only* prime ideal in R, so the noetherian ring R is in fact a local artinian ring (see *Comment* on Exercise 10.4).

Comment. (a) If R is not assumed to be noetherian, the inclusion $\mathrm{Nil}(R) \subseteq \mathfrak{m}$ in (4) above may be strict. For an example of such a commutative sub-directly irreducible ring, see McCoy's paper in Duke Math. Journal, vol. *12* (1945), pp. 382–387, on which the above exercise was based.

(b) Since any nonzero ring is a subdirect product of subdirectly irreducible rings (Birkhoff's Theorem: *FC*-(12.3)), it follows immediately from (5) above that *any commutative noetherian ring is a subdirect product of commutative local artinian rings.*

Ex. 12.3D. Let R be a commutative artinian ring. Show that R is subdirectly irreducible iff $\mathrm{soc}(R)$ is a minimal ideal.

Solution. If $\mathrm{soc}(R)$ is a minimal ideal, then any nonzero ideal contains $\mathrm{soc}(R)$, so R is subdirectly irreducible (with $\mathrm{soc}(R)$ as its little ideal). Conversely, if R is subdirectly irreducible with little ideal L, then L is the only minimal ideal of R. Hence $\mathrm{soc}(R) = L$, and it is minimal.

Comment. An explicit class of examples for the rings in this exercise are group rings R of finite (abelian) p-groups G over a field k of characteristic p. Here,

$$\mathrm{soc}(R) = L = k\sigma \quad \text{where} \quad \sigma = \sum_{g \in G} g$$

(see Ex. 12.3A). Other examples are given by $R = A/\pi^m A$, where π is a nonzero prime element in a principal ideal domain A, and $m \geq 1$. Here,

$$\mathrm{soc}(R) = L = \pi^{m-1} A / \pi^m A.$$

Ex. 12.3E. (Cf. the proof of Birkhoff's Theorem in *FC*-(12.3).) For $0 \neq a \in R$ (for any ring R), let I be any ideal of R maximal with respect to the property that $a \notin I$. Then the factor ring R/I is subdirectly irreducible (since its nonzero ideals all contain the nonzero coset $a + I$). If $R = \mathbb{Z}$, compute the ideals I arising in this manner from a given nonzero integer a, and classify the associated subdirectly irreducible rings R/I.

Solution. Write $a = \pm p_1^{r_1} \cdots p_n^{r_n}$, where the p_i's are distinct primes, and $r_i \geq 1$. Since $\mathbb{Z}$ is a PID, we have $I = b\mathbb{Z}$ where $b \nmid a$ but any proper divisor of b divides a. Clearly, $b \neq 0$. Taking b to be positive, and using the unique factorization theorem in $\mathbb{Z}$, we see that either b is a prime p distinct from the p_i's or $b = p_i^{r_i+1}$ for some i. Therefore, the subdirectly irreducible rings R/I arising from our construction are precisely the fields $\mathbb{Z}/p\mathbb{Z}$ for the primes p distinct from the p_i's, and the rings $\mathbb{Z}/p_i^{r_i+1}\mathbb{Z}$. For instance, if $a = 60 = 2^2 \cdot 3 \cdot 5$, the possibilities for R/I are: $\mathbb{Z}/p\mathbb{Z}$ for primes $p \geq 7$, and $\mathbb{Z}/8\mathbb{Z}$, $\mathbb{Z}/9\mathbb{Z}$, $\mathbb{Z}/25\mathbb{Z}$.

Comment. Certainly, the description of the rings R/I are compatible with the predictions in the Exercises 12C–D above.

Ex. 12.3F. Show that rings of the form $R[x]$ or $R[[x]]$ are never subdirectly irreducible, while a ring of the form $\mathbb{M}_n(R)$ is subdirectly irreducible iff R is.

Solution. The second conclusion is clear since the ideals of R are in one-one correspondence with those of $\mathbb{M}_n(R)$, by $I \leftrightarrow \mathbb{M}_n(I)$. (This implies that I_0 is a little ideal of R iff $\mathbb{M}_n(I_0)$ is a little ideal of $\mathbb{M}_n(R)$.)

Next, suppose a polynomial ring $A = R[x]$ is subdirectly irreducible. Take $f = r_0 + r_1 x + \cdots + r_n x^n$ in the little ideal of A, with $r_n \neq 0$. Then f lies in the nonzero ideal Ax^{n+1}. This implies that $\deg(f) \geq n+1$, a contradiction.

Finally, suppose a power series ring $B = R[[x]]$ is subdirectly irreducible. Take $g = r_n x^n + r_{n+1} x^{n+1} + \cdots$ in the little ideal of B, with $r_n \neq 0$. Then g lies in the nonzero ideal Bx^{n+1}. This implies that g does not involve an x^n-term, a contradiction.

Ex. 12.4. The following exercise appeared in an algebra text: "Let $n \geq 2$ and a, b be elements in a ring D. If

$$(*) \qquad a^n - b^n = (a - b)\left(a^{n-1} + a^{n-2}b + \cdots + ab^{n-2} + b^{n-1}\right),$$

then $ab = ba$." Show that

(1) this is true for $n = 2$;
(2) this is false for $n = 3$, even in a division ring D;
(3) if $(*)$ holds for $n = 3$ for *all* a, b in a ring D, then indeed $ab = ba$ for all a, b.

Solution. (1) For $n = 2$, the hypothesis gives

$$a^2 - b^2 = (a - b)(a + b) = a^2 - b^2 + ab - ba, \quad \text{so} \quad ab = ba.$$

(3) For $n = 3$, the hypothesis amounts to

$$a^3 - b^3 = (a - b)(a^2 + ab + b^2) = a^3 - b^3 + a(a + b)b - ba(a + b),$$

which simply means that b commutes with $a(a + b)$. *If this also holds with b replaced by $b + 1$*, then we have

$$a(a + b + 1)(b + 1) = (b + 1)a(a + b + 1).$$

Cancelling $a(a + b + 1)$ from the two sides and using $a(a + b)b = ba(a + b)$, we get $ab = ba$.

(2) By Exercise 12 below, there exists a division ring D, say of characteristic not 2, in which there are elements $a \neq c$ such that $a^2 - 2ac + c^2 = 0$. For $b := -2c$,

$$a(a + b) = a(a - 2c) = -c^2$$

certainly commutes with b, so $(*)$ is satisfied for $n = 3$. However, a, b do not commute. For, if a, b commute, then a, c also commute, and

$$0 = a^2 - 2ac + c^2 = (a - c)^2$$

implies that $a = c$, which is not the case. (We won't bother about constructing an example in characteristic 2.)

Ex. 12.5. Call a ring R *strongly* (von Neumann) *regular* if, for any $a \in R$, there exists $x \in R$ such that $a = a^2 x$. If R is strongly regular, show that R is semiprimitive, and that R is left primitive iff it is a division ring. Using this, show that any strongly regular ring is a subdirect product of division rings.

Solution. Given any element $a \in \operatorname{rad} R$, write $a = a^2 x$ for a suitable x. Then $a(1 - ax) = 0$. Since $1 - ax \in U(R)$, we have $a = 0$. This shows that $\operatorname{rad} R = 0$, so R is semiprimitive.

If R is a division ring, of course it is left primitive. Conversely, assume R is (strongly regular and) left primitive. Then it is prime. From this alone, we shall show that R is a division ring.

Step 1. R is reduced. Indeed, if $a \in R$ is such that $a^2 = 0$, then, writing $a = a^2 x$ (for some $x \in R$), we see that $a = 0$.

Step 2. R is a domain. This follows from Exercise 10.3: any reduced prime ring is a domain (and conversely).

Step 3. R is a division ring. Let $a \neq 0$ in R. Then $a = a^2 x$ for some x. Cancelling a yields $ax = 1$. By Exercise 1.2, R is a division ring.

To prove the last conclusion in the Exercise, let $R \neq 0$ be any strongly regular ring. Then R is semiprimitive by the first part of the Exercise, so (by *FC*-(12.5)) it is a subdirect product of a family of left primitive rings R_i. Since each R_i is an epimorphic image of R, R_i is also strongly regular, and hence a division ring by the above.

Comment. The *converse* of the last part of the Exercise is not true! In other words, a subdirect product of division rings (or even fields) need not be strongly regular. For instance, the ring $\mathbb{Z}$ is a subdirect product of the fields $\mathbb{F}_p$ ($p = 2, 3, 5, \ldots$); but of course $\mathbb{Z}$ is not strongly regular.

If R is von Neumann regular and left primitive, R need not be a division ring (or even an artinian ring). An immediate example is given by $\operatorname{End}(V_k)$, where V is an infinite-dimensional vector space over a division ring k. The precise relationship between "regular" and "strongly regular" will be given in the next exercise.

Ex. 12.6A. (Arens-Kaplansky, Forsythe-McCoy) Show that the following conditions on a ring R are equivalent:

(1) R is strongly regular;
(2) R is von Neumann regular and reduced;
(3) R is von Neumann regular and every idempotent in R is central;
(4) every principal right ideal of R is generated by a central idempotent.

Solution. (1) $\Rightarrow$ (2). Assume R is strongly regular. We have already observed in the last Exercise that R is reduced. For any $a \in R$, write $a = a^2 x$ where $x \in R$. Then

$$(a - axa)^2 = a^2 + axa^2xa - a^2xa - axa^2$$
$$= a^2 + axa^2 - a^2 - axa^2 = 0, \quad \text{so } a = axa.$$

(2) $\Rightarrow$ (3). We note more generally that, *in any reduced ring R, any idempotent $e \in R$ is central.* In fact, let $f = 1 - e$ and consider any $a \in R$. Then $(eaf)^2 = 0$ so $eaf = 0$, which implies that $ea = eae$. Similarly $ae = eae$, so $ea = ae$ for any $a \in R$.

(3) $\Rightarrow$ (4) is trivial in view of FC-(4.23).

(4) $\Rightarrow$ (1). Let $a \in R$. By (4), $aR = eR$ for a central idempotent $e \in R$. Write $e = ax$, $a = ey$, where $x, y \in R$. Then (1) follows since

$$a^2 x = a \cdot ax = eye = e^2 y = ey = a.$$

Comment. A ring R is called *abelian* if every idempotent in R is central. For instance, any reduced ring is abelian, as shown in (2) $\Rightarrow$ (3). The characterization (3) above shows that *an abelian regular ring is just a strongly regular ring.* For other interesting characterizations of such a ring, see Exercises 22.4B and 23.7.

Ex. 12.6B. (Jacobson, Arens-Kaplansky) Let R be a reduced algebraic algebra over a field k. Let $r \in R$, and $\varphi(x) \in k[x]$ be its minimal polynomial over k.

(1) Show that any root $a \in k$ of $\varphi(x)$ is a simple root.
(2) Show that R is strongly regular.

Solution. (1) Write $\varphi(x) = (x - a)^i \psi(x)$ with $\psi(x) \in k[x]$ and $\psi(a) \neq 0$. Then

$$[(r - a)\psi(r)]^i = (r - a)^i \psi(r)^i = \varphi(r)\psi(r)^{i-1} = 0$$

implies that $(r - a)\psi(r) = 0$. Since $\varphi(x)$ is the minimal polynomial of r, we must have $i = 1$.

(2) For $r \in R$, we would like to show that $r \in r^2 R$. Let

$$\varphi(x) = x^n + a_{n-1}x^{n-1} + \cdots + a_0 \in k[x]$$

be the minimal polynomial of r over k. If $a_0 \neq 0$, then $r \in U(R)$. In this case, the desired conclusion is trivial. If $a_0 = 0$, then by (1), $a_1 \neq 0$ and we have

$$r = -a_1^{-1}(r^n + \cdots + a_2 r^2) \in r^2 R.$$

Alternative Proof for (2). For $r \in R$, $A : = k[r]$ is a finite-dimensional reduced commutative k-algebra. By Wedderburn's Theorem, $A \cong \prod_{i=1}^{n} K_i$
where the K_i's are finite field extensions of k. Such a ring is clearly strongly
regular, so $r \in r^2 A \subseteq r^2 R$.

Comment. If the ground field k in this exercise is a *finite* field, the k-
algebra R above must in fact be commutative. This is a result of Jacobson:
see Exercise 13.11 below. In general, however, R need not be commutative.
For instance, R may be a finite-dimensional k-division algebra of dimension
> 1 over its center k.

Ex. 12.6C. (Ehrlich) Show that any strongly regular ring R is unit-regular
in the sense of Exercise 4.14B, that is, for any $a \in R$, there exists $u \in U(R)$
such that $a = aua$.

Solution. For $a \in R$, write $a = a^2 x$ where $x \in R$. By the above Exercise,
we have also $a = axa$, and $e : = ax$ is a central idempotent. Consider now
any representation of R as a subdirect product of division rings, say (φ_i) :
$R \to \prod D_i$. Since $\varphi_i(e) = \varphi_i(a)\varphi_i(x)$ is either 0 or 1 in D_i, we see that
$\varphi_i(e) = \varphi_i(x)\varphi_i(a)$ also (for every i), and hence $e = xa \in R$. Now let $f =
1 - e$, and $u = ex + f$, $v = ea + f$ in R. Since e is central, we have

$$vu = e^2 ax + f^2 = e + f = 1,$$

and similarly $uv = 1$. Therefore $u = v^{-1} \in U(R)$. From $a = a^2 x = ae$, we
get $af = 0$, so now

$$aua = a(ex + f)a = axa = a,$$

as desired. (For an alternative proof, see Exercise 22.2.)

Comment. More generally, Fisher and Snyder have shown that, *if R is any*
von Neumann regular ring for which there exists an integer $n > 0$ such
that $b^n = 0$ for any nilpotent element $b \in R$, then R is unit-regular. The
exercise above corresponds to the case when $n = 1$ (since "strongly regular"
amounts to R being reduced). For a proof of the Fisher-Snyder result, see
(7.11) in Goodearl's book "von Neumann Regular Rings," 2nd ed., Krieger
Publ. Co., Malabar, Florida, 1991. Summarizing, we have for von Neumann
regular rings R:

strongly regular $\Rightarrow$ bounded index $\Rightarrow$ unit-regular $\Rightarrow$ Dedekind-finite.

It is known that *each* of these implications is irreversible. All of these conditions should be thought of as finiteness conditions on R.

In the *Comment* on Ex. 4.17, we have briefly encountered the notion of a
strongly π-regular ring. More information on such rings can be found in the
exercises in §23. For von Neumann regular rings, the notion of strongly π-
regular rings can be added to the above display by placing it right between

"bounded index" and "unit-regular." The fact that the additional forward implications hold depends on results of Azumaya, Burgess-Stephenson, and Goodearl-Menal.

Ex. 12.7. In the solution to Exercise 6A, we have given an easy proof for the fact that any idempotent e in a reduced ring is central. Now prove the following more general fact: in any ring R, an idempotent $e \in R$ is central if (and only if) it commutes with all nilpotent elements in R.

Solution. We need to extend the earlier proof only slightly. Let $a \in R$ and write $f = 1 - e$. Since $(eaf)^2 = 0$, the hypothesis (for the "if" part) implies that

$$eaf = e(eaf) = (eaf)e = 0,$$

so $ea = eae$. Similarly, $ae = eae$, so $ea = ae$ for all $a \in R$.

Comment. For another result of the same genre, see Exercise 22.3A below.

Ex. 12.8A. Prove the following commutativity theorem without using the machinery of subdirect products: Let R be a ring such that $(ab)^2 = a^2b^2$ for any elements $a, b \in R$. Then R is commutative.

Solution. Replacing b by $1 + b$ in the equation above, we get

$$a^2(1 + 2b + b^2) = (a + ab)^2 = a^2 + (ab)^2 + a^2b + aba,$$

so we have $a^2b = aba$. Now, replacing a by $1 + a$ in this equation, we get

$$(1 + 2a + a^2)b = (1 + a)b(1 + a) = b + ab + ba + aba.$$

Cancelling b and $a^2b = aba$, we have $ab = ba$.

Comment. The proof above made substantial use of the identity element 1. Indeed, if R is a ring without 1, the conclusion in the Exercise need not hold. To construct an example, fix a prime p, and let R be the ring without identity with cardinality p^2 constructed in Exercise 1.10(b). For any $a = (w, x)$ and $b = (y, z)$ in R (where $w, x, y, z \in \mathbb{F}_p$), we have:

$$(ab)^2 = (wy, wz)^2 = (w^2y^2, w^2yz),$$
$$a^2b^2 = (w^2, wx)(y^2, yz) = (w^2y^2, w^2yz).$$

Therefore, $(ab)^2 = a^2b^2$ holds for all $a, b \in R$. But we knew that R was not commutative!

Ex. 12.8B. Let R be a ring possibly without an identity. If $a^2 - a \in Z(R)$ for every $a \in R$, show that R is a commutative ring.

Solution. We have to work a bit harder here since R may not have an identity. Noting that

$$(a+b)^2 - (a+b) - (a^2 - a) - (b^2 - b) = ab + ba,$$

we have $ab + ba \in Z(R)$ for all a, b. In particular,

$$a(ab + ba) = (ab + ba)a,$$

so $a^2 b = ba^2$. This implies that $a^2 \in Z(R)$ for all a. Therefore,

$$(a^2 - a)b = a^2 b - ab = ba^2 - ab.$$

On the other hand,

$$(a^2 - a)b = b(a^2 - a) = ba^2 - ba.$$

Comparing these two equations, we conclude that $ab = ba$.

Ex. 12.8C. Let R be a ring such that, for any $a \in R$, $a^{n(a)} = a$ for some integer $n(a) > 1$. By Jacobson's Theorem $(FC\text{-}(12.10))$, R is a commutative ring. Show that R has Krull dimension 0.

Solution. The conclusion simply means that any prime ideal $\mathfrak{p} \subset R$ is a maximal ideal. Consider the quotient ring $\overline{R} = R/\mathfrak{p}$, which is an integral domain. For any $\overline{a} \neq 0$ in $\overline{R}$, we have $\overline{a}^n = \overline{a}$ for some $n > 1$ (depending on $\overline{a}$). Cancelling $\overline{a}$, we have $\overline{a}^{n-1} = 1 \in \overline{R}$, so $\overline{a} \in U(\overline{R})$. This shows that $\overline{R}$ is a field, so $\mathfrak{p}$ is a maximal ideal.

Comment. Of course, we could have simply referred to (B) $\Longrightarrow$ (A) in Exercise 4.15.

Ex. 12.9. Give elementary proofs for the following special cases of Jacobson's Theorem $(FC\text{-}(12.10))$: Let $n \in \{2, 3, 4, 5\}$ and let R be a ring such that $a^n = a$ for all $a \in R$. Then R is commutative.

Solution. *First assume $n = 2$.* Then $-1 = (-1)^2 = 1$, so $2R = 0$. For all $a, b \in R$, we have

$$a + b = (a + b)^2 = a^2 + b^2 + ab + ba.$$

Therefore, $ab = -ba = ba$. (This simple proof for the commutativity of a Boolean ring is familiar to all beginning students in abstract algebra.)

Now assume $n = 3$. We first note that $b^2 = 0 \Rightarrow b = b^3 = 0$, so R is reduced. Next, $(a^2)^2 = a^4 = a^2$, so $a^2 \in Z(R)$ by Exercise 7. From

$$1 + a = (1 + a)^3 = 1 + 3a + 3a^2 + a^3,$$

we see that $3a \in Z(R)$. Coupling this with $2a = (a+1)^2 - a^2 - 1 \in Z(R)$, we get $a \in Z(R)$.

Next assume $n = 4$. Then $-1 = (-1)^4 = 1$, so $2R = 0$. As in the case $n = 3$, R is reduced. Since

$$(a^2 + a)^2 = a^4 + a^2 = a + a^2,$$

we have $a^2 + a \in Z(R)$ (again by Exercise 7). Replacing a by $a + b$, we see that $Z(R)$ contains

$$(a + b)^2 + a + b = a^2 + b^2 + ab + ba + a + b,$$

so it also contains $ab + ba$. In particular, $a(ab + ba) = (ab + ba)a$, so $a^2 b = ba^2$. Replacing a by a^2, we see that $a^4 b = ba^4$, which gives $ab = ba$.

Finally, assume $n = 5$. The following proof is based on suggestions of Dan Brown. First, note that R is reduced, $30R = 0$ (since $2 = 2^5 = 32 \in R$), and that $(a^4)^2 = a^8 = a^4$, so $a^4 \in Z(R)$ by Exercise 7. From

$$1 + a = (1 + a)^5 = 1 + 5a + 10a^2 + 10a^3 + 5a^4 + a^5,$$

we get $5a + 10a^2 + 10a^3 \in Z(R)$. Multiplying by 3 then yields $15R \subseteq Z(R)$. Next, note that $(a^4 + a^2)^2 = 2(a^4 + a^2)$, so $2^3(a^4 + a^2)$ is an idempotent (as $(2^3)^2 \cdot 2 = 2^7 = 2^3 \in R$). Therefore, $8(a^4 + a^2) \in Z(R)$. From this, we get $8a^2 \in Z(R)$, so

$$a^2 = 16a^2 - 15a^2 \in Z(R).$$

This yields

$$2a = (a + 1)^2 - a^2 - 1 \in Z(R),$$

so $a = 15a - 7 \cdot 2a \in Z(R)$.

Comment. Another proof for the $n = 5$ case of Jacobson's Theorem can be found on p. 145 of McCoy's Carus Monograph "Rings and Ideals." However, McCoy's proof is more complicated, and made use of the further hypothesis that $5R = 0$.

There are several different proofs of Jacobson's Theorem (FC-(12.10)) in the literature which *do not* rely on the theory of primitive rings. The shortest and the most elementary is perhaps that of J. Wamsley in J. London Math. Soc. *4* (1971), 331–332. Wamsley's proof is a modification of the one given for division rings in FC-§13 using the lemma of Herstein. However, Wamsley's proof was rather terse, so in reading his paper, the reader should perhaps be prepared to fill in a number of details along the way.

Ex. 12.10. Let R be a ring such that $a^6 = a$ for any $a \in R$. Show that $a^2 = a$ for any $a \in R$, i.e. R is a (commutative) Boolean ring.

Solution. Since $-1 = (-1)^6 = 1$, char $R = 2$. Expanding the RHS of $a + 1 = (a + 1)^6$, we get $a^2 = a^4$. Multiplying this by a^2, we get $a^4 = a^6 = a$ and therefore $a = a^2$ for all $a \in R$. (*Moral* : "$n = 6$ is trivial." Surprise!)

Ex. 12.10A. Let R be a ring of characteristic 2, and let n be a fixed 2-power. If $a^{n+1} = a$ for all $a \in R$, show that R is a Boolean ring.

Solution. (We do not assume here that R has an identity.) First note that the binomial coefficients $\binom{n}{k}$ are all even, for $1 < k < n$. (This follows, for instance, by writing n as 2^r and inducting on r.) Therefore, for any commuting pair $x, y \in R$, we have $(x + y)^n = x^n + y^n$. Thus,

$$x^2 + x = (x^2 + x)^{n+1} = (x^2 + x)^n (x^2 + x)$$
$$= (x^{2n} + x^n)(x^2 + x)$$
$$= x^{2n+2} + x^{2n+1} + x^{n+2} + x^{n+1}$$
$$= x^2 + x + x^2 + x = 0,$$

so $x^2 = x$ (for all $x \in R$).

Comment. This problem comes from the *Mathematics Magazine*. Note that the 2-power assumption on n is essential. For instance, $R = \mathbb{F}_{2^s}$ has characteristic 2, and $a^{n+1} = a$ for all $a \in R$ with $n = 2^s - 1$; but R is not Boolean if $s \geq 2$. The characteristic 2 assumption cannot be removed either: if p is a Fermat prime $2^{2^t} + 1$, then for $n = 2^{2^t}$, $a^{n+1} = a^p = a$ for all $a \in R = \mathbb{F}_p$, but R is not Boolean.

Ex. 12.11. Let p be a fixed prime. Following McCoy, define a nonzero ring R to be a *p-ring* if $a^p = a$ and $pa = 0$ for all $a \in R$. Assuming Jacobson's Theorem *FC*-(12.10), show that:

(1) A ring R is a p-ring iff it is a subdirect product of $\mathbb{F}_p$'s;
(2) A finite ring R is a p-ring iff it is a finite direct product of $\mathbb{F}_p$'s.

Solution. (1) First observe that p-rings are closed with respect to the formation of direct products and subrings. Secondly, $\mathbb{F}_p$ is obviously a p-ring. From these, it follows that any subdirect product of $\mathbb{F}_p$'s is a p-ring. Conversely, let R be a p-ring. Then R is commutative by Jacobson's Theorem. Let $\{\mathfrak{m}_i : i \in I\}$ be the maximal ideals of R. Since R is von Neumann regular and hence J-semisimple, we have $\bigcap_i \mathfrak{m}_i = 0$. Therefore, R is a subdirect product of the fields $F_i = R/\mathfrak{m}_i$, which clearly remain p-rings. But in any field the equation $x^p = x$ has at most p solutions, so $F_i \cong \mathbb{F}_p$ for all i.

(2) If R is a finite p-ring, then clearly $n = |I| < \infty$ in the above. By the Chinese Remainder Theorem, we have

$$R \cong \prod_{i=1}^{n} R/\mathfrak{m}_i \cong \prod_{i=1}^{n} \mathbb{F}_p.$$

Conversely, any finite direct product of $\mathbb{F}_p$'s is clearly a finite p-ring.

Ex. 12.12. For any field k, show that there exists a k-division algebra D with two elements $a \neq c$ such that $a^2 - 2ac + c^2 = 0$.

Solution. We take $D = K((x; \sigma))$, the division ring of twisted Laurent series in x, where K is some field extension of k, equipped with an automorphism σ. We shall choose $c = x, a = yx$, where $1 \neq y \in K$ is to be specified. Since

$$a^2 - 2ac + c^2 = yxyx - 2yxx + x^2 = (y\sigma(y) - 2y + 1)x^2,$$

we need only arrange that $y\sigma(y) - 2y + 1 = 0$. For this, we may choose K to be the rational function field $k(y)$ in the variable y. If we define $\sigma \in \mathrm{Aut}_k(K)$ by the fractional linear transformation

$$\sigma(y) = \frac{2y - 1}{y} = 2 - y^{-1},$$

then $y\sigma(y) - 2y + 1 = 0$, as desired. Note that this construction works well even in characteristic 2: in that case, we simply take $\sigma(y) = y^{-1}$, and

$$(yx)^2 + x^2 = (y\sigma(y) + 1)x^2 = 0 \quad \text{in} \quad D.$$

If k is a field of characteristic $p > 0$, we can even choose K above to be an algebraic extension of k. In fact, let K be the algebraic closure of k. Then the equation $y^{p+1} - 2y + 1 = 0$ has a root $y \in K$. If we take σ to be the Frobenius automorphism on K ($\sigma(z) = z^p$), then

$$y\sigma(y) - 2y + 1 = y^{p+1} - 2y + 1 = 0.$$

Comment. Under suitable conditions on the division ring D, $a^2 - 2ac + c^2 = 0$ may imply $a = c$. For instance this is always the case if D is a generalized quaternion division algebra over a field of characteristic not 2. For more details, see Exercise 16.15.

Ex. 12.13. Let R be a nonzero ring such that every subring of R is a division ring. Show that R must be an algebraic field extension of some $\mathbb{F}_p$.

Solution. First we cannot have char $R = 0$, for otherwise $\mathbb{Z} \subseteq R$ and we have a contradiction. Since R itself is a division ring, we must have $\mathbb{F}_p \subseteq R$, where p is a prime. Consider any nonzero element $a \in R$. Then a must be algebraic over $\mathbb{F}_p$, for otherwise $\mathbb{F}_p[a]$ is isomorphic to the polynomial ring $\mathbb{F}_p[x]$, which is not a division ring. Therefore, $\mathbb{F}_p[a]$ is a finite field, and we have $a^{n(a)} = a$ for some $n(a) \geq 2$. By Jacobson's Theorem $(FC\text{-}(12.10))$, R is commutative. Therefore, R is an algebraic field extension of $\mathbb{F}_p$.

Ex. 12.14. Show that a ring R is reduced iff, for any elements $a_1, \ldots, a_r \in R$ and any positive integers $n_1, \ldots, n_r$:

$$(*) \qquad\qquad a_1 \cdots a_r \neq 0 \Longrightarrow a_1^{n_1} \cdots a_r^{n_r} \neq 0.$$

Solution. The "if" part is trivial by using $(*)$ in the case $r = 1$. For the "only if" part, first observe that, in a reduced ring R, $xy \neq 0 \Longrightarrow yx \neq 0$. (For, if $yx = 0$, then $xy \cdot xy = 0$ and hence $xy = 0$ since R is reduced.) More generally, suppose $a_1 \cdots a_r \neq 0$. Then

$$a_1 a_2 \cdots a_r a_1 \cdots a_r \neq 0,$$

so $a_1 a_2 \cdots a_r a_1 \neq 0$, and the observation above gives $a_1^2 a_2 \cdots a_r \neq 0$. Repeating this, we get $a_1^{n_1} a_2 \cdots a_r \neq 0$. It follows that $a_2 \cdots a_r a_1^{n_1} \neq 0$, so we get similarly $a_2^{n_2} a_3 \cdots a_r a_1^{n_1} \neq 0$, and hence $a_1^{n_1} a_2^{n_2} a_3 \cdots a_r \neq 0$. Further repetition of this argument clearly gives $(*)$.

Comment. Of course, we can also prove $(*)$ easily for a reduced ring R by using a subdirect product representation $(\varphi_i) : R \to \prod_i R_i$ where the R_i's are domains. (Indeed, if $a_1^{n_1} \cdots a_r^{n_r} = 0$, then for each i, we must have $\varphi_i(a_j) = 0$ for some j. This implies that $\varphi_i(a_1 \cdots a_r) = 0$ for all i, and so $a_1 \cdots a_r = 0 \in R$.) However, it seems more fair to prove the "only if" part of this exercise in an entirely elementary fashion.

Ex. 12.15. Let a, b be elements in a reduced ring R, and r, s be two positive integers that are relatively prime. If $a^r = b^r$ and $a^s = b^s$, show that $a = b$ in R.

Solution. We may assume that $a \neq 0$. By FC-(12.7), there is a representation of R as a subdirect product of domains. Therefore, it is sufficient to prove the conclusion in the case when R is a domain. Switching r and s if necessary, we can write $1 = rp - sq$, where p, q are nonnegative integers. After replacing r, s by rp and sq, we may therefore assume that $r - s = 1$. Then

$$a^s(a - b) = a^{s+1} - a^s b = a^{s+1} - b^{s+1} = 0 \Longrightarrow a = b.$$

Comment. This exercise was proposed as a problem in the MAA Monthly *99* (1992), p. 362, by Michael Barr. Its easy solution by subdirect product representations given above is well-known to experts. Actually, it is almost just as easy to solve the exercise *without* making the reduction to the case of domains. In his solution to the problem published in the MAA Monthly *101* (1994), p. 580, Pat Stewart noted that, by a special case of Ex. 14, the equation $a^s(a - b) = 0$ in the argument above implies $a(a - b) = 0$. Similarly, $b(a - b) = 0$. Therefore, $(a - b)^2 = 0$, which yields $a = b$!

Ex. 12.16. A ring R is said to be *reversible* if $ab = 0 \in R \Longrightarrow ba = 0$. Show that:

(1) A ring R is a domain iff R is prime and reversible.
(2) R is a reduced ring iff R is semiprime and reversible.

Solution. (1) It suffices to prove the "if" part, so suppose R is prime and reversible. Then, whenever $ab = 0 \in R$, we have $a(bR) = 0$ and hence $bRa = 0$ by reversibility. Since R is prime, it follows that $a = 0$ or $b = 0$.

(2) Note that if R is reduced, of course R is semiprime, and we have observed in the solution to Ex. 12.14 that R is reversible. Conversely, assume R is semiprime and reversible. Then, whenever $a^2 = 0$, we have $a(aR) = 0$ and hence $aRa = 0$. Since R is semiprime, it follows that $a = 0$. This shows that R is reduced.

Comment. This exercise comes from P.M. Cohn's paper, "Reversible rings," Bull. London Math. Soc. *31* (1999), 641–648.

Ex. 12.17. A ring R is said to be *symmetric* if $abc = 0 \in R \Longrightarrow bac = 0$. Show that R is symmetric iff, for any n,

$$(*) \qquad a_1 \cdots a_n = 0 \in R \Longrightarrow a_{\pi(1)} \cdots a_{\pi(n)} = 0 \text{ for any permutation } \pi.$$

Solution. ("Only If") Assuming R is symmetric, we first observe that R must be reversible, since $ab = 0$ implies $(ab)(1) = 0$, which implies $0 = (ba)(1) = ba$. Since the symmetric group S_n is generated by the transposition (12) and the n-cycle $(12 \cdots n)$, the proof of $(*)$ can be reduced to the following two cases:

(A) $$a_1 \cdots a_n = 0 \Longrightarrow a_2 a_1 a_3 \cdots a_n = 0;$$

(B) $$a_1 \cdots a_n = 0 \Longrightarrow a_2 a_3 \cdots a_n a_1 = 0.$$

The first follows directly from the "symmetric" property (by taking $c = a_3 \cdots a_n$); the second follows from reversibility.

Ex. 12.18. A ring R is said to be *2-primal* if every nilpotent element of R lies in $\mathrm{Nil}_* R$. (For instance, reduced rings and commutative rings are 2-primal.) Show that, for any ring R:

$$R \text{ is reduced} \Longrightarrow R \text{ is symmetric} \Longrightarrow R \text{ is reversible} \Longrightarrow R \text{ is 2-primal},$$

but each of these implications is irreversible!

Solution. (1) Assuming R is reduced, we know already that R is reversible. From this, we see further that

$$(\dagger) \qquad xy = 0 \in R \Longrightarrow xry = 0 \quad (\forall\, r \in R),$$

since $(xry)^2 = xr(yx)ry = 0$. To check that R is *symmetric*, we start with any equation $abc = 0 \in R$. For any r, $s \in R$, we have $arbc = 0$ and hence $arbsc = 0$ (by two applications of $(\dagger)$). From this,

$$(bac)^2 = bac \cdot bac \in b(aRbRc) = 0,$$

so $bac = 0$, as desired.

Clearly, a symmetric ring need not be reduced, since any commutative ring is symmetric (but not necessarily reduced).

(2) The (trivial) fact that symmetric rings are reversible was already pointed out in the solution to Ex. 12.17. *Let us now construct a ring that is reversible but not symmetric.* The following construction is due to D.D. Anderson, V. Camillo, and independently, G. Marks.

Let K be a field, $A = K\langle x, y, z \rangle$, and $R = A/I$, where

$$I = (AxA)^2 + (AyA)^2 + (AzA)^2 + AxyzA + AyzxA + AzxyA.$$

For ease of notation, we shall use the same letters x, y, z to denote their images in R. The idea in forming the ring R is that we "kill" all monomials in A if they contain 2 factors of x, or of y, or of z, or if they contain one of the strings xyz, yzx, or zxy. A monomial K-basis for R is therefore given by $e_0 = 1$, $e_1 = x$, $e_2 = y$, $e_3 = z$ together with

$$e_4 = xy, \quad e_5 = yx, \quad e_6 = xz, \quad e_7 = zx, \quad e_8 = yz, \quad e_9 = zy,$$

$$e_{10} = xzy, \quad e_{11} = zyx, \quad e_{12} = yxz.$$

Since $xyz = 0 \neq yxz$, R is not symmetric. To prove that R is reversible, we must show that, for

$$a = \sum_{i=0}^{12} a_i e_i, \quad b = \sum_{i=0}^{12} b_i e_i \quad \text{(in } R\text{)},$$

$ab = 0$ implies $ba = 0$. Since R is a local ring[*] with unique maximal ideal spanned by $e_1, \ldots, e_{12}$, we may assume that $a_0 = b_0 = 0$ (for otherwise one of a, b is a unit and we are done). Also, since e_{10}, e_{11}, e_{12} are killed by all monomials in a, b, we can remove them from a, b to assume that $a = \sum_{i=1}^{9} a_i e_i, b = \sum_{i=1}^{9} b_i e_i$. By expanding ab, we see that $ab = 0$ amounts to the following two sets of equations:

$E_1(a,b)$: $a_1 b_2 = 0$, $a_1 b_3 = 0$, $a_2 b_3 = 0$, $a_2 b_1 = 0$, $a_3 b_1 = 0$, $a_3 b_2 = 0$;

$E_2(a,b)$: $\quad a_1 b_9 + a_6 b_2 = 0$, $\quad a_2 b_6 + a_5 b_3 = 0$, $\quad a_3 b_5 + a_9 b_1 = 0$.

Thus, reversibility amounts to the fact that these equations, taken as a whole, imply $E_1(b,a)$ and $E_2(b,a)$. The former is clear, since $E_1(b,a)$ is no different from $E_1(a,b)$. To check $E_2(b,a)$, we go into two simple cases.

Case 1. $a_1 = a_2 = a_3 = 0$. In this case, the equations in $E_2(a,b)$ involve only one term on the LHS's, and these amount to the one-term equations in $E_2(b,a)$.

Case 2. By cyclic symmetry, it remains only to treat the case $a_1 \neq 0$. In this case, $E_1(a,b)$ gives $b_2 = b_3 = 0$, and $E_2(a,b)$ gives $a_1 b_9 = 0$, $a_2 b_6 = 0$, and $a_3 b_5 + a_9 b_1 = 0$. We need only prove two equations from $E_2(b,a)$, namely, $b_5 a_3 = 0$ and $b_9 a_1 = 0$. The second is at hand, and the first needs proof only if $a_3 \neq 0$. But in this case, $E_1(a,b)$ gives $b_1 = 0$, so $a_3 b_5 + a_9 b_1 = 0$ gives the desired result.

---

[*] For the general notion of a local ring, see *FC*-§19.

In this construction, we took K to be a field only for convenience. The proof given above shows that we get a reversible ring R by using this construction if (and of course, only if) we start with a reversible ground ring K.

(3) Let R be a reversible ring, and let $a \in R$ be a nilpotent element, say with $a^n = 0$. By reversibility, $a^r x a^s = 0$ whenever $r + s = n$. Another application of reversibility shows that $a^r x a^s y a^t = 0$ whenever $r + s + t = n$. By finite induction, it follows that $x_1 \cdots x_m = 0$ as long as n of the factors x_i are equal to a. This implies that $(RaR)^n = 0$, so $RaR \subseteq \mathrm{Nil}_*(R)$. In particular, $a \in \mathrm{Nil}_*(R)$, as desired.

We finish now by constructing a 2-primal ring that is not reversible. The following construction is due to G. Bergman. Let K be any division ring that is not a field, and let R be the ring generated over K by two elements x, y commuting with K, with the relations $xy = yx = x^2 - y^2 = 0$. Then R is a local ring with (left and right) K-basis $\{1, x, y, x^2\}$, whose maximal ideal $\mathfrak{m} = Kx \oplus Ky \oplus Kx^2$ satisfies $\mathfrak{m}^3 = 0$. Clearly, such a ring is 2-primal. Now fix a pair of noncommuting elements u, $v \in K$. Noting that

(A) $$(x + uy)(x - u^{-1}y) = x^2 - y^2 = 0 \in R,$$

we have $ab = 0$ for $a = v(x + uy)$ and $b = x - u^{-1}y$. But

(B) $$ba = (x - u^{-1}y)v(x + uy) = (v - u^{-1}vu)x^2 \neq 0,$$

so R is *not* reversible.

Comment. The fact that a reduced ring satisfies the "symmetric" property $(*)$ in Ex. 12.17 is folklore in noncommutative ring theory. Kaplansky attributed this result to Herstein (unpublished), while Goodearl, in his book on von Neumann regular rings, attributed the result to Andrunakievich and Ryabukhin (see their paper in Soviet Math. Dokl. *9* (1968), 565–568). The same result was observed by Lambek in Canad. Math. Bull. *14* (1971), 359–368. In their paper in Comm. Algebra *27* (1999), 2847–2852, Anderson and Camillo proved the result more generally for reduced semigroups with 0.

The fact that a reversible ring is 2-primal is a slight improvement of Theorem 2.2 in P.M. Cohn's paper cited in the *Comment* on Ex. 12.16.

We could have also introduced the condition $ab = 0 \Longrightarrow aRb = 0$ on a ring R: this condition has been referred to as (SI) in the literature (for reasons that we shall not go into here). As we have seen, reversibility implies (SI), and the work done above is good enough to show that (SI) $\Longrightarrow$ 2-primal. Thus, we could have inserted the (SI) condition "between" reversible and 2-primal in this exercise. Note that Bergman's 2-primal example R constructed above already fails to satisfy (SI), since we have (in the notation there):

$$(x - u^{-1}y)(x + uy) = 0 \neq (x - u^{-1}y)v(x + uy) \quad (\text{in } R).$$

We leave it to the reader to find an example of a ring that satisfies (SI), but is not reversible.

A word of caution is needed on the implication "symmetric $\Longrightarrow$ reversible". In a recent monograph in ring theory, it has been claimed that this implication is an *equivalence*. The example offered in (2) above shows that this claim cannot be true. I first learned this example in 1998 from G. Marks[*], who later found the same example in the paper of Anderson and Camillo, *loc. cit.* Anderson and Camillo showed, moreover, that in any semigroup with zero, if ZC_n denotes the condition

$$a_1 \ldots a_n = 0 \Longrightarrow a_{\sigma(1)} \ldots a_{\sigma(n)} = 0 \quad (\forall \sigma \in S_n),$$

then, for $n \geq 3$, $ZC_n \Longrightarrow ZC_{n+1}$. Thus, the fact that "reversible $\not\Longrightarrow$ symmetric" is an "anomaly" to a general phenomenon. For rings with identity, clearly $ZC_{n+1} \Longrightarrow ZC_n$ for all n (in generalization of "symmetric $\Longrightarrow$ reversible"). However, for semigroups with 0 but without 1, Anderson and Camillo showed that $ZC_{n+1} \not\Longrightarrow ZC_n$ for $n \geq 2$.

Ex. 12.18*. Show that any 2-primal ring R is Dedekind-finite, but not conversely.

Solution. Given $ab = 1 \in R$, we have $a(1 - ba) = 0$, so $[(1 - ba)a]^2 = 0$. Since R is 2-primal, $(1 - ba)a \in \mathrm{Nil}_*(R)$ and hence $1 - ba \in \mathrm{Nil}_*(R)$. It follows that

$$ba \in 1 + \mathrm{Nil}_*(R) \subseteq \mathrm{U}(R),$$

so $cba = 1$ for some $c \in R$. This shows that a is also left-invertible, as desired.

Of course the converse of this exercise does not hold: for instance, for any field k, $\mathbb{M}_2(k)$ is Dedekind-finite, but not 2-primal.

Comment. It follows from Ex. 12.18 and Ex. 12.18* that *any reversible ring R is Dedekind-finite*. This is an observation of D.D. Anderson and V. Camillo in their paper referenced in the *Comment* on Ex. 12.18. They proved this fact directly as follows. Assume R is reversible. If $ab = 1$, then $(ba - 1)b = b - b = 0$ implies $b(ba - 1) = 0$, so $b^2a = b$. From this, we have $ab^2a = ab = 1$, and hence

$$ba = (ab^2a)ba = ab^2a = 1.$$

Ex. 12.19. (0) Show that a ring R is 2-primal iff the factor ring R/Nil_*R is reduced.

(1) Show that a left artinian ring R is 2-primal iff $R/\mathrm{rad}\, R$ is reduced, iff $R/\mathrm{rad}\, R$ is a finite direct product of division rings.

(2) Show that a finite direct product of 2-primal rings is 2-primal.

(3) If R is 2-primal, show that so is the polynomial ring $R[T]$ for any set of commuting variables T).

---

[*] See his paper in J. Pure and Applied Algebra *174* (2002), 311–318.

Solution. (0) Suppose R is 2-primal, and $a^n \in \mathrm{Nil}_* R$ for some $n \geq 1$. Since $\mathrm{Nil}_* R$ is a nil ideal, we have $(a^n)^m = 0$ for some $m \geq 1$. Thus, a is nilpotent and so $a \in \mathrm{Nil}_*(R)$. This shows that $R/\mathrm{Nil}_* R$ is reduced. Conversely, if $R/\mathrm{Nil}_* R$ is reduced, then for any nilpotent element $a \in R$, $\bar{a}$ is nilpotent in $R/\mathrm{Nil}_* R$. Thus, we have $a \in \mathrm{Nil}_* R$, so R is 2-primal.

(1) Since R is left artinian, rad $R = \mathrm{Nil}_* R$ by FC-(10.27). Therefore, the first "iff" follows from (0) above. The second "iff" follows from the Wedderburn-Artin Theorem, since a matrix ring $\mathbb{M}_r(D)$ over a nonzero ring D is reduced only if $r = 1$.

(2) The prime ideals in a direct product $R = R_1 \times \cdots \times R_n$ are of the form $R_1 \times \cdots \times \mathfrak{p}_i \times \cdots \times R_n$, where $\mathfrak{p}_i$ is a prime ideal in R_i. Taking the intersection of such prime ideals in R, we see that

$$\mathrm{Nil}_*(R) = \mathrm{Nil}_*(R_1) \times \cdots \times \mathrm{Nil}_*(R_n).$$

If $a = (a_1, \ldots, a_n) \in R$ is any nilpotent element, then each $a_i \in R_i$ is nilpotent. If each R_i is 2-primal, then $a_i \in \mathrm{Nil}_*(R_i)$ for all i, and so $a \in \mathrm{Nil}_*(R)$ by the equation above.

(3) Using the fact that $\mathrm{Nil}_*(R[T]) = (\mathrm{Nil}_* R)[T]$ (FC-(10.19), we get

$$\frac{R[T]}{\mathrm{Nil}_*(R[T])} = \frac{R[T]}{(\mathrm{Nil}_* R)[T]} \cong \left(\frac{R}{\mathrm{Nil}_* R} \right)[T].$$

If R is 2-primal, $R/\mathrm{Nil}_* R$ is reduced, from which we see easily that $(R/\mathrm{Nil}_* R)[T]$ is also reduced. Thus, the isomorphism above implies that $R[T]$ is 2-primal.

Comment. For (2), we should point out that an *arbitrary* direct product of 2-primal rings need not be 2-primal. Such examples have been constructed by E. Armendariz, and independently by C. Huh, H.K. Kim and Y. Lee. Even more notably, G. Marks has constructed a 2-primal ring R for which $R \times R \times \cdots$ is not 2-primal; see his paper "Direct product and power series formations over 2-primal rings," in *Advances in Ring Theory* (eds. S.K. Jain, S.T. Rizvi), pp. 239–245, Trends in Math., Birkhäuser, Boston, 1997.

Part (3) of this exercise appeared in the paper of G.F. Birkenmeier, H.E. Heatherly and E.K. Lee: "Completely prime ideals and associated radicals," in *Ring Theory* (Granville, Ohio, 1992), (eds. S.K. Jain, S.T. Rizvi), pp. 102–129, World Scientific, Singapore and River Edge, N.J., 1993. In this paper, the authors asked the related question whether R 2-primal implies that the power series ring $R[[t]]$ is 2-primal. This question was also answered negatively in G. Marks' paper cited in the last paragraph.

Ex. 12.20. If R is *right duo* in the sense that any right ideal in R is an ideal, show that R is 2-primal.

Solution. By (0) of the previous exercise, it is sufficient to show that $\bar{a}^2 = 0 \Longrightarrow \bar{a} = 0$, where $\bar{a}$ denotes the image of $a \in R$ in $\bar{R} = R/\mathrm{Nil}_*(R)$. Since aR is an ideal in R, we have $Ra \subseteq aR$. Therefore,

$$(\bar{a}\bar{R})^2 = \bar{a}\bar{R}\bar{a}\bar{R} \subseteq \bar{a}(\bar{a}\bar{R})\bar{R} = 0.$$

Since $\bar{R}$ is a semiprime ring, it follows that $\bar{a}\bar{R} = 0$, and so $\bar{a} = 0 \in \bar{R}$.

Comment. In general, a right duo ring need not be left duo: see Ex. 22.4A (as well as Ex. 22.4B) below.

Chapter 5
Introduction to Division Rings

§13. Division Rings

The study of division rings at the beginning stage is a wonderland of beautiful results. It all started with Frobenius' Theorem (c. 1877) classifying the finite-dimensional division algebras over the reals. Nowadays, we know that the theorem also works for algebraic algebras. Shortly after E.H. Moore completed his classification of finite fields, J.H.M. Wedderburn delighted the world with his "Little" Theorem (c. 1905), that all finite division rings are commutative. Over the years, this result gradually evolved into its various higher forms, of which the commutativity theorems of Jacobson, Kaplansky, and Herstein are prime examples. Then there was the Cartan-Brauer-Hua Theorem (c. 1945 or thereabout): not the result of a collaborative effort, but somehow proved at about the same time, by three famous mathematicians from three different countries. Albeit a smaller result, Herstein's "little" theorem is equally enchanting: it says that if an element in a division ring has more than one conjugate, it has infinitely many. Even one of my analyst colleagues rediscovered this result, and proposed it as a problem for the Amer. Math. Monthly. And, in the time-honored tradition of that journal, the problem was duly solved by an impressive list of fun-lovers, experts and laymen alike.

The exercises in this section explore some of the more elementary aspects of the theory of division rings. Again, many of these are folklore results. My favorite is Exercise 11, due to Jacobson, which says that reduced algebraic algebras over a finite field are always commutative. The spirit of Wedderburn certainly lives on![1]

The last three exercises deal with the theory of ranks of matrices over division rings. There are surprisingly few references for this topic. Because of this, I included a full proof for the "Kronecker Rank Theorem" (Exercise 13). After reading this proof, the reader will perhaps conclude that I have just "plagiarized" the proof for matrices over fields. And of course he/she is right! Those who are accustomed to working with matrices over fields, however, will be a bit surprised by the last exercise, which says that the rank equation $\text{rank}(A) = \text{rank}(A^t)$ fails rather miserably for matrices over a noncommutative division ring.

Exercises for §13

Ex. 13.0. Let D be a nonzero ring, not assumed to have an identity, such that for any nonzero $a \in D$, there exists a unique $b \in D$ satisfying $aba = a$. Show that D is a division ring.

Solution. We proceed in the following steps.

Step 1. D has no 0-divisors. Say $ac = 0$, where $a \neq 0$. Let $b \in D$ be such that $aba = a$. Then $a(b + c)a = aba + aca = a$ implies that $b + c = b$ (by the *uniqueness* of b), and so $c = 0$.

Step 2. For the elements a, b in Step 1, we also have $bab = b$. Indeed, since $a(bab) = (aba)b = ab$, Step 1 (and the distributive law) yields $bab = b$.

Step 3. For the elements a, b above, $e := ab$ is a multiplicative identity of D. Indeed, for any $d \in D$, we have $dea = daba = da$, so Step 1 yields $de = d$. Similarly, $bed = babd = bd$ yields $ed = d$ (since b is obviously $\neq 0$).

Now that D has an identity (which is necessarily unique), Step 3 also showed that *any* nonzero $a \in D$ has a right inverse b. Thus D is a division ring, by Ex. 1.2. (Of course, by reversing the role of a and b in Step 3, we see that ba is also the identity, so in any case it is clear that b is a 2-sided inverse of a.)

Comment. The crux of the hypothesis in this exercise is the *uniqueness* of b. Without assuming this uniqueness, D could have been any von Neumann regular ring, with or without an identity.

Ex. 13.1. Show that a nonzero ring D is a division ring iff, for any $a \neq 1$ in D, there exists an element $b \in D$ such that $a + b = ab$.

---

[1] Incidentally, Jacobson was the fifth and last Ph.D. student of J.H.M. Wedderburn.

Solution. If D is a division ring and $a \neq 1$, we can solve the equation $a + b = ab$ (uniquely) by $b = (a-1)^{-1}a$. Conversely, assume the solvability of this equation for any $a \neq 1$. Then, for any $c \neq 0$, we have $1 - c \neq 1$, so there exists $b \in D$ with

$$1 - c + b = (1 - c)b.$$

Cancelling b, we get $1 = c(1 - b)$, so c has a right inverse. By Exercise 1.2, D is a division ring.

Comment. In the terminology of Exercise 4.2, the result above says that $D \neq 0$ is a division ring iff every element $a \neq 1$ in D is right quasi-regular. In this case, $D \backslash \{1\}$ under the "$\circ$" operation will be a group isomorphic with the multiplicative group D^* of the division ring D.

Ex. 13.2. Let L be a domain and K be a division subring of L. If L is finite-dimensional as a right K-vector space, show that L is also a division ring.

Solution. Let $a \in L \backslash \{0\}$ and consider $\varphi : L \to L$ defined by $\varphi(x) = ax$ $(\forall x \in L)$. This is an injective endomorphism of the finite-dimensional K-vector space L_K, so it must be surjective. In particular, there exists $b \in L$ such that $1 = \varphi(b) = ab$. By Exercise 1.2 again, L is a division ring.

Ex. 13.3. Show that any finite prime (resp. semiprime) ring R is a matrix ring over a finite field (resp. a finite direct product of such matrix rings).

Solution. If R is semiprime, FC-(10.24) and the Artin-Wedderburn Theorem imply that $R \cong \prod_{i=1}^{n} \mathbb{M}_{n_i}(D_i)$, where the D_i's are (finite) division rings. By Wedderburn's Little Theorem FC-(13.1), each D_i is a field. The prime case also follows, since in that case we must have $n = 1$.

Ex. 13.4. For any division ring D, show that any finite abelian subgroup G of D^* is cyclic.

Solution. Let $k = Z(D)$, which is a field. Let A be the (commutative) k-subalgebra of D generated by G. Then

$$K = \{ab^{-1} \ : \ a, b \in A, \ b \neq 0\}$$

is a subfield of D. Since G is a finite subgroup of K^*, a well-known result in elementary field theory guarantees that G is cyclic.

Comment. Note that in this proof, A is a finite-dimensional commutative domain over k, so by Exer. 2 it is in fact a field, hence equal to K. However, the proof goes through without this observation.

Of course, a finite subgroup G of D^* need not be cyclic, as the case of $G = \{\pm 1, \pm i, \pm j, \pm k\}$ in the division ring of real quaternions shows.

Ex. 13.4A. Show that a division ring D is a finite field iff D^* is a cyclic group under multiplication.

Solution. We refer that reader to any textbook in abstract algebra for a proof of the standard "only if" part. For the converse, assume that D^* is cyclic, say with generator a. If $\mathrm{char}(D) = 0$, we would have $\mathbb{Q} \subseteq D$, so $\mathbb{Q}^* \subseteq D^*$ is also cyclic, which is certainly not the case. Therefore, $D \supseteq \mathbb{F}_p$ for a prime p. Since $\mathbb{F}_p(a)$ contains all powers of a, we have $\mathbb{F}_p(a) = D$. It follows that a cannot be transcendental over $\mathbb{F}_p$ (since in that case $a+1$ can't be a power of a). Thus a is algebraic over $\mathbb{F}_p$, and $D = \mathbb{F}_p(a)$ is a finite field.

Ex. 13.5. Show that an element in a division ring D commutes with all its conjugates iff it is central.

Solution. We need only prove the "only if" part. Suppose $d \in D$ commutes with every element in

$$C := \{xdx^{-1} : x \in D^*\}.$$

Assume that $d \notin Z(D)$. By FC-(13.18), D is generated as a division ring by the set C. Therefore, d commutes with every element of D, contradicting $d \notin Z(D)$.

Ex. 13.6. Let D be an algebraic division algebra over a field k. Let $a, b \in D^*$ be such that $bab^{-1} = a^n$, where $n \geq 1$. Show that a, b generate a finite-dimensional division k-subalgebra of D.

Solution. Let A be the k-subalgebra of D generated by a, b. Then A is an algebraic k-algebra. Since A is a domain, Exercise 1.13 implies that A is a division algebra. We finish by proving that $\dim_k A < \infty$. From $ba = a^n b$, we have $ba^s = a^{sn}b$ by induction on $s \geq 0$. By another induction on $t \geq 0$, we can show that

$$b^t a^s = a^{sn^t} b^t.$$

Therefore, A is spanned over k by $\{a^i b^j : i, j \geq 0\}$. It follows that

$$\dim_k A \leq (\dim_k k[a])(\dim_k k[b]) < \infty.$$

Ex. 13.7. (Brauer) Let $K \subseteq D$ be two division rings, and let $N = N_{D^*}(K^*)$, $C = C_{D^*}(K^*)$ be the group-theoretic normalizer and centralizer of the subgroup $K^* \subseteq D^*$. For any $h \in D\backslash\{0, -1\}$, show that $h, 1 + h \in N$ iff $h \in K \cup C$. Using this, give another proof for the Cartan-Brauer-Hua Theorem (FC-(13.17)).

Solution. The "if" part is clear, since K and $C \cup \{0\}$ are closed under addition. Conversely, assume that $h, 1 + h \in N$, but $h \notin C$. Then there exists an element $\delta \in K$ not commuting with h. Let $\delta_0 = h\delta h^{-1} \neq \delta$, and $\delta_1 = (h+1)\delta(h+1)^{-1}$. By hypothesis, $\delta_0, \delta_1 \in K$. Subtracting the equation $h\delta = \delta_0 h$ from $(h+1)\delta = \delta_1(h+1)$, we get

$$\delta - \delta_1 = (\delta_1 - \delta_0)h.$$

If this element is zero, we get $\delta = \delta_1 = \delta_0$, a contradiction. Therefore, $\delta_1 - \delta_0 \neq 0$, and we have

$$h = (\delta_1 - \delta_0)^{-1}(\delta - \delta_1) \in K,$$

as desired.

To derive the Cartan-Brauer-Hua Theorem, assume that $K \neq D$ and K is normal in D. Then, in the above notation, $N = D^*$. By the first part of the Exercise, we have $D = K \cup C$. Since $K \neq D$, we deduce easily that $C = D^*$, so $K \subseteq Z(D)$, as desired.

Ex. 13.8. (Cf. Herstein's Lemma in FC-(13.8)) Let D be a division ring of characteristic $p > 0$ with center F. Let $a \in D \backslash F$ be such that $a^{p^n} \in F$ for some $n \geq 1$. Show that there exists $b \in D$ such that $ab - ba = 1$, and there exists $c \in D$ such that $aca^{-1} = 1 + c$.

Solution. As in the proof of FC-(13.8), we write $\delta = \delta_a$ for the inner derivation on D defined by $\delta(x) = ax - xa$. If we can find $b \in D$ such that $\delta(b) = 1$, then, thanks to $\delta(a) = 0$, we'll have $\delta(ba) = \delta(b)a = a$. It follows that, for $c := ba$,

$$aca^{-1} = 1 + c.$$

Therefore, it is sufficient to prove the existence of b.

As in FC-(13.8), we have

$$\delta^{p^n}(x) = a^{p^n}x - xa^{p^n} = 0$$

for all $x \in D$, so $\delta^{p^n} \equiv 0$. Let $k \geq 0$ be the smallest integer such that $\delta^{k+1} \equiv 0$. Since $a \notin F$, we have $k \geq 1$. Pick $y \in D$ such that $\delta^k(y) \neq 0$, and let $z = \delta^{k-1}(y)$. Then $\delta(z) \neq 0$, and $\delta(\delta(z)) = 0$ implies that $\delta(\delta(z)^{-1}) = 0$. For $b = z\delta(z)^{-1} \in D$, we have

$$\delta(b) = \delta\left(z\delta(z)^{-1}\right) = \delta(z) \cdot \delta(z)^{-1} + z\delta\left(\delta(z)^{-1}\right) = 1,$$

as desired.

Comment. For a given element $a \in D$, $ax - xa = 1$ is known as the *metro equation* for a. In the case when a is an algebraic element over the center of D, a necessary and sufficient condition for the existence of a solution for the metro equation (in D) is given in Exercise 16.11 below.

Ex. 13.9. The following example of Amitsur shows that the Cartan-Brauer-Hua Theorem FC-(13.17) does not extend directly to simple rings. Let $F = \mathbb{Q}(x)$ and $A = F[t; \delta]$ be the ring of polynomials $\{\sum g_i(x)t^i\}$ over F with multiplication defined by the twist $tg(x) = g(x)t + \delta(g(x))$, where $\delta(g(x))$ denotes the formal derivative of $g(x)$. Show that

(1) A is a simple domain,
(2) $U(A) = F^*$, and
(3) F is invariant under all automorphisms of A, but $F \not\subseteq Z(A)$.

Solution. Since F is a simple $\mathbb{Q}$-algebra and δ is not an inner derivation on F, FC-(3.15) implies that A is a simple ring. By degree considerations, $U(A)$ is seen to be F^*, and A has no zero-divisors, so we have (1) and (2). If σ is any automorphism of A, it must take units to units, so $\sigma(F) = F$. However, $F \neq A$, and $tx = xt + 1$ shows that $x \notin Z(A)$, so $F \not\subseteq Z(A)$.

Ex. 13.10. Let k be an algebraically closed field and D be a division k-algebra. Assume that either D is an algebraic algebra over k, or $\dim_k D <$ Card k (as cardinal numbers). Show that $D = k$.

Solution. First assume D is an algebraic algebra over k. If there exists $\alpha \in D\backslash k$, $k[\alpha]$ would give a proper algebraic field extension of k. Therefore, we must have $D = k$. Next, assume $\dim_k D <$ Card k. We are done if we can show that D is an algebraic k-algebra. Consider any $\alpha \in D\backslash k$ (if it exists). The set $\{(a - \alpha)^{-1} : a \in k\}$ has cardinality Card k, so there exists a dependence relation

$$\sum_{i=1}^{r} b_i(a_i - \alpha)^{-1} = 0,$$

where $a_i \in k$, and $b_i \in k^*$. Clearing the denominators, we see (as in the proof of FC-(4.20)) that α is algebraic over k, as desired.

Ex. 13.11. (Jacobson) Let A be a reduced algebraic algebra over a finite field $\mathbb{F}_q$. Show that A is commutative.

Solution. We first prove the following fact.

Lemma. Let $\psi(x) \in \mathbb{F}_q[x]$ be such that $\psi(0) \neq 0$. Then $\psi(x)|(x^N - 1)$ for some integer N.

Proof. It suffices to prove this in the algebraic closure $\mathbb{F}$ of $\mathbb{F}_q$. Write

$$\psi(x) = a(x - a_1) \cdots (x - a_m)$$

where $a, a_i \in \mathbb{F}^*$. Each a_i is a root of unity, so there exists an integer k such that $a_i^k = 1$ for all i. Therefore, $(x - a_i)|(x^k - 1)$, and so $\psi(x)|(x^k - 1)^m$. Let $p = \text{char}(\mathbb{F})$ and pick s such that $p^s \geq m$. Then $\psi(x)$ divides

$$(x^k - 1)^{p^s} = x^{kp^s} - 1,$$

as desired.

Now let $\alpha \in A$ and let $\varphi(x) \in \mathbb{F}_q[x]$ be the minimal polynomial of α over $\mathbb{F}_q$. By Exercise 12.6B(1), we can write $\varphi(x) = x^i\psi(x)$ where $i \in \{0, 1\}$ and $\psi(0) \neq 0$. By the above lemma, there exists an integer N such that $\psi(x)|(x^N - 1)$. Therefore, $\varphi(x)|(x^{N+1} - x)$, and so $\alpha^{N+1} = \alpha$. By Jacobson's Theorem FC-(12.10), A is commutative.

For a much easier solution, we can appeal to FC-(12.7), which says that A can be expressed as a subdirect product of a family of domains $\{A_i\}$.

Since each A_i is a quotient of A, it remains algebraic over $\mathbb{F}_q$. Thus, it suffices to handle the case where A is a domain. But now for any $\alpha \in A$, $\mathbb{F}_q[\alpha]$ is a finite field, so we certainly have $\alpha^{N+1} = \alpha$ for some integer N.

Comment. The Exercise above can be further improved. According to a more general result of Arens and Kaplansky, the conclusion of the Exercise remains true if $\mathbb{F}_q$ is replaced by *any* field k satisfying the following two conditions: (a) all finite-dimensional division k-algebras are commutative, and (b) all finite separable field extensions of k are normal.[2] The proof for this more general result depends on the Skolem-Noether Theorem (for possibly infinite-dimensional division algebras).

Ex. 13.12. Let D be a division ring, and $x, y \in D^*$ be such that $\omega = xyx^{-1}y^{-1}$ lies in the center F of D.

(1) For any integers m, n, show that $x^n y^m x^{-n} y^{-m} = \omega^{mn}$.
(2) If x is algebraic over F, show that ω is a root of unity.
(3) If ω is a primitive kth root of unity, show that $(y + x)^k = y^k + x^k$.

Solution. From $xy = \omega yx$, we get $xy^m = \omega^m y^m x$ by induction on $m \geq 0$. From

$$xy^{-1} = \left(y^{-1}\omega^{-1}xy\right)y^{-1} = y^{-1}\omega^{-1}x = \omega^{-1}y^{-1}x,$$

we have similarly $xy^{-m} = \omega^{-m}y^{-m}x$ for $m \geq 0$. Therefore, the formula $xy^m = \omega^m y^m x$ is valid for *all* $m \in \mathbb{Z}$. Now $y^m x = \omega^{-m}xy^m$, so the deduction above shows that

$$y^m x^n = (\omega^{-m})^n x^n y^m$$

for all n (and all m). This gives

$$x^n y^m x^{-n} y^{-m} = \omega^{mn},$$

as desired.

Now assume x is algebraic over F. Let

$$x^r + a_1 x^{r-1} + \cdots + a_r = 0$$

be the minimal equation of x over F. Conjugating this equation by y and using $y^{-1}xy = \omega x$, we get another equation

$$(\omega x)^r + a_1(\omega x)^{r-1} + \cdots + a_r = 0.$$

Subtracting and cancelling x, we obtain

$$(\omega^r - 1)x^{r-1} + a_1(\omega^{r-1} - 1)x^{r-2} + \cdots + a_{r-1}(\omega - 1) = 0.$$

---

[2] The conditions (a), (b) are, of course, well-known properties of finite fields.

We must have $\omega^r = 1$, for otherwise x would satisfy a polynomial equation of degree $r - 1$ over F.

In order to prove the last statement in the Exercise, consider the expansion $(x + y)^k$ for any $k \geq 0$. For an indeterminate q, let

$$(k!)_q = (q^{k-1} + q^{k-2} + \cdots + 1)(q^{k-2} + \cdots + 1)\cdots(q + 1) \in \mathbb{Z}[q],$$

and define the "noncommutative q-binomial coefficients" to be

$$\begin{bmatrix} k \\ i \end{bmatrix}_q = (k!)_q / (i!)_q((k - i)!)_q \quad (0 \leq i \leq k).$$

(Note that if $q = 1$, $\begin{bmatrix} k \\ i \end{bmatrix}_1$ reduces to the usual binomial coefficients.) After showing that

$$\begin{bmatrix} k \\ i \end{bmatrix}_q = q^i \begin{bmatrix} k - 1 \\ i \end{bmatrix}_q + \begin{bmatrix} k - 1 \\ i - 1 \end{bmatrix}_q,$$

we see that $\begin{bmatrix} k \\ i \end{bmatrix}_q \in \mathbb{Z}[q]$ for all $i \leq k$. Using the relation $xy = \omega yx$, we can then show by induction on k that

$$(y + x)^k = \sum_{i=0}^{k} \begin{bmatrix} k \\ i \end{bmatrix}_\omega y^i x^{k-i} \quad (\forall k \geq 0).$$

Finally, suppose ω is a *primitive* kth root of unity. Then $\begin{bmatrix} k \\ i \end{bmatrix}_\omega = 0$ for $0 < i < k$, so the "noncommutative binomial theorem" above yields $(y + x)^k = y^k + x^k$, as desired.

Comments. The "q-binomial coefficients" $\begin{bmatrix} k \\ i \end{bmatrix}_q$ are used in various branches of combinatorics: see "The Theory of Partitions" by G. E. Andrews (p. 35), Encyclopedia of Math. and Its Applic., Vol. 2, Addison-Wesley, 1976. They also occur in the theory of norm residue symbol algebras: see "An Introduction to Algebraic K-Theory" by J. Milnor (p. 147), Princeton Univ. Press, 1976. With the advent of "quantum geometry" in recent years, the equation $xy = \omega yx$ and the q-binomial coefficients $\begin{bmatrix} k \\ i \end{bmatrix}_q$ have gained further visibility in modern mathematics.

Ex. 13.13. Let D be a division ring and $A = (a_{ij})$ be an $m \times n$ matrix over D. Define the row rank r (resp. column rank c) of A to be the *left* (resp. right) dimension of the row (resp. column) space of A as a subspace of $_D(D^n)$ (resp. $(D^m)_D$). Show that $r = c$.

Solution. Let B be an $r \times n$ matrix over D whose r rows form a basis of the (left) row space of A. Expressing the rows of A as *left* linear combinations of those of B, we get $A = B'B$ for a suitable $m \times r$ matrix B'. But the equation $A = B'B$ may also be construed as expressing the columns of A as *right* linear combinations of the r columns of B'. Hence $c \leq r$. A similar argument shows that $r \leq c$.

Ex. 13.14. Keep the notations in Exercise 13. The common value $r = c$ is called the *rank* of the matrix A. Show that the following statements are equivalent:

(a) rank $A = s$.
(b) s is the largest integer such that A has an $s \times s$ invertible submatrix.
(c) A has an $s \times s$ invertible submatrix M such that any $(s+1) \times (s+1)$ submatrix of A containing M is not invertible.

(This is called "Kronecker's Rank Theorem.")

Solution. We first treat a special case of this theorem. In the case when $A \in \mathbb{M}_n(D)$, we claim that A *has rank n iff it is invertible.* Indeed, if rank $A = n$, we can express the unit vectors as left linear combinations of the row vectors of A, so $A'A = I_n$ for a suitable matrix A', and this implies that A is invertible since $\mathbb{M}_n(D)$ is Dedekind-finite. The converse can be proved by reversing the argument.

$(a) \Longrightarrow (b)$. If $t > s =$ rank A, any $t \times t$ submatrix of A has rank $< t$, for otherwise A would have t rows which are left linearly independent. Therefore, any such submatrix is not invertible. Furthermore, since rank $A = s$ and A has size $m \times n$, there is a submatrix A_0 of size $s \times n$ with rank $A_0 = s$. Taking s columns of A_0 which are right linearly independent, we arrive at an $s \times s$ submatrix M of A_0 with rank $M = s$. By the paragraph above, M is invertible.

$(b) \Longrightarrow (c)$ is a tautology.

$(c) \Longrightarrow (a)$. The s rows of A supporting the matrix M in (c) are clearly left linearly independent, so rank $A \geq s$. If rank $A > s$, we can add one more row to the above set to produce $s + 1$ left linearly independent row vectors. Let A_1 be the submatrix of A consisting of these rows. Then rank $A_1 = s + 1$. Let A_2 be the unique $(s+1) \times s$ submatrix of A_1 containing M. The s columns of A_2 are right linearly independent (since rank $M = s$), so we can add one more column of A_2 to get a set of $s + 1$ right linearly independent columns. This leads to an $(s+1) \times (s+1)$ submatrix M' of A_2 containing M, with rank $M' = s + 1$. This contradicts the choice of M in (c).

Ex. 13.15. Show that $\operatorname{rank}(A) = \operatorname{rank}(A^t)$ for all matrices A over a division ring D iff D is a field.

Solution. If D is a field, the rank equation is true since there is no need to distinguish the left row space from the right row space of A. Now assume D is not a field. Fix two elements $a, b \in D$ such that $ab \neq ba$. Then the matrix

$$A = \begin{pmatrix} 1 & b \\ a & ab \end{pmatrix}$$

has rank 1 since $\begin{pmatrix} b \\ ab \end{pmatrix} = \begin{pmatrix} 1 \\ a \end{pmatrix} b$, and $\begin{pmatrix} 1 \\ a \end{pmatrix} \neq 0$. But

$$A^t = \begin{pmatrix} 1 & a \\ b & ab \end{pmatrix}$$

has rank 2. Indeed, if we have an equation

$$c(1, a) + d(b, ab) = 0,$$

then $db = -c$ and $dab = -ca = dba$, so $d = 0$ and $c = 0$. We have therefore $\mathrm{rank}(A) \neq \mathrm{rank}(A^t)$.

Comment. For the elements a, b above, the matrix $B = \begin{pmatrix} 1 & a \\ b & ab \end{pmatrix}$ is invertible, but its transpose $B^t = \begin{pmatrix} 1 & b \\ a & ab \end{pmatrix}$ is not. This is *not* a contradiction, since over a noncommutative ring $(XY)^t$ is in general not equal to $Y^t X^t$.

  The apparently defective situation $\mathrm{rank}(A) \neq \mathrm{rank}(A^t)$, however, should not cause undue alarm. In fact, if we define a new transpose A^T by $(A^T)_{ij} = A_{ji}^{\mathrm{op}}$, so that A^T is a matrix over the *opposite* division ring D^{op}, it is then easy to verify that $\mathrm{rank}(A) = \mathrm{rank}(A^T)$ for any rectangular matrix A over D. Moreover, $(XY)^T = Y^T X^T$, so X is invertible over D iff X^T is invertible over D^{op}.

Ex. 13.16. Let G be the group of order 21 generated by two elements a, b with the relations $a^7 = 1$, $b^3 = 1$, and $bab^{-1} = a^2$. Using the Wedderburn decomposition of $\mathbb{Q}G$ obtained in Ex. (8.28)(2), show that G cannot be embedded in the multiplicative group of *any* division ring.

Solution. Suppose, instead, $G \subseteq \mathrm{U}(D)$, where D is a division ring. Since G is not cyclic, we must have $\mathrm{char}(D) = 0$ by FC-(13.3). Let D_0 be the $\mathbb{Q}$-span of G in D. Then D_0 is a finite-dimensional $\mathbb{Q}$-domain, so D_0 is a division ring. Since D_0 is a homomorphic image of $\mathbb{Q}G$, it must be isomorphic to a simple component of the semisimple ring $\mathbb{Q}G$ (by Ex. (3.3)(c)). This is a contradiction, since $\mathbb{Q}G$ *does not* have such a simple component, according to our solution to Ex. 8.28(2)!

Ex. 13.17. Give a ring-theoretic proof for the conclusion of the above exercise (without using the representation theory of groups).

Solution. Using the notation in the last exercise, let us assume that $G \subseteq U(D)$, where D is a division ring (or just a domain). From $ba = a^2b$, we get $b^2a = a^4b^2$. Now

$$0 = b^3 - 1 = (b-1)(b^2 + b + 1) \in D$$

shows that $b^2 + b + 1 = 0$. Right multiplying this by a, we get

$$0 = (b^2 + b + 1)a = a^4b^2 + a^2b + a$$
$$= a^4(-b - 1) + a^2b + a$$
$$= a^2(1 - a^2)b - a^4 + a \in D.$$

Since $a^2 \neq 1$, this implies that b *commutes* with a, which contradicts $bab^{-1} = a^2$.

Comment. The above argument suggests that certain classes of finite groups can be shown to be not embeddable in the multiplicative group of any division ring by completely elementary ring-theoretic calculations. For more details, see the author's short note "Finite groups embeddable in division rings" in Proc. Amer. Math. Soc. *129* (2001), 3161–3166.

§14. Some Classical Constructions

In principle, it is not difficult to "construct" all fields. Since any field K has a transcendence base $\{x_i : i \in I\}$ over its prime field, we can obtain K as an algebraic extension of $\mathbb{Q}(\{x_i\})$ or of $\mathbb{F}_p(\{x_i\})$. For division rings, however, there is no such simple procedure. As a result, we have to try much harder in order to come up with good constructions of noncommutative division rings.

This section deals with two of the best known classical constructions of division rings, due to L.E. Dickson, A.I. Mal'cev, and B.H. Neumann. In order to set up the notations to be used in the exercises, a brief summary of these two constructions is given below.

Given a cyclic field extension K/F with Galois group $\langle \sigma \rangle$ of order s, Dickson's cyclic algebra $D = (K/F, \sigma, a)$ (with respect to an element $a \in F$) is a formal direct sum

$$D = K \oplus Kx \oplus \cdots \oplus Kx^{s-1},$$

made into an F-algebra with the multiplication rules $x^s = a$ and $xb = \sigma(b)x$ for every $b \in K$. If $a \neq 0$, D is always a simple F-algebra with $Z(D) = F$, and with K as a maximal subfield. Under suitable assumptions on a, D will be a division algebra: cf. *FC*-(14.8), (14.9). From the perspective of skew polynomial rings, $(K/F, \sigma, a)$ is just the quotient ring $K[t; \sigma]/(t^s - a)$. For instance, nonsplit quaternion algebras are cyclic algebras.

While Dickson's cyclic algebra constructions give only centrally finite division rings, the Mal'cev-Neumann formal Laurent series constructions lead to many examples of centrally infinite division rings. Let $(G, <)$ be a multiplicatively written ordered group, R be a ring, and $\omega : G \to \text{Aut}(R)$ $(g \mapsto \omega_g)$ be a fixed group homomorphism. The Mal'cev-Neumann ring $A = R((G, \omega))$ is defined to be the ring of formal sums $\alpha = \sum_{g \in G} a_g g$ with well-ordered supports. Addition and multiplication are defined formally, the latter being given the "twist":

$$gr = \omega_g(r)g, \quad \text{for } g \in G \text{ and } r \in R.$$

Substantial work is needed in verifying that A is a well-defined ring (see FC-pp. 130–131). Once this is known, it can further be checked that, if R is a division ring, then so is $A = R((G, \omega))$ (see FC-(14.21)). The construction of A has various useful applications. For instance, it leads to the Mal'cev-Neumann-Moufang Theorem on the embeddability of free algebras over division rings into division rings (as in FC-(14.25)).

The exercises in this section deal with numerous examples of cyclic algebras, their maximal subfields, and the Mal'cev-Neumann rings. Some of these exercises are of a number-theoretic flavor. Other exercises deal with dimension questions (Exercises 1, 2), skew group rings (Exercises 13A, 13B, 13C), and quadratic extensions of division rings (Exercise 12).

Exercises for §14

Ex. 14.1. Let D be a ring containing a division ring K. Let $[D : K]_\ell$ (resp. $[D{:}K]_r$) denote the dimension (as a cardinal number) of D as a left (resp. right) K-vector space. The following construction shows that $[D : K]_\ell \neq [D : K]_r$ in general. Let k be a field and $K = k(t)$ where t is an indeterminate. Let $D = K \oplus K \cdot x$, made into a ring with the multiplication rules $x^2 = 0$ and $x \cdot b(t) = b(t^2)x$ for any $b(t) \in K$. Show that $K \cdot x = xK \oplus txK$, and so $[D : K]_\ell = 2$, $[D : K]_r = 3$. If we set $K_n = k(t^n)$, then the transitivity formulas for left and right dimensions show that $[D : K_n]_\ell = 2n$, $[D : K_n]_r = 3n$.

Solution. Let $E = k(t^2)$, so $K = E \oplus tE$. We have

$$K \cdot x = (E \oplus tE)x = Ex \oplus tEx = xK \oplus txK,$$

in view of $x \cdot b(t) = b(t^2)x$ for $b(t) \in K$. Therefore, $[K \cdot x : K]_r = 2$ and

$$[D : K]_r = [K \oplus Kx : K]_r = 1 + 2 = 3,$$

in contrast to $[D : K]_\ell = 2$. Since $[K : K_n] = n$, we have, by the transitivity formulas for left and right dimensions:

$$[D : K_n]_l = [D : K]_l[K : K_n] = 2n, \quad \text{and}$$
$$[D : K_n]_r = [D : K]_r[K : K_n] = 3n.$$

Ex. 14.2. The following question was raised by E. Artin: *if $D \supseteq K$ are a pair of division rings, is $[D : K]_\ell = [D : K]_r$?* Show that this holds if D is centrally finite.

Solution. Let $F = Z(D)$, and assume $[D : F] < \infty$. Let $A = F \cdot K$ be the subring of D generated by F and K. Since A is a domain with $[A : F] < \infty$, A is a division ring. Since $A = F \cdot K$, we can choose a basis of A_K consisting of elements of F. Therefore, $[A : K]_\ell = [A : K]_r$. On the other hand

$$[D : F] = [D : A]_\ell[A : F] = [D : A]_r[A : F] < \infty$$

implies that $[D : A]_\ell = [D : A]_r$. We have then

$$[D : K]_\ell = [D : A]_\ell[A : K]_\ell = [D : A]_r[A : K]_r = [D : K]_r,$$

as desired.

Comment. The answer to Artin's original question is "no" in general. P.M. Cohn has found examples of division rings $D \supseteq K$ such that $[D : K]_\ell$ is finite but $[D : K]_r$ is infinite: see p.126 of his book "Skew Field Constructions," Cambridge University Press, Cambridge, 1977. More recently, A.H. Schofield has constructed examples of division rings $D \supseteq K$ such that $[D : K]_\ell$ and $[D : K]_r$ are arbitrarily prescribed integers ≥ 2: see his paper "Artin's problem for skew field extensions," Proc. Cambridge Phil. Soc. *97* (1985), 1–6.

Ex. 14.3. For K/F a cyclic extension with $\mathrm{Gal}(K/F) = \langle \sigma \rangle$, let $D = (K/F, \sigma, a)$ be a cyclic algebra, where $a \in F$. Show that, if we allow a to be zero, the resulting algebra D is no longer simple (unless $K = F$).

Solution. By definition,

$$D = K \oplus Kx \oplus \cdots \oplus Kx^{s-1},$$

where $s = [K : F]$, and $x^s = a$, $xb = \sigma(b)x$ for all $b \in K$. We claim that, if $a = 0$,

$$\mathfrak{A} = Kx \oplus \cdots \oplus Kx^{s-1}$$

is an ideal of D. In fact, let $\beta = bx^i$ where $b \in K$ and $i \geq 1$. For any $\gamma = cx^j$ with $c \in K$ and $j \geq 0$, we have

$$\gamma\beta = cx^j bx^i = c\sigma^j(b)x^{i+j} \quad \text{and} \quad \beta\gamma = bx^i cx^j = b\sigma^i(c)x^{i+j}.$$

Since $i + j \geq 1$ and $x^{i+j} = 0$ if $i + j \geq s$, both products above belong to $\mathfrak{A}$. This shows that $\mathfrak{A}$ is an ideal of D, so D is not a simple algebra, unless $s = 1$.

Ex. 14.4. Let $D = (K/F, \sigma, a)$ be a cyclic algebra for which there exists $c \in K^*$ with $N_{K/F}(c) = a$. Give an alternative proof for the splitting of D by constructing an explicit isomorphism L from D to $\mathrm{End}_F(K)$.

Solution. For $b \in K$, let $\lambda(b) \in \operatorname{End}_F(K)$ denote the left multiplication by b on K. Define $L : D \to \operatorname{End}_F(K)$ by $L(b) = \lambda(b)$ $(b \in K)$, and $L(x) = \lambda(c)\sigma$. To check that L is a well-defined F-algebra homomorphism, we must check that L respects the defining relations for the cyclic algebra. More specifically, we must check that

$$[\lambda(c)\sigma]^s = \lambda(a), \quad \text{and} \quad \lambda(c)\sigma\lambda(b) = \lambda(\sigma b)\lambda(c)\sigma \quad \text{for all } b \in K.$$

By direct computations, we have, for any $d \in K$:

$$[\lambda(c)\sigma]^i(d) = c \cdot \sigma c \cdots \sigma^{i-1}c \cdot \sigma^i(d),$$
$$[\lambda(c)\sigma]^s(d) = N_{K/F}(c)\sigma^s(d) = \lambda(a)(d), \quad \text{and}$$
$$[\lambda(c)\sigma\lambda(b)](d) = c\sigma(bd) = \sigma(b)c \cdot \sigma(d) = [\lambda(\sigma b)\lambda(c)\sigma](d),$$

as desired. Since D is a simple algebra, L must be injective. It follows that L is an isomorphism, since $\dim_F \operatorname{End}_F(K) = s^2 = \dim_F D$.

Ex. 14.5. Let $K = \mathbb{Q}(v)$ be the cyclic cubic field defined by

$$v^3 + v^2 - 2v - 1 = 0,$$

with $\operatorname{Gal}(K/\mathbb{Q}) = \langle \sigma \rangle$ where $\sigma(v) = v^2 - 2$. Let D be the cyclic division algebra

$$(K/\mathbb{Q}, \sigma, 2) = K \oplus Kx \oplus Kx^2$$

considered by Dickson (cf. *FC*-p. 227). Show that, for any nonzero $\alpha \in K$, we have

$$(*) \qquad \mathbb{Q}(\alpha x) \cong \mathbb{Q}\left(\sqrt[3]{2N(\alpha)}\right) \quad \text{and} \quad \mathbb{Q}(\alpha x^2) \cong \mathbb{Q}\left(\sqrt[3]{4N(\alpha)}\right)$$

(where $N = N_{K/\mathbb{Q}}$), and that these are maximal subfields of D. Note that $K = \mathbb{Q}(v)$ is Galois over $\mathbb{Q}$, but these new maximal subfields are not.

Solution. Let $y = \alpha x$ and $z = \alpha x^2$ in D. Then

$$y^3 = \alpha x \cdot \alpha x \cdot \alpha x = \alpha\sigma(\alpha)x^2\alpha x = \alpha\sigma(\alpha)\sigma^2(\alpha)x^3 = 2N(\alpha),$$
$$z^3 = \alpha x^2 \cdot \alpha x^2 \cdot \alpha x^2 = \alpha\sigma^2(\alpha)\sigma^4(\alpha)x^6 = \alpha\sigma(\alpha)\sigma^2(\alpha)x^6 = 4N(\alpha).$$

Since $2 \notin N(\dot{K})$ (see *FC*-p. 227), $2N(\alpha)$ and $4N(\alpha)$ are not in $\dot{\mathbb{Q}}^3$, so the minimal polynomials of y and z over $\mathbb{Q}$ are respectively $t^3 - 2N(\alpha)$ and $t^3 - 4N(\alpha)$, and $(*)$ follows. To see that $K' = \mathbb{Q}(y)$ is a maximal subfield of D, let $E \subseteq D$ be a subfield containing K'. Then $\dim_\mathbb{Q} E \,|\, 9$ and $3 \,|\, \dim_\mathbb{Q} E$ by the transitivity formulas for dimensions. These force $\dim_\mathbb{Q} E$ to be 3, so $E = K'$. Similarly, $K'' = \mathbb{Q}(z)$ is also a maximal subfield of D. However, neither K' nor K'' is Galois over $\mathbb{Q}$, since K' (resp. K'') does not contain the nonreal roots of $t^3 - 2N(\alpha)$ (resp. $t^3 - 4N(\alpha)$).

Comment. If one assumes the results in FC-(15.4) about dimensions of centralizers in a division ring, it will be immediate that the K', K'' above are maximal subfields of D. This observation applies also to the two exercises below.

Ex. 14.6. Let k be a field, with an automorphism σ of order s. Let $D = k((x, \sigma))$ (division ring of formal Laurent series in x with $x^i a = \sigma^i(a)x^i$), which may be viewed as the cyclic algebra $(K/F, \sigma, x^s)$ where $F = k_0((x^s))$, $K = k((x^s))$, and $k_0 = k^\sigma$ (fixed field of σ).[3] Show that $K' = k_0((x))$ as well as K are maximal subfields of D. If $s > 1$, show that K and K' are not isomorphic over F.

Solution. Of course, K is a maximal subfield of D, since D is a cyclic algebra constructed from the cyclic extension K/F (cf. FC-(14.6)(3)). To show that $K' = k_0((x))$ is also a maximal subfield, it suffices to check that $C_D(K') \subseteq K'$. Let $\beta \in C_D(K')$. We can write

$$\beta = b_0 + b_1 x + \cdots + b_{s-1} x^{s-1}$$

where $b_i \in K$. From $\beta x = x\beta$, we get

$$b_0 x + \cdots + b_{s-2} x^{s-1} + b_{s-1} x^s = \sigma(b_0)x + \cdots + \sigma(b_{s-2})x^{s-1} + \sigma(b_{s-1})x^s,$$

and so $b_i = \sigma(b_i)$ for all i. This yields $b_i \in K^\sigma = F$, whence

$$\beta \in \sum_{i=0}^{s-1} F x^i = k_0((x)) = K'.$$

Now assume $s > 1$. Then the two maximal subfields K, K' of D cannot be isomorphic over the center F. In fact, the element $x^s \in F$ is an sth power in K'; however, a straightforward lowest-degree-term argument shows that $x^s \in F$ is *not* an sth power in K. This clearly implies that K and K' are not F-isomorphic.

Ex. 14.7. Let $\zeta \in \mathbb{C} \backslash \{1\}$ be an sth root of unity, where s is a prime. Let $k = \mathbb{Q}(\zeta), K = k(y)$, and $F = k(y^s)$, where y is an indeterminate commuting with k.

(1) Show that K/F is a cyclic (Kummer) extension with $\mathrm{Gal}(K/F) = \langle \sigma \rangle$, where σ is defined by $\sigma(y) = \zeta y$.

(2) Show that $\zeta \notin N_{K/F}(K^*)$. Therefore, $D = (K/F, \sigma, \zeta)$ is an s^2-dimensional division algebra over its center F. (The cyclic algebra relations boil down to $x^s = \zeta$, $xy = \zeta yx$. Note that the former relation already implies that $x\zeta = \zeta x$.)

(3) Show that K and $K' = F(x)$ are both maximal subfields of D. (Note that $K = \mathbb{Q}(\zeta, y)$ and $K' = k(x)(y^s) \cong \mathbb{Q}(\zeta', y)$, where ζ' is a primitive s^2th root of unity. In particular, K and K' are nonisomorphic fields.)

---

[3] We view σ as an automorphism on K by extending $\sigma \in \mathrm{Aut}(k)$ to K via $\sigma(x^s) = x^s$.

Solution. (1) We have $[K : F] = s$, and K is obtained by adjoining an sth root of an element $y^s \notin F^s$. Since F contains the primitive sth root of unity ζ, K/F is a cyclic Kummer extension. The conjugates of $y \in K$ are $y, \zeta y, \ldots, \zeta^{s-1} y$, so $\mathrm{Gal}(K/F) = \langle \sigma \rangle$, where σ is the identity on F and $\sigma(y) = \zeta y$.

(2) Assume, instead, $\zeta = N_{K/F}(f(y)/g(y))$, where $f, g \in k[y], g \neq 0$. Then

$$(*) \qquad f(y)f(\zeta y) \cdots f(\zeta^{s-1} y) = \zeta g(y) g(\zeta y) \cdots g(\zeta^{s-1} y).$$

Let

$$f(y) = a_n y^n + a_{n+1} y^{n+1} + \cdots,$$
$$g(y) = b_m y^m + b_{m+1} y^{m+1} + \cdots,$$

where $a_i, b_j \in k$, $a_n \neq 0 \neq b_m$. Comparing the lowest degree terms in $(*)$, we see that $n = m$. Therefore, $f/g = f_1/g_1$ where

$$f_1 = a_n + a_{n+1} y + \cdots, \quad \text{and}$$
$$g_1 = b_n + b_{n+1} y + \cdots.$$

After replacing f, g by f_1, g_1, we may then assume $f(0) \neq 0 \neq g(0)$. Setting $y = 0$ in $(*)$, we see that $\zeta = f(0)^s / g(0)^s \in k^s$. This is, of course, impossible, so we have shown that $\zeta \notin N_{K/F}(K^*)$. It follows from FC-(14.8) that $D = (K/F, \sigma, \zeta)$ is an s^2-dimensional division algebra over its center F.

(3) As usual, K is a maximal subfield of D. The argument used in the solution of Exercise 6 shows that $K' = F(x)$ is also a maximal subfield of D. (This is valid in any cyclic division algebra $D = K \oplus Kx \oplus \cdots \oplus Kx^{s-1}$.) The two fields K, K' are not isomorphic since K' contains a primitive s^2th root of unity (namely x), but K does not.

Ex. 14.8. Let a be an integer which is not a norm from the extension $E = \mathbb{Q}(\sqrt[3]{2})$ of $\mathbb{Q}$. Show that the $\mathbb{Q}$-algebra with generators ω, α, x and relations

$$(*) \quad \omega^2 + \omega + 1 = 0, \quad \alpha^3 = 2, \quad x^3 = a, \quad \omega\alpha = \alpha\omega, \quad \omega x = x\omega, \quad x\alpha = \omega\alpha x$$

is an 18-dimensional $\mathbb{Q}$-division algebra with center $\mathbb{Q}(\omega)$.

Solution. Let $F = \mathbb{Q}(\omega)$ where ω is a primitive cubic root of unity, and let $\alpha = \sqrt[3]{2}$. Then
$$K = E \cdot F = \mathbb{Q}(\omega, \alpha)$$
is a cyclic cubic extension of F, with $\mathrm{Gal}(K/F) = \langle \sigma \rangle$, where $\sigma(\alpha) = \omega\alpha$. The cyclic algebra

$$D = (K/F, \sigma, a) = K \oplus Kx \oplus Kx^2 = \mathbb{Q} \langle \omega, \alpha, x \rangle$$

is therefore defined by the relations

$$x^3 = a, \quad xb = \sigma(b)x \quad (\forall b \in K),$$

together with the relations in K, namely,

$$\omega^2 + \omega + 1 = 0, \quad \alpha^3 = 2, \quad \text{and} \quad \omega\alpha = \alpha\omega.$$

To spell out the relations $xb = \sigma(b)x$ more explicitly, note that $K = \mathbb{Q}(\omega, \alpha)$ simplifies these relations to the following two:

$$x\omega = \sigma(\omega)x = \omega x, \quad x\alpha = \sigma(\alpha)x = \omega\alpha x.$$

Therefore, $D = \mathbb{Q}\langle\omega, \alpha, x\rangle$ is precisely defined by the relations $(*)$. Moreover,

$$\dim_F D = (\dim_F D)(\dim_\mathbb{Q} F) = 9 \cdot 2 = 18,$$

and $Z(D) = F = \mathbb{Q}(\omega)$ by $FC\text{-}(14.6)(1)$. To see that D is a division algebra, it suffices to show that $a \notin N_{K/F}(\dot{K})$ (by $FC\text{-}(14.8)$). Assume, instead, that $a = N_{K/F}(b)$ for some $b \in K$. Then

$$N_{K/\mathbb{Q}}(b) = N_{F/\mathbb{Q}}(a) = a^2,$$

so $a^{-2} \in N_{E/\mathbb{Q}}(E)$ and $a = a^3 a^{-2} \in N_{E/\mathbb{Q}}(E)$, a contradiction.

Ex. 14.9. Let R be a ring, $(G, <)$ be an ordered group, and $\omega : G \to \text{Aut}(R)$ be a group homomorphism. Let $A = R((G, \omega))$ be the Mal'cev-Neumann ring constructed from the data $\{R, G, <, \omega\}$ (cf. $FC\text{-}(14.18)$). For any $\alpha \in A$ such that

$$\text{supp}(\alpha) \subseteq P := \{g \in G : g > 1\}$$

and any positive integer n which is invertible in R, show that $1 + \alpha$ has an nth root in A.

Solution. The main trick is to use the following

Lemma. *For any integers $k \geq 0$, $n \geq 1$, and $m \in \mathbb{Z}$, the generalized binomial coefficient $\binom{m/n}{k} \in \mathbb{Z}[1/n]$.*

Proof. For a fixed m, n, let $a_k = \binom{m/n}{k} \in \mathbb{Q}$. Following the suggestion of D. Shapiro, we proceed by induction on k. Since $a_0 = 1$ and $a_1 = m/n$, there is no problem in starting the induction. By Newton's generalized binomial theorem,

$$(*) \qquad (1 + a_1 x + a_2 x^2 + \cdots)^n = (1 + x)^m \in \mathbb{Z}[[x]].$$

For $k \geq 2$, comparison of the coefficients of x^k on the two sides of this equation yields $na_k \in \mathbb{Z}[a_1, a_2, \ldots, a_{k-1}]$. By the inductive hypothesis, $a_i \in \mathbb{Z}[1/n]$ for all $i \leq k - 1$. Therefore,

$$a_k \in (1/n)\,\mathbb{Z}[1/n] \subseteq \mathbb{Z}[1/n].$$

Alternatively, there is a p-adic proof of the Lemma, which I learned from H. Lenstra. Since

$$\mathbb{Z}[1/n] = \bigcap_{p \nmid n} \mathbb{Z}_{(p)} \qquad (\mathbb{Z}_{(p)} = \text{localization of } \mathbb{Z} \text{ at the prime } (p)),$$

it suffices to show that $\binom{m/n}{k} \in \mathbb{Z}_{(p)}$ for any prime $p \nmid n$. Given such a prime p, consider the polynomial function

$$f(x) = \binom{x}{k} = x(x-1)\ldots(x-k+1)/k!$$

from $\hat{\mathbb{Q}}_p$ to itself (where $\hat{\mathbb{Q}}_p$ denotes the field of p-adic numbers). As we have noted already in (*), $f(\mathbb{Z}) \subseteq \mathbb{Z}$. Since f is a continuous function (with respect to the p-adic topology), we have $f\left(\hat{\mathbb{Z}}_{(p)}\right) \subseteq \hat{\mathbb{Z}}_{(p)}$, where $\hat{\mathbb{Z}}_{(p)}$ is the ring of p-adic integers. Therefore,

$$\binom{m/n}{k} = f(m/n) \in \mathbb{Q} \cap \hat{\mathbb{Z}}_{(p)} = \mathbb{Z}_{(p)}.$$

The third time being the charm, we record here also a purely arithmetic proof of the Lemma given by Pólya and Szegő (Aufgaben und Lehrsätze aus der Analysis II, Problem 138 on p. 353). Write $k! = rs$, where r is divisible only by the primes dividing n, and $(s,n) = 1$. Since

$$\binom{m/n}{k} = \frac{m(m-n)\cdots(m-(k-1)n)}{(n^k r)s},$$

our job is to show that $N := m(m-n)\cdots(m-(k-1)n)$ is divisible by s. Let $t \in \mathbb{Z}$ be such that $tn \equiv 1 \pmod{s}$. Then

$$t^k N = (tm)(tm - tn)\cdots(tm - (k-1)tn)$$
$$\equiv (tm)(tm-1)\cdots(tm-(k-1)) \pmod{s}.$$

Since a product of k consecutive integers is always divisible by $k! = rs$, we see that $t^k N \equiv 0 \pmod{s}$, and hence $N \equiv 0 \pmod{s}$, as desired!

Now, coming back to the Exercise, let n be a fixed natural number which is invertible in R. Then we have a ring homomorphism from $\mathbb{Z}[1/n]$ to R. In particular, by the Lemma (applied with $m = 1$), all binomial coefficients $\binom{1/n}{k}$ make sense in R. Let $\alpha \in A$ be given as in the exercise. According to FC-(14.23), the infinite sum

$$1 + \binom{1/n}{1}\alpha + \binom{1/n}{2}\alpha^2 + \cdots + \binom{1/n}{k}\alpha^k + \cdots$$

gives a *well-defined* element β of A (thanks to $\text{supp}(\alpha) \subseteq P$). Since the above series is a formal nth root of $1 + \alpha$, we have $\beta^n = 1 + \alpha$, as desired.

Ex. 14.10. Let $A = R((G))$ (with trivial ω), where R is a division ring and $(G, <)$ is a (not necessarily abelian) ordered group. Define a map φ : $A^* \to G$ by $\varphi(\alpha) = \min(\text{supp } \alpha)$ for any $\alpha \in A^*$. Show that φ is a Krull valuation of the division ring A with value group G in the sense that it has the following properties:

(1) $\varphi(\alpha\beta) = \varphi(\alpha)\varphi(\beta)$ for any $\alpha, \beta \in A^*$, and
(2) $\varphi(\alpha + \beta) \geq \min \{\varphi(\alpha), \varphi(\beta)\}$ for any $\alpha, \beta, \alpha + \beta \in A^*$.

Solution. Let

$$\alpha = ag + \cdots , \qquad \beta = bh + \cdots ,$$

where $a, b \in R^*$, $g = \min(\text{supp } \alpha)$, and $h = \min(\text{supp } \beta)$. Then $\alpha\beta = (ab)gh + \cdots$, and by the ordered group axioms, $gh = \min(\text{supp } \alpha\beta)$. Therefore

$$\varphi(\alpha\beta) = \min(\text{supp } \alpha\beta) = gh = \varphi(\alpha)\varphi(\beta).$$

For the sum, we have

$$\alpha + \beta = ag + bh + \cdots .$$

The "smallest" group element occurring here is clearly no smaller than $\min \{g, h\}$. Therefore

$$\varphi(\alpha + \beta) = \min(\text{supp}(\alpha + \beta)) \geq \min\{g, h\} = \min\{\varphi(\alpha), \varphi(\beta)\},$$

if $\alpha + \beta \in A^*$.

Ex. 14.11. Let $A = R((G, \omega))$ be as in Exercise 9, where R is a division ring. In case G is an additively written ordered abelian group, it is convenient to introduce a symbol x and write the elements of G "exponentially" as $\{x^g : g \in G\}$. The elements of A are then written as $a = \sum_{g \in G} a_g x^g$. Now let G be the additive group $\mathbb{R}$ with the usual ordering. Let A_1 be the subset of A consisting of $\sum_{n=1}^{\infty} a_n x^{g_n}$ where $a_n \in R$ and $\{g_1, g_2, \ldots\}$ is any strictly increasing sequence in $\mathbb{R}$ with $\lim_{n \to \infty} g_n = +\infty$. Show that A_1 is a proper division subring of A. If

$$A_2 = \left\{ \sum_{n=1}^{\infty} a_n x^{g_n} : a_n \in R \quad \text{and} \quad g_1 < g_2 < \cdots \text{ in } \mathbb{R} \right\},$$

is A_2 a subring of A?

Solution. Consider any two elements

$$\alpha = \sum a_n x^{g_n}, \qquad \beta = \sum b_n x^{h_n}$$

where $g_1 < g_2 < \cdots$, $h_1 < h_2 < \cdots$, with $\lim g_n = \infty = \lim h_n$. For any real number M, we have $g_n, h_n \geq M$ for sufficiently large n, so there are only finitely many g_i, h_j with $g_i < M$ and $h_j < M$. Thus, we can rearrange the elements $\{g_1, g_2, \ldots, h_1, h_2, \ldots\}$ into

$$k_1 \leq k_2 \leq k_3 \leq \cdots$$

where no three consecutive elements are equal, and $\lim k_n = \infty$. Since $\alpha \pm \beta = \sum c_n x^{k_n}$ for suitable elements $c_n \in R$, we have $\alpha \pm \beta \in A_1$. To handle the product $\alpha\beta$, we observe similarly that, for any real number M', there are only finitely many g_i, h_j such that $g_i + h_j < M'$. This enables us to "rearrange" the elements $\{g_i + h_j\}$ in a nondecreasing sequence as before, so that we can conclude that $\alpha\beta \in A_1$. To show that any

$$\alpha = \sum a_n x^{g_n} \in A_1 \backslash \{0\}$$

has its inverse in A_1, we may assume (after extracting the factor $a_1 x^{g_1}$ from α) that $a_1 = 1$ and $g_1 = 0$. Let

$$\gamma = -(a_2 x^{g_2} + a_3 x^{g_3} + \cdots),$$

with $\operatorname{supp}(\gamma) > 0$. Then

(1) $$\alpha^{-1} = (1 - \gamma)^{-1} = 1 + \gamma + \gamma^2 + \cdots,$$

where the sum on the RHS makes sense according to FC-(14.23). Consider the set of positive real numbers

(2) $$\{g_i, \ g_i + g_j, \ g_i + g_j + g_k, \ \ldots\} \quad (i, j, k, \ldots \geq 2).$$

For any real number N, there are only a finite number of elements of the above set which are $< N$. Therefore, we can again "rearrange" the elements in (2) above in a nondecreasing sequence (which tends to ∞). It follows that the RHS of (1) is an element of A_1, so A_1 is a (proper) division subring of A, as claimed. On the other hand, A_2 is not a subring of A. To see this, look at

$$\alpha = x^{1/2} + x^{3/4} + x^{7/8} + \cdots \quad \text{and} \quad \beta = x^1,$$

both of which are elements of A_2. However, $\alpha + \beta$ with support $\{1/2, 3/4, 7/8, \ldots, 1\}$ is not an element of A_2. Therefore, A_2 is not even an additive subgroup of A, let alone a subring.

Ex. 14.12. Let K be a division ring and σ be an automorphism of K such that $\sigma^2(b) = aba^{-1}$ for all $b \in K$, where a is an element of K fixed by σ. Let $D = K \oplus Kx$ where $x^2 = a$ and $xb = \sigma(b)x$ for all $b \in K$.

(1) Show that D is a division ring iff there does not exist $c \in K$ such that $a = \sigma(c)c$.

(2) Compute $Z(D)$.

(3) Show that D is centrally finite iff K is.

Solution. (1) Let R be the skew polynomial ring $K[x;\sigma]$ (in which $xb = \sigma(b)x$ for all $x \in K$), and let $\mathfrak{A} = R \cdot (x^2 - a)$. Since

$$(x^2 - a)x = x^3 - x\sigma^{-1}(a) = x(x^2 - a), \quad \text{and}$$
$$(x^2 - a)b = \sigma^2(b)x - ab = aba^{-1}x - ab = aba^{-1}(x - a),$$

$\mathfrak{A}$ is an ideal in R. The given ring D may be identified with the quotient $R/\mathfrak{A}$. If $a = \sigma(c)c$ for some $c \in K$, then

$$(x + \sigma(c))(x - c) = x^2 - xc + \sigma(c)x - \sigma(c)c = x^2 - a.$$

In this case $R/\mathfrak{A}$ has a proper left ideal $R \cdot (x - c)/\mathfrak{A}$, so $R/\mathfrak{A}$ is not a division ring. Conversely, if $R/\mathfrak{A}$ is not a division ring, it will have a proper left ideal $\mathfrak{B}/\mathfrak{A}$. Since R is a principal left ideal domain (see FC-(1.25)), $\mathfrak{B} = Rf$ for a (say monic) polynomial f. We have $x^2 - a = gf$ for some $g \in R$, so clearly $\deg f = \deg g = 1$, and g is also monic. Writing $f = x - c$ and $g = x + d$ $(c, d \in K)$, we have

$$x^2 - a = (x + d)(x - c) = x^2 + (d - \sigma(c))x - dc \in R.$$

Therefore, $d = \sigma(c)$, and $a = dc = \sigma(c)c$, as desired.

(2) Clearly, σ induces an automorphism of order ≤ 2 on $Z(K)$ (the center of K). We write

$$K^\sigma = \{k \in K : \sigma(k) = k\} \quad \text{and} \quad F = Z(K)^\sigma = Z(K) \cap K^\sigma.$$

Then $[Z(K) : F] \leq 2$. For $b, b' \in K$, we can work out the necessary and sufficient conditions for $b + b'x \in Z(D)$. These conditions are:

$$(b + b'x)c = c(b + b'x) \ (\forall c \in K), \quad \text{and} \quad (b + b'x)x = x(b + b'x),$$

which translate into: $b \in F$, $b' \in K^\sigma$, and $b'\sigma(c) = cb'$ $(\forall c \in K)$. The computation of $Z(D)$ can therefore be divided into two cases:

(a) There does not exist $b_0 \in K^\sigma$ such that $\sigma(c) = b_0^{-1}cb_0$ $(\forall c \in K)$. In this case, $Z(D) = F$ from the above.
(b) There exists $b_0 \in K^\sigma$ such that $\sigma(c) = b_0^{-1}cb_0$ $(\forall c \in K)$. In this case, $b_0x \in Z(D)$, and the above considerations easily lead to $Z(D) = F \oplus F \cdot b_0x$.

(3) First assume $[D : Z(D)] = m < \infty$. Then $[D : F] \leq 2m$, so

$$[K : Z(K)] \leq [K : F] = [D : F]/2 \leq m < \infty.$$

Conversely, assume $[K : Z(K)] = n < \infty$. Then

$$[D : Z(D)] \leq [D : F] = 2[K : F] = 2[K : Z(K)][Z(K) : F] \leq 4n < \infty.$$

Ex. 14.13A. Let G be a group of automorphisms of a field K, and let F be the fixed field of G. Let $A = K * G$ be the skew group ring of G over K, with respect to the natural action of G on K. (The elements of A are formal combinations $\sum_{\sigma \in G} a_\sigma \sigma$ ($a_\sigma \in K$), with multiplication induced by $(a_\sigma \sigma)(b_\tau \tau) = a_\sigma \sigma(b_\tau) \sigma\tau$: see FC-(1.11).) Show that

(1) A is a simple ring with center F.
(2) $K(= K \cdot 1)$ is a maximal subfield of A.

Solution. The arguments here are almost the same as those used to prove the properties of the cyclic algebra in FC-p. 219 (except that here G may be infinite). To prove (1), let $\mathfrak{A} \neq 0$ be an ideal in A. Choose a nonzero element

$$z = b_{i_1} \sigma_{i_1} + \cdots + b_{i_r} \sigma_{i_r} \in \mathfrak{A} \quad (b_{i_j} \in K; \quad \sigma_{i_j} \text{ distinct in } G),$$

with r minimal. If we can show that $r = 1$, then $z \in U(A)$ and $\mathfrak{A} = A$. Assume, instead, $r \geq 2$. Then $\sigma_{i_1}(b) \neq \sigma_{i_r}(b)$ for some $b \in K$. Subtracting zb from $\sigma_{i_1}(b)z$ as in FC-p. 219, we obtain a "shorter" nonzero element in $\mathfrak{A}$, a contradiction. Next, we show that $C_A(K) \subseteq K$. Let

$$z = \sum b_\sigma \sigma \in C_A(K).$$

Then, for any $a \in K$, $za = az$ leads to $b_\sigma \sigma(a) = ab_\sigma$. Therefore, $b_\sigma \neq 0 \Rightarrow \sigma = 1$, and hence $z \in K$. From $C_A(K) \subseteq K$, (2) follows immediately. It is easy to see that $F \subseteq Z(A)$. To show the reverse inclusion, let $b \in Z(A)$. Then $b \in C_A(K) \subseteq K$, and

$$b\sigma = \sigma b = \sigma(b)\sigma \quad (\forall \sigma \in G)$$

shows that $b \in K^G = F$.

Ex. 14.13B. (W. Sinnott) Keep the notations in Exercise 13A.

(1) Show that K is a left A-module under the action

$$\left(\sum a_\sigma \sigma\right) \cdot c = \sum a_\sigma \sigma(c).$$

(2) Show that $_A K$ is faithful and simple, with $\operatorname{End}(_A K) \cong F$.
(3) Using the Wedderburn-Artin Theorem (but without assuming *any* facts from Galois Theory), show that $|G| < \infty$ iff $[K : F] < \infty$, and that, in this case, $|G| = [K : F]$.

Solution. (1) It suffices to show that, for any $c \in K$:

$$(a_\sigma \sigma) \cdot [(b_\tau \tau) \cdot c] = [(a_\sigma \sigma)(b_\tau \tau)] \cdot c.$$

This follows since

$$(a_\sigma \sigma) \cdot [(b_\tau \tau) \cdot c] = (a_\sigma \sigma) \cdot (b_\tau \tau(c)) = a_\sigma \sigma(b_\tau) \sigma\tau(c),$$

$$[(a_\sigma \sigma)(b_\tau \tau)] \cdot c = [a_\sigma \sigma(b_\tau) \sigma\tau] \cdot c = a_\sigma \sigma(b_\tau) \sigma\tau(c).$$

(2) Since $_KK$ is a simple K-module, of course $_AK$ is a simple A-module. By Exercise 13, A is a simple ring, so $_AK$ is automatically faithful. Consider now any $f \in \text{End}(_AK)$. Since we also have $f \in \text{End}(_KK)$, f is the multiplication on K by some element $a \in K$. The endomorphism law $f(\sigma \cdot c) = \sigma \cdot f(c)$ amounts to

$$a\sigma(c) = \sigma(ac) \quad (\forall \sigma \in G, \ c \in K).$$

Therefore, multiplication by a gives an A-endomorphism iff $a \in K^G = F$. This shows that $\text{End}(_AK) \cong F$.

(3) *First assume the simple ring A is not artinian.* In this case, the Wedderburn-Artin Theorem implies that K is infinite-dimensional over $\text{End}(_AK) = F$. Also, G must be an infinite group, for otherwise $\dim_K A < \infty$ would imply A is artinian. *Finally, assume A is artinian.* In this case, the Wedderburn-Artin Theorem implies that $n := [K : F] < \infty$, and that $A \cong \mathbb{M}_n(F)$. Computing F-dimensions, we find that

$$n^2 = \dim_F A = (\dim_K A) \, [K : F] = |G| \cdot n,$$

and therefore $|G| = n = [K : F]$.

Comment. (3) above is a somewhat surprising application of the Wedderburn-Artin Theory. The fact that $|G| = [K : F]$ in case G is finite is a basic fact in Galois Theory, known as "Artin's Theorem" (cf. p. 264 in Lang's "Algebra," 3rd edition). This theorem also asserts that when $|G| < \infty$, K/F is a Galois extension: Artin used this theorem to develop Galois Theory in his famous Notre Dame lecture notes.

The faithfulness of the module $_AK$ (in the general case), which we deduced from the simplicity of the ring A, means the following: *if $\sum_{\sigma \in G} a_\sigma \sigma$ ($a_\sigma \in K$) acts as the zero operator on K, then all $a_\sigma = 0$.* This property, called the "K-linear independence of automorphisms," was first discovered by Dedekind, and is often referred to as the "Dedekind-Artin Lemma." This lemma can be formulated and proved already for *monoids* of automorphisms of K.

Ex. 14.13C. (Berger-Reiner) Keep the notations in Exercise 13A, and assume that $n = |G| < \infty$. The field K has the natural structure of a left FG-module. Noether's *Normal Basis Theorem* in Galois Theory states that the FG-module K is free of rank 1, i.e. there exists an element $c \in K$ such that $K = \bigoplus_{\sigma \in G} F \cdot \sigma(c)$.

(1) Deduce this theorem from the Krull-Schmidt Theorem for finite-dimensional FG-modules (FC-(19.22)).

(2) Deduce from (1) that, for any subgroup $H \subseteq G$, the fixed field $L = K^H$ has a primitive element over F.

Solution. (1) Since G acts trivially on F, we may (and shall) view $R = FG$ as an F-subalgebra of $A = K * G$. The module structure on $_R K$ is just that on $_A K$ restricted to the subring R. *We claim that $_R A$ is a free R-module of rank n.* To see this, fix any F-basis $\{c_i : 1 \le i \le n\}$ on K. Calculating in A, we have

$$A = \bigoplus_\sigma K\sigma = \bigoplus_\sigma \sigma \cdot K = \bigoplus_\sigma \sigma \cdot \left(\bigoplus_i F c_i \right)$$

$$= \bigoplus_i \left(\bigoplus_\sigma F\sigma \right) c_i = \bigoplus_{i=1}^n R \cdot c_i.$$

Since the c_i's are units in A, each $R \cdot c_i$ is R-free of rank 1, so $_R A \cong {_R R^n}$. On the other hand, by Exercise 13B, $_A K$ is the unique simple module over the simple F-algebra $A \cong \mathbb{M}_n(F)$, so we have $_A A \cong {_A K^n}$. Restricting this isomorphism to the subring $R \subseteq A$, we get

$$_R K^n \cong {_R A} \cong {_R R^n},$$

so the Krull-Schmidt Theorem yields $_R K \cong {_R R}$!

(2) For $c \in K$ as in (1), let $d = \sum_{\tau \in H} \tau(c)$, which clearly lies in $K^H = L$. We fix a coset decomposition

$$G = \bigcup_{i=1}^m \sigma_i H,$$

where $m = [G : H]$. Since

$$\{\sigma_i \tau(c) : \ 1 \le i \le m, \ \tau \in H\}$$

is an F-basis on K, $\sigma_1 d, \ldots, \sigma_m d$ are linearly independent over F (and in particular distinct). Thus, $d \in L$ has at least m distinct conjugates over F, so $[F(d) : F] \ge m$. On the other hand, $[K : L] = |H|$ by Exercise 13A, so

$$[L : F] = [K : F]/[K : L] = |G|/|H| = m.$$

Since $d \in L$, we must have $L = F(d)$, as desired.

Comment. Noether's Normal Basis Theorem is proved in many books in Galois theory and algebraic number theory, usually by calculations with determinants. By Galois theory, part (2) above amounts to the existence of primitive elements for any finite separable extension of F. This theorem is usually proved by considering separately the case of finite fields and infinite fields. The nice approach to both of these theorems in the present exercise is taken from the paper of T.R. Berger and I. Reiner in the MAA Monthly, vol. *82* (1975), 915–918. This paper of the two authors, however, did not originate as a joint work. At about the same time and independently of each other, Berger and Reiner submitted papers to the MAA Monthly on new proofs of Noether's Normal Basis Theorem. Since the two papers used similar ideas, they were combined and reorganized into a joint paper (*loc. cit.*), at the suggestion of the Monthly's editor, A. Rosenberg. This interesting piece of anecdotal information was supplied to me by Irma Reiner. For a "unified version" of the Primitive Element Theorem and the Normal Basis Theorem, see the recent article of W. C. Waterhouse in Comm. Algebra *22* (1994), 2305–2308.

Ex. 14.14. ("Dickson's Example") Let $K = \mathbb{Q}(v)$, where $v^3 + v^2 - 2v - 1 = 0$. This is a cyclic Galois extension of $\mathbb{Q}$ with $\mathrm{Gal}(K/F) = \langle \sigma \rangle$, where $\sigma(v) = v^2 - 2$. As in Exer. 5, let

$$D = (K/\mathbb{Q}, \sigma, 2) = K \oplus Kx \oplus Kx^2 = \mathbb{Q}\langle v, x \rangle,$$

with the usual cyclic algebra relations. It is known that D is a division algebra. Find the inverse for $v + x$ in D.

Solution. We try to find $b, c, d \in K$ such that

$$d = (b + cx + x^2)(v + x) = (bv + 2) + (b + c\sigma v)x + (c + \sigma^2 v)x^2.$$

Clearly,

$$c = -\sigma^2(v) = v^2 + v - 1,$$
$$b = -c\sigma(v) = \sigma^2(v)\sigma(v).$$

Since $\sigma^2(v)\sigma(v)v = 1$, we have $b = v^{-1} = v^2 + v - 2$. Finally

$$d = (v^2 + v - 2)v + 2 = v^3 + v^2 - 2v + 2 = 3.$$

Therefore, $(v + x)^{-1}$ is given by

$$[x^2 + (v^2 + v - 1)x + (v^2 + v - 2)]/3 \in D.$$

Ex. 14.15. Let $K = \mathbb{Q}(v)$ be the cubic field defined by the minimal equation $v^3 + av + b = 0$ where a, b are odd integers. Show that for any element $\alpha = p + qv + rv^2 \in K$ where $p, q, r \in \mathbb{Z}$, we have

$$N_{K/\mathbb{Q}}(\alpha) \equiv 1 + (p+1)(q+1)(r+1) \pmod{2}.$$

Using this, show that, for any even integer n, if $n \in N_{K/\mathbb{Q}}(\dot{K})$, then $8 \mid n$.

Solution. Since $\alpha v = -rb + (p - ra)v + qv^2$, and

$$\alpha v^2 = pv^2 + q(-av - b) + r(-av - b)v$$
$$= -qb - (qa + rb)v + (p - ra)v^2,$$

the matrix of left multiplication by α on K (with respect to the basis $\{1, v, v^2\}$) is given by:

$$\begin{pmatrix} p & -rb & -qb \\ q & p-ra & -(qa+rb) \\ r & q & p-ra \end{pmatrix}.$$

Computing the determinant of this matrix modulo 2, we have:

$$N_{K/\mathbb{Q}}(\alpha) \equiv p(p^2 + r^2a^2 + q^2a + qrb) + rb(pq + r^2b) + qb(q^2 + pr + r^2a).$$

Since $s^n \equiv s \pmod 2$ for any $s \in \mathbb{Z}$, and $a \equiv b \equiv 1 \pmod 2$,

$$N_{K/\mathbb{Q}}(\alpha) \equiv p + pr + pq + pqr + pqr + r + q + pqr + qr$$

$$\equiv p + pr + pq + r + q + pqr + qr$$

$$\equiv 1 + (p+1)(q+1)(r+1) \pmod 2.$$

Now let $n \in 2\mathbb{Z}$, and assume $n \in N(\dot{K})$, where $N = N_{K/\mathbb{Q}}$. Write $n = N(\alpha/m)$ where

$$\alpha = p + qv + rv^2 \quad (p, q, r \in \mathbb{Z}),$$

and $m \in \mathbb{N}$ is chosen as small as possible. By the above, we have

$$m^3 n \equiv 1 + (p+1)(q+1)(r+1) \pmod 2.$$

Since n is even, p, q, r must all be even. But then m must be odd (by its minimal choice). Writing $p = 2p_0, q = 2q_0$ and $r = 2r_0$, we have

$$m^3 n = 8\,N(p_0 + q_0 v + r_0 v^2) \in 8\mathbb{Z},$$

so $n \in 8\mathbb{Z}$, as desired. (The last part of the proof here is just a repetition of the argument given in *FC*-p. 227.)

Ex. 14.16. Show that $K = \mathbb{Q}(v)$ with $v^3 - 3v + 1 = 0$ is a cyclic cubic field, and find the conjugates of v. Then, proceeding as in Dickson's example (cf. *FC*-pp. 226–227), show that $D = \mathbb{Q}\langle v, x\rangle$ defined by the relations

$$(*) \qquad v^3 - 3v + 1 = 0, \quad x^3 = 2, \quad \text{and} \quad xv = (v^2 - 2)x$$

is a 9-dimensional central division algebra over $\mathbb{Q}$.

Solution. We start more generally with a cubic field $K = \mathbb{Q}(v)$ defined by a minimal equation $v^3 + av + b = 0$. By a known formula in field theory (see Jacobson's "Basic Algebra I," p. 259), the discriminant for $K/\mathbb{Q}$ with respect to the basis $\{1, v, v^2\}$ is $d = -(4a^3 + 27b^2)$. Furthermore, $K/\mathbb{Q}$ is Galois iff $d \in \dot{\mathbb{Q}}^2$. In our case, $v^3 - 3v + 1$ being clearly irreducible over $\mathbb{Q}$, the discriminant is

$$d = -(27 - 4 \cdot 3^3) = 3^4,$$

so $K/\mathbb{Q}$ is Galois.

More directly, in the polynomial ring $K[t]$, we can use long division to obtain the factorization

$$t^3 - 3t + 1 = (t - v)\left(t^2 + vt + (v^2 - 3)\right).$$

By brute-force computation, one can check that the discriminant $12 - 3v^2$ of the quadratic factor above is a perfect square $(2v^2 + v - 4)^2$ in K, so we can further factor

$$t^3 - 3t + 1 = (t - v)\left(t - (v^2 - 2)\right)\left(t + (v^2 + v - 2)\right).$$

Therefore, the conjugates of v are

$$v, \quad v^2 - 2 \quad \text{and} \quad 2 - v - v^2$$

in K, and K is Galois. We can write $\text{Gal}\,(K/\mathbb{Q}) = \langle\sigma\rangle$ where $\sigma(v) = v^2 - 2$. The cyclic algebra

$$D = (K/\mathbb{Q}, \sigma, 2) = K \oplus Kx \oplus Kx^2 = \mathbb{Q}\langle v, x\rangle$$

is, therefore, defined exactly by the relations (∗). By Exercise 15, $2 \notin N_{K/\mathbb{Q}}(\dot{K})$, so by FC-(14.8), D is a (9-dimensional central) division algebra over $\mathbb{Q}$.

Ex. 14.17. Do the same for $K = \mathbb{Q}(v)$ with $v^3 - 7v + 7 = 0$.

Solution. After testing for roots, we see that $f(t) = t^3 - 7t + 7$ is irreducible over $\mathbb{Q}$. The discriminant of K with respect to $\{1, v, v^2\}$ is

$$d = -4(27 \cdot 7^2 - 4 \cdot 7^3) = 2^2 7^2 \in \dot{\mathbb{Q}}^2,$$

so K is again Galois over $\mathbb{Q}$. After some lengthy computations, we arrive at the following factorization of $f(t)$ over K:

$$f(t) = (t - v)\left(t^2 + vt + (v^2 - 7)\right)$$
$$= (t - v)\left[t - (3v^2 + 4v - 14)\right]\left[t + (3v^2 + 5v - 14)\right].$$

Therefore, the conjugates of v are

$$v, \quad 3v^2 + 4v - 14 \quad \text{and} \quad -(3v^2 + 5v - 14).$$

Writing $\text{Gal}(K/\mathbb{Q}) = \langle\sigma\rangle$ where $\sigma(v) = 3v^2 + 4v - 14$, we can construct $D = (K/\mathbb{Q}, \sigma, 2) = \mathbb{Q}\langle v, x\rangle$ with the relations

$$v^3 - 7v + 7 = 0, \quad x^3 = 2 \quad \text{and} \quad xv = (3v^2 + 4v - 14)x.$$

Again, Exercise 15 gives $2 \notin N_{K/\mathbb{Q}}(\dot{K})$, so D is a 9-dimensional central division algebra over $\mathbb{Q}$.

Comment. There is a field-theoretic theorem behind Exercises 16 and 17. It is known that every cyclic cubic field K over $\mathbb{Q}$ is defined by a minimal equation

$$v^3 - 3(r^2 + 3s^2)v + 2r(r^2 + 3s^2) = 0,$$

where $r, s \in \mathbb{Q}$ and $s \neq 0$. Conversely, if the polynomial on the left-hand side above is irreducible over $\mathbb{Q}$, then it defines a cyclic cubic field over $\mathbb{Q}$. In this case, the two other conjugates of v are

$$\pm\frac{1}{s}(r^2 + 3s^2) \mp \frac{r \pm s}{2s}v \mp \frac{1}{2s}v^2 \in K.$$

For a proof of these facts, see C. MacDuffee's "An Introduction to Abstract Algebra," p. 90, J. Wiley, 1940. The polynomial $t^3 - 3t + 1$ in Exercise 16 corresponds to $r = s = 1/2$; the polynomial $t^3 - 7t + 7$ in Exercise 17 corresponds to $r = 3/2$ and $s = 1/6$.

§15. Tensor Products and Maximal Subfields

A subfield K of a division ring D is a maximal subfield iff it is self-centralizing, in the sense that the centralizer $C_D(K)$ of K in D is just K. Such a maximal subfield always contains $F = Z(D)$, the center of D. The study of the maximal subfields of D is of great importance in understanding the structure of D, especially in the case when D is a centrally finite division ring.

One of the principal tools for studying division subalgebras $K \subseteq D$ in general is the formation of the tensor product $R = D \otimes_F K^{\mathrm{op}}$, where K^{op} denotes the opposite ring of K. Under the action

$$(d \otimes a^{\mathrm{op}})(v) = dva \quad \text{for} \quad d, \ v \in D \quad \text{and} \quad a \in K,$$

D becomes a faithful simple left R-module (cf. *FC*-(15.3)). The F-algebra R is always simple, and acts densely on D_L, where $L = \mathrm{End}(_RD)$ is essentially the centralizer $C_D(K)$. In the case when $\dim_F K < \infty$, R is even a simple *artinian* ring. Here, the density result mentioned above leads directly to the so-called Double Centralizer Theorem for $K \subseteq D$ (cf. *FC*-(15.4)). As it turns out, the division ring D is centrally finite iff it has a maximal subfield that is finite-dimensional over F. In this case, the maximal subfields of D are precisely those subfields $K \supseteq F$ such that $(\dim_F K)^2 = \dim_F D$, or such that $D \otimes_F K$ is a matrix algebra over K. Moreover, classical results of Noether, Köthe and Jacobson guarantee that D always admits a maximal subfield that is *separable* over F (cf. *FC*-(15.12)).

The full text for §15 in *FC* offers an assortment of other results, e.g. Kaplansky's Theorem on division rings D which are "radical" over their centers F (every $d \in D$ has a power in F), the Gerstenhaber-Yang Theorem on division rings that have finite right dimension over a real-closed subfield, and the Brauer-Albert Theorem on the existence of special F-bases for centrally finite division rings, etc.

The exercise set for this section is relatively modest, offering a few examples for the general theory of maximal subfields, and some counterexamples (for instance for the Double Centralizer Theorem for centrally infinite division rings). The first two exercises, due to Hua, Kaplansky and Jacobson, are taken directly from the classical literature (see, for instance, Jacobson's "Structure of Rings").

Exercises for §15

Ex. 15.1. (Hua, Kaplansky) Suppose, in a noncommutative division ring D, an integer $n(x) > 0$ is assigned to every element $x \in D$ in such a way that $n(axa^{-1}) = n(x)$ for every $a \in D^*$. Show that $\{x^{n(x)} : x \in D^*\}$ generates D as a division ring.

Solution. Let K be the division ring generated by $\{x^{n(x)} : x \in D^*\}$ in D. Clearly, K is invariant under all inner automorphisms of D. Assume, for the moment that $K \neq D$. By the Cartan-Brauer-Hua Theorem $(FC\text{-}(13.17))$, $K \subseteq Z(D)$. But then D is radical over $Z(D)$, so by $FC\text{-}(15.15)$, we must have $D = Z(D)$, which contradicts $K \neq D$.

Ex. 15.2. (Jacobson) Let D be an algebraic division algebra of infinite dimension over its center F. Show that there exist elements in D of arbitrarily high degree over F.

Solution. Assume otherwise and let $K \supseteq F$ be a finite separable field extension of F in D with the largest degree. Let $L = C_D(K) \supseteq K$, which is also a division ring. Since $[K : F] < \infty$, $FC\text{-}(15.4)$ implies that $Z(L) = K$. If $K \neq L$, there would exist an element $\alpha \in L\backslash K$ which is separably algebraic over K, by $FC\text{-}(15.11)$. This would contradict the choice of K. Therefore, we must have $K = L$, in which case K is a maximal subfield of D. But then $FC\text{-}(15.8)$ implies that $[D : F] < \infty$, again a contradiction.

Ex. 15.3. Let R be a field, $(G, <)$ be a nontrivial ordered group, and $\omega : G \to \text{Aut } R$ be an injective group homomorphism. Show that R is a maximal subfield of the Mal'cev-Neumann division ring $D = R((G, \omega))$.

Solution. It suffices to show that $C_D(R) \subseteq R$ (cf. $FC\text{-}(15.7)$). Consider any element

$$\alpha = \sum \alpha_g g \in C_D(R).$$

(Recall that the formal sum may be infinite in general.) Then, for any $r \in R$, we have

$$\sum r\alpha_g g = r\alpha = \alpha r = \left(\sum \alpha_g g\right) r = \sum \alpha_g \omega_g(r) g.$$

Suppose $g \in G$ is such that $\alpha_g \neq 0$. Then $r\alpha_g = \alpha_g w_g(r)$ implies that $w_g(r) = r$ for every $r \in R$. This means that $w_g = I$ and hence $g = 1$ (by the injectivity of w). We have thus shown that $\alpha = \alpha_1 \in R$.

Ex. 15.4. Let $R = \mathbb{Q}(y)$ and let σ be the $\mathbb{Q}$-automorphism of R defined by $\sigma(y) = 2y$. Let D be the division ring $R((x,\sigma))$. Show that $K_1 = \mathbb{Q}(y)$ and $K_2 = \mathbb{Q}((x))$ are both maximal subfields of D. Let $L_1 = \mathbb{Q}(y^2)$ and $L_2 = \mathbb{Q}((x^2))$, both of which contain $Z(D) = \mathbb{Q}$. Show that $C_D(L_i) = K_i$ and so $C_D(C_D(L_i)) = K_i \supsetneq L_i$ for $i = 1, 2$. (This provides counterexamples to the Double Centralizer Theorem for subfields of D infinite-dimensional over the center.)

Solution. We can apply the previous exercise with $G = \langle x \rangle$ and

$$w : G \to \text{Aut } R \quad \text{given by} \quad w(x) = \sigma.$$

Since σ is of infinite order, w is an injection. It follows that K_1 is a maximal subfield of D. Alternatively, let us show directly that $C_D(L_i) \subseteq K_i$ ($i = 1, 2$). (Note that this also implies $C_D(K_i) = K_i$, so each K_i is a maximal subfield of D.) Consider any

$$\alpha = \sum_{j \geq m} a_j x^j \in C_D(L_i), \quad \text{where} \quad a_j \in R.$$

Let $i = 1$. Commuting α with $b \in L_1$, we have $a_j \sigma^j(b) = ba_j$ for $j \geq m$. Therefore, whenever $a_j \neq 0$, we have $\sigma^j(b) = b$ for all $b \in L_1$. The latter is possible only when $j = 0$, so we see that $\alpha = a_0 \in K_1$.

Now let $i = 2$. Commuting α with x^2, we get $a_j = \sigma^2(a_j)$ for $j \geq m$. Since the fixed field of σ^2 is $\mathbb{Q}$, this gives $a_j \in \mathbb{Q}$ for all $j \geq m$, and so $\alpha \in \mathbb{Q}((x)) = K_2$.

Ex. 15.5. Let D be a centrally finite division ring with center k. Let $[D, D]$ be the additive subgroup generated by all additive commutators $ab - ba$, where $a, b \in D$. Show that $[D, D] \subsetneq D$.

Solution. Let K be a maximal subfield of D. Consider an element of the form $(ab - ba) \otimes \alpha \in D^K$, where $a, b \in D$ and $\alpha \in K$. Since

$$(ab - ba) \otimes \alpha = (a \otimes \alpha)(b \otimes 1) - (b \otimes 1)(a \otimes \alpha),$$

we see that $[D, D]^K \subseteq [D^K, D^K]$. Thus, it suffices to check that

$$[D^K, D^K] \subsetneq D^K.$$

Now $D^K \cong \mathbb{M}_r(K)$ for some r, and under this isomorphism, $[D^K, D^K]$ maps into the subspace

$$S = \{M \in \mathbb{M}_r(K) : \text{tr}(M) = 0\}.$$

Since $S \subsetneq \mathbb{M}_r(K)$, we see that $[D^K, D^K] \subsetneq D^K$, as desired.

Ex. 15.6. Let D be a centrally finite division algebra with center k, and let K be a subfield of D containing k. Show that K is a splitting field for D iff K is a maximal subfield of D.

Solution. Let $L = C_D(K)$, and recall from FC-(15.4) that, if $r = \dim_F K$, then $\dim(D_L) = r$, and $D \otimes_F K \cong \mathbb{M}_r(L)$. Therefore, K is a splitting field for D iff $D \otimes_F K$ is a matrix ring over K, iff $K = L = C_D(K)$, iff K is a maximal subfield of D.

§16. Polynomials over Division Rings

In the polynomial ring $D[t]$ over a division ring D, the indeterminate t commutes with the scalars. However, if we "evaluate" $f(t) = \sum a_i t^i$ at $d \in D$ (to get $f(d) = \sum a_i d^i$), the element d need no longer commute with the a_i's. What this means is that evaluation at d is no longer a ring homomorphism from $D[t]$ to D. Nevertheless, a "weak" homomorphism property survives, and a suitable form of the usual Remainder Theorem remains true: see FC-(16.2) and (16.3). Thanks to these facts, one can still develop a substantial amount of polynomial theory, at least in one variable, over the division ring D.

In this new polynomial theory, various new phenomena come to light. One has Dickson's Theorem on the conjugacy of algebraic elements over the center, and Wedderburn's Theorem on the factorization of minimal polynomials of such algebraic elements. For the theory of roots of polynomials in one variable, one has the theorems of Gordon-Motzkin, Bray-Whaples, as well as the more classical results of Niven, Jacobson, and Baer. By and large, this theory seems to have a life of its own, and bears little resemblance to the theory of polynomials in the commutative case.

The exercises in this section deal with various aspects of polynomials over division rings D, including: ideals in $D[t]$ and central polynomials, primitivity questions for $D[t]$, interpolation and Vandermonde matrices, and equations of the type $ax - xb = c$, etc. A few explicit polynomial equations are solved over the real quaternions in Exercises 3 and 15, with somewhat surprising results!

Exercises for §16

In the following exercises, D denotes a division ring with center F.

Ex. 16.1. Let K be a division subring of D and let $f(t) = h(t)k(t) \neq 0$ in $D[t]$. If $f(t), k(t) \in K[t]$, show that $h(t) \in K[t]$.

Solution. We induct on $\deg h$, the case $\deg h = 0$ being trivial. For the inductive step, write

$$h(t) = a_n t^n + h'(t)$$

where $a_n \neq 0$, and $h' = 0$ or deg $h' < n$. By comparing leading coefficients in $f(t) = h(t)k(t)$, we see that $a_n \in K^*$. Since

$$(a_n t^n + h'(t))k(t) \in K[t] \Longrightarrow h'(t)k(t) \in K[t],$$

the inductive hypothesis gives $h'(t) \in K[t]$. Therefore, $h(t) \in K[t]$, as desired.

Ex. 16.2. Let K be a division subring of D. An element $d \in D$ is said to be (right) algebraic over K if $f(d) = 0$ for some polynomial $f(t) \in K[t] \backslash \{0\}$. Among these polynomials, there is a unique monic $f_0(t) \in K[t]$ of the least degree, which is called the minimal polynomial of d over K. Is $f_0(t)$ always irreducible in $K[t]$? In case $f_0(t)$ is indeed irreducible in $K[t]$, are two roots of $f_0(t)$ in D always conjugate by an element of $C_D(K)$?

Solution. In general, $f_0(t)$ need not be irreducible in $K[t]$. For instance, let D be the division ring of rational quaternions, and let $K = \mathbb{Q}[i]$. The element $d = j$ satisfies the polynomial $t^2 + 1 \in K[t]$, and does not satisfy any linear polynomial in $K[t]$ (since $j \notin K$). Hence, the minimal polynomial $f_0(t)$ of d over K is $t^2 + 1$. This factors into $(t + i)(t - i)$ in $K[t]$! The answer to the second question in the exercise is also in the negative. Here, we consider the two elements

$$d_1 = j + i, \quad d_2 = j - i.$$

Both of these have minimal polynomial

$$f_0(t) = t^2 + 2 \in K[t],$$

which is irreducible over K. However, d_1, d_2 are not conjugate by an element of $C_D(K) = K$. Indeed, if $x + yi \in K^*$ is such that

$$(x + yi)(j + i) = (j - i)(x + yi),$$

where $x, y \in \mathbb{Q}$, we get

$$xj + xi + yk - y = xj - yk - xi + y.$$

This yields $x = y = 0$, a contradiction.

Ex. 16.3. In the division ring $\mathbb{H}$ of real quaternions, find all roots of the quadratic polynomials

$$f_1(t) = (t - j)(t - i), \quad f_2(t) = (t - (j - i))(t - i),$$

and of the cubic polynomial $f_3(t) = (t - k)(t - j)(t - i)$.

Solution. (1) Of course $f_1(i) = 0$. We claim that i is the only root. Indeed, suppose $f_1(z) = 0$ where $z \neq i$ in $\mathbb{H}$. Expanding $f_1(t)$ into $t^2 - (i + j)t + ji$, we have

$$z^2 - (i + j)z = -ji,$$

so $(z - i)z = j(z - i)$. This shows that z is conjugate to j, so in particular $z^2 = j^2 = -1$. Using this, we have

$$ji = (i + j)z - z^2 = (i + j)z + 1.$$

Since this is a linear equation in z, there is a unique solution in $\mathbb{H}$. But i is obviously a solution, so no other solution can exist.

(2) We have $f_2(t) = t^2 - jt + (1 - k)$ by expansion. Aside from the root i, we also have another root $i + j$, since

$$(i + j)^2 - j(i + j) + 1 - k = -2 + ij + ji - ji - j^2 + 1 - k = 0.$$

Now the two roots $i, i + j$ are not conjugate in $\mathbb{H}$, since $(i + j)^2 \neq -1$. The theorem of Bray and Whaples in FC-(16.13) then implies that i and $i + j$ are the *only* roots of f_2 in $\mathbb{H}$.

Aside from this solution, the following direct transformations are of interest:

$$\begin{aligned}
f_2(t) &= (t - (j - i))(t - i) \\
&= t^2 - (j - i + i)t + (j - i)i \\
&= t^2 - (i + j - i)t - i(i + j) \\
&= (t + i)(t - (i + j)).
\end{aligned}$$

(3) Since i, j and k are conjugate in $\mathbb{H}$, FC-(16.4) implies that any root of f_3 is conjugate to i. We claim that i is the only root of f_3. Indeed, assume f_3 has another root (in the conjugacy class of i). By FC-(16.17),

$$f_3(t) = (t - a)(t^2 + 1) \quad \text{for some} \quad a \in \mathbb{H}.$$

(Note that $t^2 + 1$ is the minimal polynomial of i over $\mathbb{R}$). Comparing constant coefficients, we have $a = kji = 1$. Comparing coefficients of t^2, we get $-1 = -(i + j + k)$, a contradiction.

Ex. 16.4. Let $c_0, \ldots, c_n \in D$ be pairwise nonconjugate elements. Show that the $(n + 1) \times (n + 1)$ Vandermonde matrix

$$V = V(c_0, \ldots, c_n) = \begin{pmatrix} 1 & \cdots & 1 \\ c_0 & \cdots & c_n \\ c_0^2 & \cdots & c_n^2 \\ \vdots & & \vdots \\ c_0^n & \cdots & c_n^n \end{pmatrix}$$

is invertible over D. If $d_0, \ldots, d_n$ are any given elements in D, show that there exists a unique polynomial

$$f(t) = a_n t^n + \cdots + a_0 \in D[t]$$

such that $f(c_i) = d_i$ for $0 \leq i \leq n$.

Solution. We view V as a D-linear transformation acting on the right of the left D-space $\{(b_0, \ldots, b_n)\}$. It suffices to show that $\ker(V) = 0$. Let

$$g(t) = b_0 + b_1 t + \cdots + b_n t^n \in D[t].$$

For $(b_0, \ldots, b_n) \in \ker(V)$, we have

$$0 = (b_0, \ldots, b_n)V = (g(c_0), \ldots, g(c_n)),$$

so g has roots in $n+1$ different conjugacy classes of D. By the Gordon-Motzkin Theorem (FC-(16.4)), $g(t)$ must be the zero polynomial, so $(b_0, \ldots, b_n) = 0$. For the second conclusion of the exercise, let $d_0, \ldots, d_n$ be given. To find the desired polynomial f amounts to solving the equation

$$(a_0, \ldots, a_n)V = (d_0, \ldots, d_n).$$

Since V is invertible, a solution $(a_0, \ldots, a_n)$ exists and is unique.

Ex. 16.5. Let a, b, c be three *distinct* elements in D.

(1) Show that the 3×3 Vandermonde matrix $V = V(a, b, c)$ is not invertible if and only if
$$(b - a)b(b - a)^{-1} = (c - a)c(c - a)^{-1}.$$

(2) Show that V is invertible if a, b, c do not lie in a single conjugacy class of D.

Solution. (2) follows from (1) and from symmetry. In the following, we offer two proofs for (1).

For the first proof, note that the noninvertibility of V amounts to the existence of a nonzero monic polynomial f of degree ≤ 2 such that

$$f(a) = f(b) = f(c) = 0.$$

Thus, we seek a polynomial $f(t) = (t - d)(t - a)$ which vanishes on b and c. By FC-(16.3),

$$f(b) = \left((b - a)b(b - a)^{-1} - d\right)(b - a),$$
$$f(c) = \left((c - a)c(c - a)^{-1} - d\right)(c - a).$$

These are both zero for some d iff

$$(b - a)b(b - a)^{-1} = (c - a)c(c - a)^{-1}.$$

The latter is, therefore, the necessary and sufficient condition for the existence of the desired polynomial f.

The second proof is independent of the theory of polynomials over division rings. It uses, instead, the basic technique of elementary row transformations. First, we left-multiply V by a suitable matrix (which corresponds to two successive row operations):

$$\begin{pmatrix} 1 & 0 & 0 \\ -a & 1 & 0 \\ 0 & -a & 1 \end{pmatrix} V = \begin{pmatrix} 1 & 1 & 1 \\ 0 & b-a & c-a \\ 0 & (b-a)b & (c-a)c \end{pmatrix}.$$

We are now reduced to looking at the 2×2 matrix

$$V' = \begin{pmatrix} b-a & c-a \\ (b-a)b & (c-a)c \end{pmatrix},$$

and finding a criterion for its noninvertibility. A second left-multiplication yields

$$\begin{pmatrix} (c-a)^{-1} & 0 \\ 0 & (c-a)^{-1} \end{pmatrix} V' = \begin{pmatrix} (c-a)^{-1}(b-a) & 1 \\ (c-a)^{-1}(b-a)b & c \end{pmatrix},$$

and a third and last one yields

$$\begin{pmatrix} (c-a)^{-1}(b-a) & 1 \\ (c-a)^{-1}(b-a)b - c(c-a)^{-1}(b-a) & 0 \end{pmatrix}.$$

Therefore, V is not invertible iff

$$(c-a)^{-1}(b-a)b = c(c-a)^{-1}(b-a),$$

or equivalently,

$$(b-a)b(b-a)^{-1} = (c-a)c(c-a)^{-1}.$$

For instance, if $b, c \in Z(D)$, or $b-a, c-a \in Z(D)$, then

$$b \neq c \Longrightarrow (b-a)b(b-a)^{-1} \neq (c-a)c(c-a)^{-1},$$

so we can conclude that V is invertible.

Comment. The converse of (2) above is not true in general. However, if D is the division ring $\mathbb{H}$ of real quaternions, the converse is true. For, if a, b, c are elements in a given conjugacy class A of $\mathbb{H}$, then A satisfies a monic quadratic polynomial over $\mathbb{R}$, and our earlier work in Exercise 4 shows that V is *not* invertible. The simplest examples illustrating this are

$$V(i, j, k) \quad \text{and} \quad V(j+k, k+i, i+j).$$

Of course these Vandermonde matrices are not invertible since their first and third rows are obviously linearly dependent! The general theory of Vandermonde matrices over division rings was developed by the author in an article in Expos. Math. *4* (1986), 193–215.

Ex. 16.6. If a conjugacy class A of D is not algebraic over F, show that no $h(t) \in D[t]\backslash\{0\}$ can vanish identically on A.

Solution. Suppose

$$h(t) = \sum_{i=0}^{n} b_i t^i \in D[t]\backslash\{0\}$$

vanishes identically on A. We may assume that n is chosen minimal, and that $b_n = 1$. Fix $a \in A$, and use the notation a^z for the conjugate $z^{-1}az$. For any $x, y \in D^*$, we have

$$0 = \sum b_i \left(a^{xy^{-1}}\right)^i = \sum b_i(a^i)^{xy^{-1}}, \quad \text{so}$$

$$0 = \sum b_i^y(a^i)^x = \sum b_i^y(a^x)^i.$$

This means that $h'(t) = \sum b_i^y t^i$ also vanishes identically on A. Since $h(t)$ and $h'(t)$ are both monic of degree n, we must have $b_i = b_i^y$ for all $y \in D^*$. This implies that all $b_i \in F$, so A is now algebraic over F, a contradiction.

Ex. 16.7. Let $f(t) = g(t)h(t)$ where $g \in F[t]$ and $f, h \in D[t]$. Show that $d \in D$ is a root of f iff d is a root of g or a root of h.

Solution. Since $f = gh = hg$, if d is a root of g or h, then d is a root of f by the Remainder Theorem (see FC-(16.2)). Now assume d is a root of f, but not a root of h. Let $a = h(d) \neq 0$. By FC-(16.3),

$$0 = f(d) = g(ada^{-1})h(d),$$

so $g(ada^{-1}) = 0$. Since $g(t) \in F[t]$, $g(ada^{-1})$ is just $ag(d)a^{-1}$. Therefore, $g(d) = 0$.

Ex. 16.8. For a polynomial $f(t) \in D[t]$, write (f) for the ideal generated by f. By a central factor of f, we mean a polynomial $h \in F[t]$ that divides f. Assume $f \neq 0$.

(a) Let $f(t) = g(t)f_0(t)$, where $f_0(t)$ is a central factor of $f(t)$ of the largest possible degree. Show that $(f) = (f_0)$.
(b) Show that $(f) = D[t]$ iff f has no nonconstant central factor.
(c) If f vanishes on some conjugacy class A of D, then $(f) \subsetneq D[t]$.

Solution. (a) Let h be a monic polynomial of the least degree in (f). By the solution to Exercise 1.16, $(f) = D[t]h$ and h is central. Let f_0 and g be as in (a), and assume (for convenience) that f_0 is monic. We are done if we can show that $f_0 = h$. Write $h = \sum g_i f g_i'$, where $g_i, g_i' \in D[t]$. Then

$$h = \left(\sum g_i g g_i'\right) f_0,$$

so $\deg h \geq \deg f_0$. By the choice of f_0, we must have $\deg h = \deg f_0$. Since h, f_0 are both monic, it follows that $h = f_0$.

(b) follows immediately from (a).

(c) Let $k(t)$ be a monic polynomial of the smallest degree vanishing on A. By the solution to Exercise 6, we have $k(t) \in F[t]$. The usual Euclidean algorithm argument shows that $f(t) = q(t)k(t)$ for some $q(t) \in D[t]$. Therefore, f has the central factor k, so $(f) \subsetneq D[t]$ by (b).

Ex. 16.9. (Jacobson) Assume that $d = \dim_F D < \infty$. For any polynomial $f(t) = \sum_{i=0}^{n} a_i t^i \in R = D[t]$ of degree n, show that there exists a nonzero polynomial $g(t) \in R$ of degree $n(d-1)$ such that $f(t)g(t) = g(t)f(t) \in F[t]$. As a consequence, show that R is neither left primitive nor right primitive.

Solution. The second conclusion follows from the first, in view of Exercise 11.20. To construct the polynomial g in the first conclusion, we may assume $a_n = 1$. Let
$$V = De_1 \oplus \cdots \oplus De_n$$
be a left D-vector space with basis $e_1, \ldots, e_n$, and let A be the D-endomorphism defined by: $e_i A = e_{i+1}$ for $i < n$, and

(1) $$e_n A = -a_0 e_1 - a_1 e_2 - \cdots - a_{n-1} e_n.$$

The matrix of A with respect to $\{e_i : 1 \le i \le n\}$ is the companion matrix of f, with last row
$$(-a_0, -a_1, \ldots, -a_{n-1}).$$

We shall identify $E = \mathrm{End}(_D V)$ with $\mathbb{M}_n(D)$ via the basis $\{e_i : 1 \le i \le n\}$, and view $\mathbb{M}_n(D)$ as a (D,D)-bimodule.[4] For any $i < n$, we have

(2) $$e_1(A^i a) = (e_1 A^i)(Ia) = e_{i+1}(Ia) = ae_{i+1} = a(e_1 A^i).$$

We claim that
$$f(A) = \sum_{j=0}^{n} a_j A^j = 0.$$

For this, it suffices to show that $e_k f(A) = 0$ for all k. Write $k = i + 1$, where $0 \le i < n$. By (2) and (1):

$$e_k f(A) = (e_1 A^i)\left(\sum_{j=0}^{n} a_j A^j\right) = \sum_{j=0}^{n} (e_1 A^i a_j)A^j$$

$$= \sum_{j=0}^{n} (a_j e_1 A^i)A^j = \sum_{j=0}^{n} \left(a_j e_1 A^j\right) A^i$$

$$= (a_0 e_1 + a_1 e_2 + \cdots + a_{n-1} e_n + e_n A)A^i = 0,$$

---

[4] Note that E itself does not have a natural D-structure on either side. We get a (D,D)-bimodule structure on E only by identifying E with $\mathbb{M}_n(D)$ using a basis.

as desired. On the other hand, if $r(t) = \sum_{j=0}^{n-1} r_j t^j$ has degree $< n$ and $r(A) = 0$, the same calculation shows that

$$0 = e_1 r(A) = r_0 e_1 + r_1 e_2 + \cdots + r_{n-1} e_n,$$

so we must have $r(t) = 0$. This implies, by the usual Euclidean algorithm argument, that for any polynomial h,

$$h(A) = 0 \Longrightarrow h \in Rf.$$

Now, if we view V as an F-vector space (of dimension nd), the F-linear transformation A satisfies its characteristic polynomial, say $h(t) \in F[t]$. Therefore $h = gf$ for some $g \in R$ with

$$\deg g = \deg h - \deg f = nd - n = n(d-1).$$

Moreover, $gh = hg = gfg$ shows that $h = fg$ as well.

Comment. The second conclusion of the above exercise states that if $D[t]$ is left primitive, then $\dim_F D$ is infinite. The converse of this is however not true, although, if D contains a *transcendental* element over F, then indeed $D[t]$ is left primitive: see *FC*-(11.14). In general, $D[t]$ is left primitive iff *some* matrix ring $\mathbb{M}_r(D)$ contains a transcendental element over F: see the *Comment* on Exercise 11.20.

Ex. 16.10. Let a, b be elements of D that have the same minimal polynomial $f(t)$ over the center F of D. For $c \in D^*$, P.M. Cohn has shown that the equation $ax - xb = c$ has a solution $x \in D$ iff $f(t)$ has a right factor $(t - cbc^{-1})(t - a)$ in $D[t]$. Give a proof for the "only if" part.

Solution. Fix $x \in D^*$ such that $ax - xb = c$ and write

$$y = xbx^{-1} - a = -cx^{-1} \neq 0.$$

Since $f(a) = 0$, $f(t) = g(t)(t - a)$ for some $g(t) \in D[t]$. Evaluating this polynomial at $t = xbx^{-1}$, we have

$$0 = f(xbx^{-1}) = g(yxbx^{-1}y^{-1})y$$

by *FC*-(16.3). Therefore,

$$yxbx^{-1}y^{-1} = (-c)b(-c)^{-1} = cbc^{-1}$$

is a root of $g(t)$. It follows that

$$g(t) = h(t)(t - cbc^{-1})$$

for some $h(t)$, and hence

$$f(t) = h(t)(t - cbc^{-1})(t - a).$$

Comment. The proof of the "if" part above can be found in P.M. Cohn's paper "The range of derivations on a skew field and the equation $ax - xb = c$," J. Indian Math. Soc. *37* (1973), 61–69.

Ex. 16.11. Let $a \in D$ be algebraic over F. Show that the "metro equation" $ax - xa = 1$ has a solution $x \in D$ iff a is not separable over F. (You may use Cohn's result mentioned in Exercise 10.)

Solution. Let $f(t) \in F[t]$ be the minimal polynomial of a over F, and let $E = F[a] \subseteq D$. If a is not separable over F, then, in $E[t]$, $f(t)$ is divisible by $(t - a)^p$, where $p = \operatorname{char} F > 0$. In particular, in $D[t]$, $f(t)$ has a quadratic right factor $(t - a)^2$. By Cohn's result (applied with $b = a$ and $c = 1$), $ax - xa = 1$ has a solution in D. Conversely, if $ax - xa = 1$ has a solution in D, then

$$f(t) = h(t)(t - a)^2 \quad \text{for some} \quad h(t) \in D[t].$$

Since $f(t), (t - a)^2 \in E[t]$, Exercise 1 shows that $h(t) \in E[t]$. Now $f(t)$ has a nonsimple root a in the algebraic closure of F, so a is *not* separable over F.

Alternative Solution (not using the result of Cohn). First assume there exists $x \in D$ such that $ax - xa = 1$. By induction on $n \geq 1$, we have $a^n x - xa^n = na^{n-1}$. Therefore, if $f(t) \in F[t]$ is the minimal polynomial of a over F, then

$$f'(a) = f(a)x - xf(a) = 0,$$

so a is not separable over F. For the converse, we follow an argument of J.-P. Tignol. Assume a is not separable over F. Then $a \notin K : = F[a^p]$ where $p = \operatorname{char} F > 0$. Let $L = C_D(K)$ (the centralizer of K in D). By FC-(15.4), $Z(L) = K$, so $a \in L \backslash Z(L)$ and $a^p \in Z(L)$. By Exercise 13.8, $ax - xa = 1$ has a solution in $L \subseteq D$.

Comment. The term "metro equation" for $ax - xa = 1$ was coined in the early 1970's by P.M. Cohn. According to reliable sources, Cohn and Amitsur were traveling in the Paris subway system (presumably to get to some conference) when Cohn asked Amitsur whether he knew of any equation in a division ring D which did not have solutions in any division ring extension of D. Amitsur reportedly suggested looking at the equation $ax - xa = 1$ where $a \in D$ is a separable algebraic element over $Z(D)$. In his subsequent papers, Cohn fondly referred to $ax - xa = 1$ as the "metro equation." The same equation was also considered at about the same time by M. Boffa and P. van Praag. (The first papers dealing with the equation $ax - xb = c$ were written in 1944 by N. Jacobson and R.E. Johnson.)

Ex. 16.12. (Cohn) Let $a, b \in A$, where A is an algebra over a commutative ring k. Suppose there exists $f(t) \in k[t]$ with $f(b) = 0$ and $f(a) \in U(A)$. For any $c \in A$, show that the equation $ax - xb = c$ has a unique solution $x \in A$.

Solution. Let L denote left multiplication by a and R denote right multiplication by b, on A. Then $LR = RL$ (by associativity), $f(R) = 0$, while $f(L)$ is a bijection. (The latter is just left multiplication by $f(a) \in \mathrm{U}(A)$.) Given $c \in A$, we wish to solve the equation $(L - R)(x) = c$ for $x \in A$. Note that

$$f(L) = f(L) - f(R) = (L - R) \cdot g(L, R) = g(L, R) \cdot (L - R)$$

for some polynomial $g \in k[t, s]$. Since $f(L)$ is a bijection, it follows from this operator equation that $L - R$ is also a bijection. Therefore, $(L - R)(x) = c$ has a unique solution $x \in A$.

Comment. Cohn's formulation of the above result is quite a bit more general. Let A, B be k-algebras (where k is as above), and let C be an (A, B)-bimodule such that $\alpha c = c\alpha$ for every $\alpha \in k$ and $c \in C$. For given elements $a \in A$ and $b \in B$, suppose there exists $f(t) \in k[t]$ with $f(b) = 0$ and $f(a) \in \mathrm{U}(A)$. Then, for any $c \in C$, the equation $ax - xb = c$ has a unique solution $x \in C$. The proof above works verbatim in this case. (In our exercise, $A = B = C$.)

Ex. 16.13. Let A be a k-algebra where k is a field, and let $a, b \in A$ be algebraic elements over k with minimal polynomials $m_a(t), m_b(t) \in k[t]$. Consider the following conditions:

(1) $m_a(t), m_b(t)$ are relatively prime in $k[t]$;
(2) for any $c \in A$, the equation $ax - xb = c$ has a unique solution $x \in A$;
(3) for $x \in A$, $ax = xb \Longrightarrow x = 0$.

Show that $(1) \Rightarrow (2) \Rightarrow (3)$, and that $(3) \Rightarrow (1)$ if A is a simple algebra with $\dim_k A < \infty$ and $Z(A) = k$. Does $(3) \Rightarrow (1)$ hold in general for any finite-dimensional central k-algebra?

Solution. $(1) \Longrightarrow (2)$. Fix an equation

$$\alpha(t)m_a(t) + \beta(t)m_b(t) = 1$$

in $k[t]$. Evaluating this on a, we have

$$1 = \beta(a)m_b(a) = m_b(a)\beta(a), \quad \text{so} \quad m_b(a) \in \mathrm{U}(A).$$

Since we also have $m_b(b) = 0$, Exercise 12 gives (2).

$(2) \Longrightarrow (3)$ follows by setting $c = 0$ in (2).

$(3) \Longrightarrow (1)$ (assuming here that A is simple, $\dim_k A < \infty$ and $Z(A) = k$). In order to shorten the proof, we shall assume several facts about A when we extend scalars from k to $\overline{k}$, the algebraic closure of k.

(a) The $\overline{k}$-algebra $\overline{A} : = A \otimes_k \overline{k}$ is isomorphic to $\mathbb{M}_n(\overline{k})$ (for some n).
(b) Any $a \in A$ has the same minimal polynomial as $a \otimes 1 \in \overline{A}$.

The fact (b) implies that the condition (1) is unchanged if we replace the k-algebra A by the $\overline{k}$-algebra $\overline{A}$. Next, notice that, upon this scalar extension, the condition (3) is also unchanged. In fact, if we express $x \in A$ in (3) in terms of a fixed k-basis of A, the equation $ax = xb$ in A amounts to a homogeneous system of linear equations in the "coordinates" of x. As is well-known, such a system admits a nontrivial solution in k iff it does in $\overline{k}$.

In view of the above remarks, it is sufficient to prove (3) $\Longrightarrow$ (1) when k is algebraically closed and $A = \mathbb{M}_n(k)$. Assume that (1) does not hold. Then there exists $\lambda \in k$ which is a common eigenvalue of the matrices a, b. If we can find a nonzero $x \in A$ such that

$$(a - \lambda)x = x(b - \lambda),$$

then $ax = xb$ and we are done. Thus, after replacing a, b by $a - \lambda$ and $b - \lambda$, we may assume that a, b are both *singular* matrices. We shall think of a, b as linear endomorphisms of k^n. Fix a nonzero vector $u \in \ker(a)$ and a nonzero linear functional f on k^n vanishing on $\mathrm{im}(b)$. Then $x \in \mathrm{End}_k(k^n) = A$ defined by

$$x(v) = f(v)u \quad (v \in k^n)$$

is clearly nonzero, with $ax = 0$ and $xb = 0$. In particular, $ax = xb$, as desired.

In general, (3) $\Rightarrow$ (1) does not hold for finite-dimensional central k-algebras. To construct a counterexample, consider the $\mathbb{C}$-algebra A consisting of all matrices of the form $\begin{pmatrix} r & t & u \\ 0 & r & v \\ 0 & 0 & s \end{pmatrix}$ in $\mathbb{M}_3(\mathbb{C})$. A straightforward computation shows that $Z(A) = \mathbb{C}$. Let

$$a = \begin{pmatrix} 1 & 1 & 1 \\ 0 & 1 & 1 \\ 0 & 0 & 0 \end{pmatrix} \quad \text{and} \quad b = \begin{pmatrix} 0 & 0 & 0 \\ 0 & 0 & 1 \\ 0 & 0 & 2 \end{pmatrix}$$

in A. Since a, b are singular matrices, $m_a(t)$ and $m_b(t)$ are both divisible by t, so (1) does not hold.[5] However, (3) holds. In fact, if

$$\begin{pmatrix} 1 & 1 & 1 \\ 0 & 1 & 1 \\ 0 & 0 & 0 \end{pmatrix} \begin{pmatrix} r & t & u \\ 0 & r & v \\ 0 & 0 & s \end{pmatrix} = \begin{pmatrix} r & t & u \\ 0 & r & v \\ 0 & 0 & s \end{pmatrix} \begin{pmatrix} 0 & 0 & 0 \\ 0 & 0 & 1 \\ 0 & 0 & 2 \end{pmatrix},$$

a quick computation shows that $r = s = t = u = v = 0$.

Comment. Note that, for finite-dimensional algebras A, (3) $\Longrightarrow$ (2) does hold. For, if (3) holds, then the k-linear map $x \mapsto ax - xb$ on A is a monomorphism, and hence an isomorphism. The counterexample given

---

[5] An explicit computation shows that $m_a(t) = t(t-1)^2$, and $m_b(t) = t(t-2)$.

above shows, therefore, that $(2) \Longrightarrow (1)$ also fails in general for finite-dimensional central k-algebras.

Ex. 16.14. Let $a \in D$ where D is a division ring. If a is the only root of the polynomial $(t - a)^2$ in D, show that a is the only root of $(t - a)^n$ in D for any $n \geq 2$.

Solution. Suppose $c \neq a$ is another root of

$$f(t) = (t - a)^n = (t - a)^{n-1}(t - a), \quad \text{where } n \geq 3.$$

By FC-(16.3), $(c - a)c(c - a)^{-1}$ must be a root of $(t - a)^{n-1}$. Invoking an inductive hypothesis, we have then

$$(c - a)c(c - a)^{-1} = a.$$

This amounts to $(c - a)c = a(c - a)$, or equivalently $a^2 - 2ac + c^2 = 0$. This means that c is a root of $(t - a)^2$, so $c = a$, a contradiction!

Ex. 16.15. Let $a \in D$ where D is a division algebra over a field F. Suppose $F(a)$ is a separable quadratic extension of F. Show that the polynomial

$$f(t) = (t - a)^n = \sum_{i=0}^{n} (-1)^i \binom{n}{i} a^i t^{n-i}$$

has a unique root (namely a) in D. (In particular, this is always the case for every $a \in D$ if D is a generalized quaternion division algebra over a field F of characteristic not 2.)

Solution. By Exercise 14, it suffices to prove the conclusion in the case $n = 2$. Let $a' \neq a$ be the Galois conjugate of a in $F(a)$, and let

$$g(t) = (t - a')(t - a) \in F[t]$$

be the minimal polynomial of a over F. Consider any $c \in D$ which is a root of $f(t) = (t - a)^2$. By the Gordon-Motzkin Theorem, c is conjugate to a in D. Since a is a root of $g(t) \in F[t]$ and $F \subseteq Z(D)$, so is c. Therefore, c is a root of

$$f(t) - g(t) = (t - a)^2 - (t - a')(t - a) = (a' - a)(t - a).$$

Since $a' \neq a$, this implies that $c = a$.

Comment. Recall that, in the solution to Exercise 12.12, we have constructed division algebras D with an element $a \in D$ such that the quadratic polynomial

$$f(t) = (t - a)^2 = t^2 - 2at + a^2$$

has two *distinct* roots $a, c \in D$.

Ex. 16.16. Let $\mathbb{H}$ be the division ring of real quaternions. For $\alpha = w + xi + yj + zk \in \mathbb{H}$, write $\bar{\alpha} = w - xi - yj - zk$, and define the trace and norm of α by

$$T(\alpha) = \alpha + \bar{\alpha} = 2w, \quad N(\alpha) = \alpha\bar{\alpha} = \bar{\alpha}\alpha = w^2 + x^2 + y^2 + z^2.$$

Using Dickson's Theorem $(FC\text{-}(16.8))$, show that two quaternions $\alpha, \beta \in \mathbb{H}$ are conjugate iff $T(\alpha) = T(\beta)$ and $N(\alpha) = N(\beta)$. Does this conclusion also hold in the division ring of rational quaternions?

Solution. First assume $\beta = \gamma\alpha\gamma^{-1}$, where $0 \neq \gamma \in \mathbb{H}$. Since $N(\sigma\tau) = N(\sigma)N(\tau)$ and $T(\sigma\tau) = T(\tau\sigma)$, we have

$$N(\gamma\alpha\gamma^{-1}) = N(\gamma)N(\alpha)N(\gamma^{-1}) = N(\alpha), \quad \text{and}$$
$$T(\gamma\alpha\gamma^{-1}) = T(\alpha\gamma^{-1}\gamma) = T(\alpha).$$

Conversely, assume $T(\alpha) = T(\beta)$ and $N(\alpha) = N(\beta)$. If $\alpha = w \in \mathbb{R}$ and $\beta = w_1 + x_1 i + y_1 j + z_1 k$, then we have

$$2w = 2w_1, \quad \text{and} \quad w^2 = w_1^2 + x_1^2 + y_1^2 + z_1^2.$$

This implies that $x_1 = y_1 = z_1 = 0$, and so $\alpha = w = w_1 = \beta$. Now we may assume $\alpha, \beta \notin \mathbb{R}$. Since

$$\alpha^2 - T(\alpha)\alpha + N(\alpha) = \alpha^2 - (\alpha + \bar{\alpha})\,\alpha + \bar{\alpha}\,\alpha = 0,$$

we see that α is quadratic over $\mathbb{R}$, with minimal polynomial $f(t) = t^2 - T(\alpha)t + N(\alpha)$. But then β has the *same* minimal polynomial $f(t)$, so Dickson's Theorem implies that α and β are conjugate in $\mathbb{H}$. The same proof, of course, also works over the rational quaternions.

Comment. Essentially the same proof would work in any "generalized" quaternion division algebra D over a field F of characteristic $\neq 2$; see my book "The Algebraic Theory of Quadratic Forms", Revised Edition, W.A. Benjamin, 1982. However, in the case of a "split" quaternion algebra, the "if" part of this exercise does not work. For instance, if there exist $(x_1, y_1, z_1) \in F^3 \backslash \{0\}$ such that $x_1^2 + y_1^2 + z_1^2 = 0$, then $\alpha = 1$ and $\beta = 1 + x_1 i + y_1 j + z_1 k$ both have trace 2 and norm 1. But of course, α and β are *not* conjugate. Here, the quaternion algebra $F \oplus Fi \oplus Fj \oplus Fk$ (with $i^2 = j^2 = -1$ and $k = ij = -ji$) splits.

Ex. 16.17. Use Dickson's Theorem $(FC\text{-}(16.8))$ to show that Herstein's Lemma $(FC\text{-}(13.8))$ holds in a division ring D of *any* characteristic.

Solution. The Lemma in question states that, if $a \in D^*$ is a noncentral, torsion element, then there exists $y \in D^*$ such that $yay^{-1} = a^i \neq a$, where i is some positive integer. Moreover, y may be chosen to be an additive commutator. In $FC\text{-}(13.8)$, this result was proved for division rings D of characteristic $p > 0$. In the following, we shall give a characteristic-free proof of the result for *all* division rings D.

Let $n > 1$ be the order of $a \in D^*$, and write

$$t^n - 1 = f_1(t) \cdots f_r(t) \in F[t] \qquad (F = Z(D)),$$

where each f_j is monic irreducible in $F[t]$. Working in the field $F(a)$, we also have a factorization

$$t^n - 1 = (t - 1)(t - a)(t - a^2) \cdots (t - a^{n-1}).$$

Applying the unique factorization theorem in $F(a)[t]$, we see that, for a suitable index j, $f_j(t) = (t - a) \cdot (\cdots)$. Since $f_j(t) \in F[t]$ but $a \notin F$, $f_j(t)$ must have another factor $t - a^i$ where $i \in \{2, \ldots, n - 1\}$. Thus, $f_j(t)$ is the minimal polynomial of *both* a and a^i over F. By Dickson's Theorem (*FC*-(16.8)), we have $xax^{-1} = a^i \neq a$ for some $x \in D^*$. As was pointed out in the proof of *FC*-(13.8), a simple replacement of x by $y := ax - xa \neq 0$ leads to $yay^{-1} = a^i \neq a$, where y is now an additive commutator.

Comment. Since the original proof of Herstein's Lemma made heavy use of techniques in characteristic p, it is remarkable that the same result actually holds in all characteristics.

Ex. 16.18. (Lam-Leroy) Let $f(t) = (t - d)g(t) \in D[t]$. If D is a centrally finite division ring, show that $f(d') = 0$ for a suitable conjugate d' of d.

Solution. We offer a solution here that depends on the existence of a nonzero F-linear function $T : D \to F = Z(D)$ with the property that $T(rs) = T(sr)$ for any $r, s \in D$. Let us first assume the existence of such a function T.

Writing $g(t) = a_n t^n + \cdots + a_0$ $(a_i \in D)$, we have

$$f(t) = (t - d)g(t)$$
$$= a_n t^{n+1} + (a_{n-1} - da_n)t^n + \cdots + (a_0 - da_1)t - da_0.$$

We try to find a conjugate $d' = rdr^{-1}$ $(r \in D^*)$ such that

(1)
$$0 = f(d')$$
$$= a_n rd^{n+1}r^{-1} + (a_{n-1} - da_n)rd^n r^{-1} + \cdots + (a_0 - da_1)rdr^{-1} - da_0.$$

Consider the map $\varphi : D \to D$ defined by

(2) $\varphi(r) = a_n rd^{n+1} + (a_{n-1} - da_n)rd^n + \cdots + (a_0 - da_1)rd - da_0 r$

for any $r \in D$. This map is clearly F-linear. Moreover, since

$$T(\varphi(r)) = T(a_n rd^{n+1}) - da_n rd^n) + \cdots + T(a_0 rd - da_0 r) = 0$$

for all $r \in D$, we have $\varphi(D) \subseteq \ker(T) \subsetneq D$. Invoking now the assumption that $\dim_F D < \infty$, we see that $\ker(\varphi) \neq (0)$. Thus $\varphi(r) = 0$ for some $r \in D^*$, and a comparison of (1) and (2) shows that $f(rdr^{-1}) = 0$, as desired.

The existence of the nonzero functional T above is well-known in the theory of finite-dimensional division algebras. In fact, T may be taken to be the "reduced trace" function on D. We'll give only a quick sketch of its construction here. By FC-(15.8), if $K \supseteq F$ is a maximal subfield of D, then

$$D \otimes_F K \cong \mathbb{M}_r(K), \quad \text{where} \quad r = (\dim_F D)^{1/2}.$$

Viewing the isomorphism here as an identification, we get a function

$$D \otimes_F K \to K$$

by taking the trace of matrices. It is easy to see that this restricts to a nonzero. F-linear map $T : D \to F$, and we clearly have $T(rs) = T(sr)$ for all r, $s \in D$ since this equation holds for traces of matrices over K.

Comment. As the reader might suspect, the conclusion of this exercise no longer holds if D is not assumed to be centrally finite. An example to this effect can be found in T.Y. Lam and A. Leroy: "Wedderburn polynomials over division rings, I," J. Pure and Applied Algebra, to appear.

Chapter 6
Ordered Structures in Rings

§17. Orderings and Preorderings in Rings

From our first course in abstract algebra, we learned that $\mathbb{Z}$, the set of all integers, is not only a ring, but an *ordered* ring, in that we have the ordering relation

$$\cdots < -2 < -1 < 0 < 1 < 2 < \cdots$$

between its elements. For arbitrary rings, it is also of significance to study their orderings if they exist.

We say $(R, <)$ is an *ordered ring* if R is a ring and "$<$" is a transitive total ordering of its elements such that, for all $a, b, c \in R$, we have

$$a < b \Longrightarrow a + c < b + c,$$

$$0 < a, \ 0 < b \Longrightarrow 0 < ab.$$

The "positive cone" $P = \{c \in R : 0 < c\}$ has the properties:

$$P + P \subseteq P, \quad P \cdot P \subseteq P, \quad \text{and} \quad P \cup (-P) = R \backslash \{0\}.$$

Conversely, if we are given a nonempty set $P \subseteq R$ satisfying these three properties, then, *defining* $a < b$ by $b - a \in P$, we obtain an ordered ring $(R, <)$. By abuse of terminology, we shall often refer to P itself (rather than "$<$") as an ordering on R.

If P is any ordering on R as above, then $1 \in P$, $P \cap (-P)$ is empty, and R is a domain of characteristic zero $(FC\text{-}(17.4))$. Thus, rings with 0-divisors cannot be ordered in our sense. (If we want to consider ordering structures in rings with 0-divisors, our definition of an ordering must be modified first; we shall not go into this generality in the present section.)

In order to deal effectively with existence questions about orderings, we need the notion of a preordering. By definition, a *preordering* in R is a nonempty subset $T \subseteq R\backslash\{0\}$ such that $T + T \subseteq T$, and for any

$$a_1, \ldots, a_m \in R\backslash\{0\} \quad \text{and} \quad t_1, \ldots, t_n \in T,$$

the product of

$$a_1, a_1, \ldots, a_m, a_m, t_1, \ldots, t_n,$$

taken in any order, lies in T. It turns out that *such a T is an ordering on R iff T is maximal as a preordering:* see FC-(17.10). If T is not yet maximal, we can always "enlarge" T to a maximal preordering by Zorn's Lemma. Therefore, *R has a preordering iff R has an ordering.*

The above approach to orderings via preorderings leads quickly to a criterion for a nonzero ring R to admit an ordering. To wit, *R can be ordered iff R is formally real* in the sense that there is no equation $\sum t_i = 0$ where each t_i is a product of

$$a_{i1}, a_{i1}, \ldots, a_{in}, a_{in},$$

taken in some order, where $a_{ij} \in R\backslash\{0\}$. In the case of fields, this is a famous theorem of Artin and Schreier. The extension of this result to division rings is due to Szele (and Pickert), while the extension to arbitrary rings is due to R.E. Johnson.

If R is formally real, there is a unique smallest preordering $T(R) \subseteq R$ consisting of all (finite) sums $\sum t_i$, where the t_i's are as described in the last paragraph.[1] This $T(R)$ is known as the *weak preordering* of R. It can be used, for instance, to describe the so-called totally positive elements of R: a nonzero element $a \in R$ is positive in all orderings of R *iff* there exists $b \in R\backslash\{0\}$ such that $ab^2 \in T(R)$ (FC-(17.15)).

An ordered ring $(R, <)$ is said to be *archimedean* if, for any $a, b > 0$ in R, $na > b$ for a suitable positive integer n. Such an ordered ring is always commutative, and is in fact order-isomorphic to a unique subring of $\mathbb{R}$ (with the induced ordering). This classical result is proved by using Dedekind cuts in FC-(17.21). Another proof appears below in the solution to Exercise 10.

The exercises in this section offer examples of ordered rings and expand on the various themes discussed in this introduction. While most exercises are oriented toward noncommutative rings, a couple of them are in the commutative setting. The most memorable one is Artin's result in Exercise 9A which states that, for a formally real field k, any totally positive element in the polynomial ring $R = k[t]$ (in one variable) is a sum of squares in R.

---

[1] The "formally real" assumption guarantees that $0 \notin T(R)$.

Exercises for §17

Ex. 17.1. Let $(R, <)$ be an ordered ring and $(G, <)$ be a multiplicative ordered group. In the group ring $A = RG$, define P to be the set of $\sum_{i=1}^{n} r_i g_i$ where $r_1 > 0$ in R and $g_1 < g_2 < \cdots < g_n$ in G. Show that (A, P) is an ordered ring.

Solution. The two properties

$$P + P \subseteq P, \quad P \cup (-P) = A \backslash \{0\}$$

for an ordering cone are both clear. To check the last property, $P \cdot P \subseteq P$, take $\alpha = \sum_{i=1}^{n} r_i g_i$ as above, and another element $\alpha' = \sum_{j=1}^{m} r_j' g_j'$ where $r_1' > 0$ in R and $g_1' < g_2' < \cdots < g_m'$ in G. Then

$$\alpha \alpha' = r_1 r_1' g_1 g_1' + s_2 h_2 + s_3 h_3 + \cdots \quad (s_i \in R, \ h_i \in G)$$

with $r_1 r_1' > 0$ in R and

$$g_1 g_1' < h_2 < h_3 < \cdots \quad \text{in } G.$$

Therefore, $\alpha \alpha' \in P$, as desired.

Comment. The same conclusion holds already if $(G, <)$ is an *ordered semigroup*, since the inverse operation plays no role in the above construction (and proof).

Ex. 17.2. Show that any free algebra F over a formally real field k can be ordered.

Solution. Say F is freely generated over k by $\{x_i : i \in I\}$. Let G (resp. H) be the free group (resp. semigroup) generated by $\{x_i : i \in I\}$. By the Artin-Schreier Theorem (cf. *FC*-(17.11)), k can be ordered, and by the Birkhoff-Iwasawa-Neumann Theorem (see *FC*-(6.31)), G can be ordered. Therefore, by Exercise 1, the group ring $A = kG$ can be ordered. Since $F = kH$ may be viewed as a subring of $A = kG$, we can get an ordering on F by restricting an ordering on A to F. Alternatively, we can apply the *Comment* on Exercise 1 to get directly an ordering on F, using an ordering on k and an ordering on H (the latter obtained by restricting an ordering on G).

Ex. 17.3. Let R by the Weyl algebra generated over $\mathbb{R}$ by x and y, with the relation $xy - yx = 1$. Elements of R have the canonical form

$$r = r_0(x) + r_1(x)y + \cdots + r_n(x)y^n,$$

where each $r_i(x) \in \mathbb{R}[x]$, $r_n(x) \neq 0$ (if $r \neq 0$). Let $P \subset R$ be the set of all nonzero elements $r \in R$ above for which $r_n(x)$ has a positive leading coefficient. Show that P defines an ordering "$<$" on R on which

(*) $\quad \mathbb{R} < x < x^2 < \cdots < y < xy < x^2y < \cdots < y^2 < xy^2 < x^2y^2 < \cdots$.

Solution. As in Exercise 1, the two properties

$$P + P \subseteq P, \quad P \cup (-P) = R \backslash \{0\}$$

are clear. To show that $P \cdot P \subseteq P$, consider $r \in P$ as above, and another element

$$s = s_0(x) + s_1(x)y + \cdots + s_m(x)y^m \in P.$$

Upon multiplying r with s and writing the product in the "canonical form," the highest degree term in y comes from the product $r_n(x)y^n s_m(x)y^m$. If

$$s_m(x) = a_0 + a_1 x + \cdots + a_k x^k \quad \text{(where } a_k > 0),$$

we have

$$y^n s_m(x) = y^n \sum_{i=0}^{k} a_i x^i = \left(\sum_{i=0}^{k} a_i x^i \right) y^n + \cdots$$

where the suppressed terms have y-degree $< n$. Therefore,

$$r_n(x)y^n s_m(x)y^m = r_n(x)s_m(x)y^{n+m} + \text{(terms with } y\text{-degree} < n + m).$$

Since the leading coefficient of $r_n(x)s_m(x)$ is the product of those of $r_n(x)$ and $s_m(x)$, it is positive. This checks that $rs \in P$. The inequalities in $(*)$ can be checked easily by forming differences and writing them in canonical form. For instance,

$$x^2 y^i - x y^i = (x^2 - x)y^i$$

shows that $xy^i < x^2 y^i$, etc.

Ex. 17.4. In $R = \mathbb{Q}(t)$, define P to be the set of $f(t)/g(t)$ where f, g are polynomials with positive leading coefficients. Show that (R, P) is a nonarchimedean ordered field, and that all the order-automorphisms of (R, P) are induced by $t \mapsto at + b$ where $a, b \in \mathbb{Q}$, $a > 0$.

Solution. The three properties for an ordering cone:

$$P + P \subseteq P, \quad P \cdot P \subseteq P, \quad P \cup (-P) = R \backslash \{0\}$$

are all clear for the P defined in this exercise. Therefore, P defines an ordering "<" on R. For any $c \in \mathbb{Q}$, we have $t - c \in P$. This shows that $c < t$ for any $c \in \mathbb{Q}$, so "<" is a *nonarchimedean* ordering. It is well known that any automorphism φ of R is the identity on $\mathbb{Q}$, with

$$\varphi(t) = \frac{at + b}{ct + d}$$

where $a, b, c, d \in \mathbb{Q}$, and $ad - bc \neq 0$. Now assume φ is an *order-automorphism*, i.e. with $\varphi(P) \subseteq P$. We claim that $c = 0$. Indeed, assume $c \neq 0$. Then

$$\varphi\left(t - \frac{a}{c}\right) = \frac{at + b}{ct + d} - \frac{a}{c} = \frac{cb - ad}{c(ct + d)},$$

$$\varphi\left(t - \frac{a}{c} - 1\right) = \frac{-c^2 t + (cb - ad - cd)}{c(ct + d)}.$$

In the second equation, the RHS is in $-P$, but the LHS is in $\varphi(P) \subseteq P$, a contradiction. Therefore, we must have $c = 0$. We may assume, without loss of generality, that $d = 1$, so now $\varphi(t) = at + b$, where $a \in \mathbb{Q}$ is necessarily > 0 since $\varphi(t) \in \varphi(P) \subseteq P$. Conversely, it is easy to see that, for any $a > 0$, $\varphi(t) = at + b$ induces a unique order-automorphism of R.

Ex. 17.5. Let $k \subseteq \mathbb{R}$ be a field of real algebraic numbers, and let π be any transcendental real number. Let $P \subseteq k[t]$ be the set of polynomials $f(t)$ such that $f(\pi) > 0$. Show that P is an archimedean ordering in $A = k[t]$.

Solution. Since $f(\pi) \neq 0$ for $f \in A \backslash \{0\}$, we have either $f(\pi) > 0$ or $f(\pi) < 0$. Therefore, $P \cup (-P) = A \backslash \{0\}$. The two properties

$$P + P \subseteq P, \quad P \cdot P \subseteq P$$

are both clear. Therefore, P defines an ordering "$<$" on the integral domain A. To show that P is an archimedean ordering, consider any $f, g \in P$. Fix a positive integer n such that $f(\pi)/g(\pi) < n$. Then

$$(ng - f)(\pi) = ng(\pi) - f(\pi) > 0,$$

so $ng - f \in P$, i.e. $ng > f$ with respect to the ordering "$<$" defined by P.

Comment. Of course, this construction works also for the rational function field $F = k(t)$, and defines an ordering $\tilde{P}$ on F extending the ordering P on A. We can think of F as a subfield of $\mathbb{R}$ by using the embedding $f(t) \mapsto f(\pi)$ (for any $f(t) \in F$); the ordering $\tilde{P}$ is then obtained simply by restricting to F the usual ordering on $\mathbb{R}$.

Ex. 17.6. If a commutative domain R is not formally real, must -1 be a sum of squares in R?

Solution. To say that the commutative domain R is not formally real means that there exists an equation $a_1^2 + \cdots + a_n^2 = 0$ in R where the a_i's are not all zero. This need not imply that that -1 is a sum of squares in R. For instance, consider the ring

$$R = \mathbb{R}[x_1, \ldots, x_n]/(x_1^2 + \cdots + x_n^2),$$

in which $\bar{x}_1^2 + \cdots + \bar{x}_n^2 = 0$ and $\bar{x}_i \neq 0$ for all i. We have an $\mathbb{R}$-algebra homomorphism $\varphi : R \to \mathbb{R}$ induced by $\varphi(\bar{x}_i) = 0$ for all i. If -1 was a sum of squares in R, an application of φ would show that -1 is a sum of squares in $\mathbb{R}$, which is absurd. Therefore, -1 cannot be a sum of squares in R.

Ex. 17.7. Show by a direct calculation that, in any ring R, the division closure $\tilde{T}$ of any preordering T is also a preordering.

Solution. Recall from FC-(17.12) that $\tilde{T}$ is defined to be any one of the four equal sets below:

$$\{a \in R : at \in T \text{ for some } t \in T\}$$
$$= \{a \in R : t'a \in T \text{ for some } t' \in T\}$$
$$= \{a \in R : ab^2 \in T \text{ for some } b \neq 0\}$$
$$= \{a \in R : b'^2 a \in T \text{ for some } b' \neq 0\}.$$

Since $0 \notin T$, we have also $0 \notin \tilde{T}$. To show that $\tilde{T}$ is a preordering involves checking the following two properties:

(1) $\tilde{T} + \tilde{T} \subseteq \tilde{T}$.
(2) For $a_1, \ldots, a_n \in \tilde{T}$ and $c_1, \ldots, c_m \in R \backslash \{0\}$, per $(c_1^2 \cdots c_m^2 a_1 \cdots a_n) \in \tilde{T}$.

Here, as in FC-p. 263, per $(c_1^2 \cdots c_m^2 a_1 \cdots a_n)$ means any product of the following factors

$$c_1, c_1, \ldots, c_m, c_m, a_1, \ldots, a_n,$$

taken in any order. To check (2), fix $t_i \in T$ such that $a_i t_i \in T$, and let $t = (a_1 t_1) \cdots (a_n t_n) \in T$. For any

$$a = \text{per } (c_1^2 \cdots c_m^2 a_1 \cdots a_n),$$

at is a product of the form

$$\text{per } (c_1^2 \cdots c_m^2 a_1^2 \cdots a_n^2 t_1 \cdots t_n),$$

which is in T. Therefore, by definition, $a \in \tilde{T}$. To check (1), consider $a_1, a_2 \in \tilde{T}$; say $a_i t_i \in T$, where $t_i \in T$. Then, since T is a preordering,

$$(a_1 + a_2) t_1 a_2 t_2 = (a_1 t_1)(a_2 t_2) + a_2 t_1 a_2 t_2 \in T.$$

Noting that $t_1 a_2 t_2 = t_1 (a_2 t_2) \in T$, we conclude that $a_1 + a_2 \in \tilde{T}$.

Comment. There is a much more conceptual approach to the fact that $\tilde{T}$ is a preordering. In fact, in FC-(17.13), it is proved that $\tilde{T}$ is precisely the intersection of all the orderings of R containing T. This implies the result in the exercise since any intersection of orderings (or preorderings) is always a preordering. However, it is nice to see a direct check of the preordering axioms for $\tilde{T}$.

Ex. 17.8. Show that

$$R = \mathbb{R}[x_1, \ldots, x_n, y, z]/(x_1^2 + \cdots + x_n^2 - y^2 z)$$

is formally real, and that its weak preordering $T = T(R)$ is not division-closed, i.e. $T \neq \tilde{T}$.

Solution. Recall that $T(R)$ denotes all sums of terms of the form

$$\text{per } (a_1^2 \ldots a_n^2) \quad \text{where} \quad a_i \in R\backslash\{0\}.$$

By definition, R being formally real means that $0 \notin T(R)$. In the commutative case, $T(R)$ is simply the set of all sums $\sum a_i^2$ where the a_i's are nonzero, and R being formally real means that $\sum a_i^2 \neq 0$ whenever all $a_i \neq 0$.

Consider now the commutative ring R in this exercise. Since $x_1^2 + \cdots + x_n^2 - y^2 z$ is an irreducible polynomial, the ideal

$$(x_1^2 + \cdots + x_n^2 - y^2 z)$$

is prime. Therefore, R is an integral domain. The quotient field F of R is easily seen to be the rational function field $\mathbb{R}(\bar{x}_1, \ldots, \bar{x}_n, \bar{y})$. Since F is formally real (why?), so is R. Thus, $T = T(R)$ is a preordering in R. To see that T is not division-closed, consider the $\mathbb{R}$-homomorphism $f : R \to \mathbb{R}$ defined by

$$f(\bar{x}_1) = \cdots = f(\bar{x}_n) = f(\bar{y}) = 0 \quad \text{and} \quad f(\bar{z}) = -1.$$

Since $f(T) \geq 0$ in $\mathbb{R}$ and $f(\bar{z}) = -1$, we see that $\bar{z} \notin T$. On the other hand,

$$\bar{y}^2 \bar{z} = \bar{x}_1^2 + \cdots + \bar{x}_n^2 \in T$$

implies that $\bar{z} \in \tilde{T}$, so $\tilde{T} \neq T$.

Ex. 17.9A. (E. Artin) For a formally real field k, show that the weak preordering T of $R = k[t]$ is division-closed.

Solution. We present here Artin's original solution to this problem. For the sake of authenticity, we shall also use Artin's own notations.

Initial Observation. If $h, f \neq 0$ are sums of squares in $k(t)$, then so is h/f. (This follows easily from the equation $h/f = fh/f^2$.)

To prove that T is division-closed, we must show that, if

$$(1) \qquad \varphi(t)^2 F(t) = \sum_{i=1}^{N} g_i(t)^2 \neq 0 \quad (F, \varphi, g_i \in R),$$

then $F(t) \in T$. This will be done by induction on $n = \deg F$, the case $n = 0$ being clear (by comparing leading coefficients). We may also assume that

$F(t)$ is not divisible by the square of any nonconstant polynomial, and that φ is chosen with deg φ minimal in (1).

Write $g_i = q_i F + h_i$ where deg $h_i < n$. We claim that the h_i's are not all 0. In fact, if all $h_i = 0$, then (1) gives $\varphi^2 = F \sum_{i=1}^{N} q_i^2$. Since F has no square factors, we have $F|\varphi$ in R, and $(\varphi/F)^2 F = \sum q_i^2$ contradicts the minimal choice of deg φ. Plugging $g_i = q_i F + h_i$ into (1) and squaring, we get

$$(2) \qquad fF = \sum_{i=0}^{N} h_i^2 \neq 0 \quad \text{for some } f \in R.$$

Here, deg $f \leq 2(n-1) - n = n - 2$. Among *all* identities of the form (2), we may assume that the above one is chosen with $m = \deg f$ minimal (and hence $\leq n - 2$). By our "Initial Observation" above, f is a sum of squares in $k(t)$, so by the inductive hypothesis, $f \in T$. Assume, for the moment, that $m > 0$. Write $h_i = p_i f + r_i$ where $p_i, r_i \in R$, with deg $r_i < m$. If all $r_i = 0$, then (2) yields

$$F = f \left(\sum p_i^2 \right) \in T \cdot T \subseteq T.$$

Hence we may assume some $r_i \neq 0$, in which case (2) yields $\sum r_i^2 = f f_1 \neq 0$ for some $f_1 \in R$, with

$$\deg f_1 \leq 2(m-1) - m = m - 2.$$

Now by Lagrange's Identity:

$$(3) \qquad \left(\sum_{i=1}^{N} h_i^2 \right) \left(\sum_{i=1}^{N} r_i^2 \right) = \left(\sum_{i=1}^{N} h_i r_i \right)^2 + \sum_{i<j} (h_i r_j - h_j r_i)^2.$$

Here, the LHS is $F f^2 f_1 \neq 0$. On the RHS, since $h_i \equiv r_i \pmod{f}$, we have

$$\sum_{i=1}^{N} h_i r_i \equiv \sum_{i=1}^{N} h_i^2 \equiv 0 \pmod{f}, \quad \text{and}$$

$$h_i r_j - h_j r_i \equiv h_i h_j - h_j h_i \equiv 0 \pmod{f}.$$

Therefore, the RHS of (3) has the form $f^2 G$ for some $G \in T$. Cancelling f^2 from (3) now yields $f_1 F = G \in T$, which contradicts the minimal choice of m. Thus, we must have $m = 0$, i.e. $f = a \in k\backslash\{0\}$. Since $f \in T$, a is a sum of squares in k, and so

$$F = a \sum (h_i/a)^2 \in T,$$

as desired.

Comment. The above result and its proof appeared in Artin's famous paper "Über die Zerlegung definiter Funktionen in Quadraten," Hamburg Abh. *5* (1927), 100–115, in which he solved Hilbert's 17th Problem ("Every positive semidefinite polynomial $F \in \mathbb{R}[x_1, \ldots, x_r]$ is a sum of squares in $\mathbb{R}(x_1, \ldots, x_r)$"). Much later (namely in 1964), Cassels modified Artin's argument to get the following quantitative result: *If $\varphi^2 F$ is a sum of N squares in $k[t]$ (where $0 \neq \varphi \in k[t]$ and k is any field), then F is also a sum of N squares in $k[t]$.* Shortly thereafter, Pfister improved this version further: *If $\varphi^2 F = a_1 g_1^2 + \cdots + a_N g_N^2$ where $0 \neq \varphi \in k[t]$, $g_i \in k[t]$, and $a_i \in k$, then $F = a_1 f_1^2 + \cdots + a_N f_N^2$ for suitable $f_i \in k[t]$.* Now known as the Cassels-Pfister Theorem, this result plays a crucial role in the modern algebraic theory of quadratic forms over fields. For more details, see Chapter 9 of the author's book "The Algebraic Theory of Quadratic Forms," Benjamin/Addison-Wesley, 1973. Although the Cassels-Pfister argument yields a stronger conclusion valid in the more general setting of quadratic forms, historians of algebra are not likely to forget Artin's original theorem (c. 1927), or his very elegant proof thereof, reproduced in this Exercise.

Artin's Theorem does not extend to sums of higher even powers: for every $d \geq 2$, there exists $F \in \mathbb{R}[t]$ which is a sum of $2d$ powers in $\mathbb{R}(t)$, but not in $\mathbb{R}[t]$. Finally, Artin's Theorem also fails in the multivariate case: For $r \geq 2$ and k a formally real field, the weak preordering T of $k[x_1, \ldots, x_r]$ is no longer division-closed. This is the content of the next exercise, which essentially goes back to Hilbert (c. 1888).

Ex. 17.9B. Let k be any formally real field and let $r \geq 2$. Show that there exist polynomials in $R = k[x_1, \ldots, x_r]$ which are sums of squares in $k(x_1, \ldots, x_r)$, but not in R.

Solution. It suffices to prove this in the $r = 2$ case, where we write $R = k[x, y]$. Consider the sextic polynomial

(∗) $$F(x, y) = x^4 y^2 + y^4 + x^2 - 3x^2 y^2 \in R.$$

By direct computation, one can check that

$$(x^2 + y^2)^2 F(x, y) = x^4 y^2 (x^2 + y^2 - 2)^2 + (x^2 - y^2)^2 (x^2 y^2 + x^2 + 1),$$

so F is a sum of four squares in $k(x, y)$. *We claim that F is not a sum of squares in R.* Indeed, assume it is, say $F = \sum f_i^2$, where $f_i \in R$. Then

$$f_i = a_i x^3 + b_i y^3 + c_i x^2 y + d_i xy^2 + r_i x^2 + s_i y^2 + t_i xy + p_i x + q_i y + e_i,$$

where the coefficients are in k. Comparing coefficients of x^6, y^6 and the constant terms in $F = \sum f_i^2$, we see that

$$a_i = b_i = e_i = 0 \quad \text{for all } i.$$

Using these, and comparing coefficients of x^2y^4, x^4 and y^2, we see further that

$$d_i = r_i = q_i = 0 \quad \text{for all } i.$$

But then

$$-3 = (\text{coefficient of } x^2y^2) = \sum_i t_i^2 \in k,$$

a contradiction.

Comment. The above example and proof are taken from the author's joint paper with M. D. Choi in Math. Ann. *231* (1977), 1–18. The polynomial F is a modification of Motzkin's polynomial

$$M(x,y) = x^4y^2 + x^2y^4 + 1 - 3x^2y^2,$$

which was the first explicit polynomial in two variables known to have the properties in this exercise. Note that, by the Arithmetic-Geometric Inequality, both M and F are positive semidefinite polynomials, with respect to any ordering on k.

Ex. 17.10. Without using Dedekind cuts, show that an archimedean ordered ring $(R, <)$ is commutative, and that the only order-automorphism of $(R, <)$ is the identity.

Solution. For the first part, it is sufficient to show that any two positive elements a, $b \in R$ commute. For any $m \in \mathbb{N}$, there exists $n \in \mathbb{N}$ such that $(n-1)a \le mb < na$. Thus,

$$m(ab - ba) = a(mb) - (mb)a < a(na) - (n-1)aa = a^2.$$

Since this holds for all $m \in \mathbb{N}$ (and $(R, <)$ is archimedean), we must have $ab - ba \le 0$. By symmetry, we also have $ba - ab \le 0$; hence $ab = ba$.

For the second part, consider any order-automorphism φ of $(R, <)$. Let $b > 0$ in R. For any $m \in \mathbb{N}$ again, there exists $n \in \mathbb{N}$ such that

$$n - 1 \le mb < n.$$

Writing $b' = \varphi(b)$ and applying φ to these inequalities, we get $n - 1 \le mb' < n$. Subtraction now yields

$$-1 < m(b - b') < 1,$$

that is,[2] $m|b - b'| < 1$ for any $m \in \mathbb{N}$. Since $(R, <)$ is archimedean, this is possible only when $b = b' = \varphi(b)$. This implies also that

$$\varphi(-b) = -\varphi(b) = -b,$$

so φ is the identity map on R.

---

[2] The absolute value (with respect to the ordering "$<$") is defined in the usual way: $|a| = a$ if $a \ge 0$, and $|a| = -a$ if $a < 0$.

Comment. The result in this exercise is due to Hilbert, who first proved it for ordered division rings. Compare *FC*-(17.21) for the treatment using Dedekind cuts. See also the proof of the analogous "Hölder's Theorem" (Exercise 6.11) for archimedean ordered groups. Note that the proof in the ring-theoretic case is quite a bit easier since $\mathbb{Z} \subseteq R$ for any ordered ring $(R, <)$.

Ex. 17.11. Give an example of an ordered ring R which has infinitely large elements but no infinitely small elements. Give an example of an ordered ring other than $\mathbb{Z}$ for which there exists no element between 0 and 1.

Solution. Let $R = \mathbb{Z}[t]$, with the ordering given by the set P of all polynomials with positive leading coefficients. As in Exercise 4, P is not archimedean: for instance, the element $t \in P$ is infinitely large. Consider any

$$f(t) = a_n t^n + \cdots + a_0 \in P$$

where necessarily $a_n > 0$. If $n > 0$, then f is also infinitely large. If $n = 0$, then $f = a_0 \geq 1$. Therefore, there are no elements in R between 0 and 1. In particular, there are no infinitely small (positive) elements.

Ex. 17.12. Let $(R, <)$ be any ordered ring that is algebraic over a subfield $F \subseteq Z(R)$. If F is archimedean with respect to the induced ordering, show that $(R, <)$ is an archimedean ordered field.

Solution. First, since R has no 0-divisors, it is clear that R is a division ring (see Exercise 1.13(c)). Next, we claim that, for any $r \in R$, $|r| < N$ for a suitable integer $N > 0$. Fix a monic equation satisfied by r over F, say

$$r^n + a_1 r^{n-1} + \cdots + a_n = 0 \quad (a_i \in F).$$

We claim that

$$|r| \leq 1 + |a_1| + \cdots + |a_n|.$$

Since $(F, <)$ is archimedean, this clearly gives what we want. To prove the inequality, we may assume, of course, that $|r| > 1$. Then

$$|r|^n = |a_1 r^{n-1} + \cdots + a_n| \leq |a_1| \cdot |r|^{n-1} + |a_2| \cdot |r|^{n-2} + \cdots + |a_n|$$
$$\leq |a_1| \cdot |r|^{n-1} + |a_2| \cdot |r|^{n-1} + \cdots + |a_n| \cdot |r|^{n-1}$$
$$= (|a_1| + |a_2| + \cdots + |a_n|) \cdot |r|^{n-1}.$$

Therefore,

$$|r| \leq |a_1| + \cdots + |a_n| \leq 1 + |a_1| + \cdots + |a_n|,$$

as claimed.

Since $(R, <)$ is now archimedean, Exercise 10 implies that R is commutative. Therefore, R is a field.

Ex. 17.13. Let (R, P) be an ordered ring, and $\sigma : R \to R$ by any endomorphism of the additive group of R such that $\sigma(P) \subseteq P$. Assume that, for any $r \in R$, there exists an integer $n \geq 1$ such that $\sigma^n(r) = r$. Show that σ is the identity map.

Solution. Note that σ respects the total ordering "<" defined by P. In fact, if $a < b$, then $b - a \in P$, so $\sigma(b - a) \in P$. Since σ is an additive endomorphism,

$$\sigma(b - a) = \sigma(b) - \sigma(a).$$

Therefore, we have $\sigma(a) < \sigma(b)$.

Assume, for the moment, that $\sigma \neq id$. Then $\sigma(r) \neq r$ for some $r \in R$. We have either $r < \sigma(r)$ or $\sigma(r) < r$. Say $r < \sigma(r)$. Fix an integer $n \geq 1$ such that $\sigma^n(r) = r$. By the first paragraph, we have

$$r < \sigma(r) < \sigma^2(r) < \cdots < \sigma^n(r) = r,$$

a contradiction. If $\sigma(r) < r$, a similar contradiction results.

Ex. 17.14. Let (R, P) be an ordered ring for which P is a well-ordered set, i.e. every nonempty subset of P has a smallest element. Show that (R, P) is order-isomorphic to $\mathbb{Z}$ with its usual ordering.

Solution. Clearly, $P \supseteq \mathbb{N}$, the set of positive integers. We are done if we can show that $P \subseteq \mathbb{N}$. Assume, for the moment, that this is not the case. Let x be the smallest element of $P \backslash \mathbb{N}$. If $x > 1$, then $x - 1 \in P \backslash \mathbb{N}$ is smaller than x, a contradiction. Hence we must have $0 < x < 1$. But then

$$x - x^2 = x(1 - x) \in P$$

implies that $0 < x^2 < x < 1$, which is again a contradiction.

§18. Ordered Division Rings

In this section, we focus our study of ordered rings on the case of division rings. Therefore, in the following, D always denotes a division ring.

Due to the fact that any nonzero element of D has an inverse, the definition of a preordering on D can be recast in a somewhat simpler form. As is shown in *FC*-(18.1), *a set $T \subseteq D^*$ is a preordering iff $T + T \subseteq T$, $T \cdot T \subseteq T$, and $d^2 \in T$ for every $d \in D^*$.* (Here, we no longer need to consider products of

$$a_1, a_1, \ldots, a_m, a_m, t_1, \ldots, t_n,$$

taken in some order.) If T is such a preordering, then T is a normal subgroup of D^*, and D^*/T is an abelian group of exponent 2. Finally, *this T is an ordering of D iff $|D^*/T| = 2$.* In this case, T is itself a multiplicative ordered group, with the ordering cone $\{a \in T : a - 1 \in T\}$.

The gist of the matter is that, in a division ring, any product of $a_1, a_1, \ldots, a_n, a_n$ (in any order, where $a_i \in D^*$) is always a "square-product," i.e. of the form $b_1^2 \cdots b_m^2$. Therefore, the definition for D to be formally real requires here that a sum of nonzero square-products be again nonzero. The criterion for the existence of an ordering on D then takes on the following simpler form: *D has an ordering iff* -1 *is not a sum of square-products in D (FC-(18.2))*. In this case, $T(D)$, the weak preordering on D, is the set of all sums of nonzero square-products, and *an element $a \in D^*$ is positive in all orderings of D iff $a \in T(D)$ (FC-(18.4))*.

The first noncommutative ordered division ring was constructed by Hilbert, in connection with his study of the foundations of geometry. Many other examples can be constructed by using the Mal'cev-Neumann division rings of Laurent series associated with ordered groups (*FC-*(18.5)). However, Albert has shown that, *for any orderable division ring D, the center F of D is always algebraically closed in D (FC-(18.10))*. In particular, if D is a noncommutative algebraic algebra over its center F, then D can never be ordered. (Stated in a positive form: *any formally real algebraic division algebra over its center is commutative*. In this form, Albert's Theorem may be thought of as another commutativity result.)

Ordered division rings sometimes exhibit rather unexpected properties: see, for instance, *FC-*(18.12) and Exercise 7 below. The other exercises in this section deal with the natural Krull valuation on the Mal'cev-Neumann Laurent series ring $R((G))$ over a given ordered division ring R. Exercise 6 offers an easy, but nevertheless surprising, example of a noncommutative division ring D that is archimedean over its center.

Exercises for §18

Ex. 18.1. (This exercise, due to A. Wadsworth, provides an example of a centrally finite division algebra D in which -1 is a square-product, but is *not* a sum of squares.) Let k be a field with at least two orderings, P, P', and let $e \in P \backslash P'$. Let $F = k((y))$ and let D be the F-algebra with generators i, j and relations

$$i^2 = e, \; j^2 = y, \quad \text{and} \quad k := ij = -ji.$$

Then D is a 4-dimensional (generalized quaternion) division algebra over F. Show that -1 is a square-product in D (so D has level 1 in the terminology of *FC-*§18), but $\sum H_\alpha^2 = 0$ in D implies that $H_\alpha = 0$ for all α.

Solution. The fact that D is a division F-algebra is not obvious, but the proof is fairly elementary, using a quadratic form theoretic criterion for a generalized quaternion algebra to be a division algebra. In any case, we are not required to prove this part in detail in this exercise.

Note that (in any generalized quaternion algebra)

$$k^2 = (ij)^2 = ijij = -i^2 j^2,$$

so $-1 = (i^{-1})^2 k^2 (j^{-1})^2$ is always a product of 3 squares. This checks that D has level 1. Now consider an equation $\sum H_\alpha^2 = 0$ where

$$H_\alpha = a_\alpha + b_\alpha i + c_\alpha j + d_\alpha k.$$

Upon squaring and comparing constant terms, we get an equation

(*) $$\sum a_\alpha^2 + e \sum b_\alpha^2 = y \left(e \sum d_\alpha^2 - \sum c_\alpha^2 \right) \in F.$$

Suppose the Laurent series $a_\alpha, b_\alpha, c_\alpha, d_\alpha$ are not all zero, and that the lowest degree terms in them are of the form $y^r (r \in \mathbb{Z})$. We write

$$a_\alpha = y^r (a_{\alpha 0} + \cdots) \quad (a_{\alpha,0} \in k),$$

and similarly for b_α, c_α and d_α. Then the $a_{\alpha 0}, b_{\alpha 0}, c_{\alpha 0}, d_{\alpha 0} \in k$ are not all zero. Comparing terms of degree y^{2r} in (*), we get

$$\sum a_{\alpha 0}^2 + e \sum b_{\alpha 0}^2 = 0.$$

If some $b_{\alpha 0} \neq 0$, then $\sum b_{\alpha 0}^2 \neq 0$ and this equation shows that $-e$ is a sum of squares in k (cf. "Initial Observation" in the solution to Exercise 17.9A), contradicting $e \in P \backslash P'$. Therefore, we must have all $b_{\alpha 0} = 0$ and hence all $a_{\alpha 0} = 0$. Going back to (*) and comparing terms of degree y^{2r+1}, we get

$$e \sum d_{\alpha 0}^2 = \sum c_{\alpha 0}^2.$$

Repeating the argument above, we see that all $d_{\alpha 0} = 0$ and hence all $c_{\alpha 0} = 0$, a contradiction!

Ex. 18.2. Let (R, P_o) be an ordered division ring, $(G, <)$ be an ordered group, and $A = R((G))$ be the division ring of (untwisted) Laurent series as defined in FC-§14. Let $v : A^* \to G$ be the Krull valuation constructed in Exercise 14.10, and let $\mathfrak{m}$ be the Jacobson-radical of the valuation ring of v. Show that the ordering $P \subset A$ constructed in FC-(18.5):

$$P = \left\{ \alpha = \sum_{g \in G} \alpha_g g : \ \alpha_{g_o} \in P_o \ \text{ for } \ g_o = \min(\text{supp}(\alpha)) \right\}$$

has the following properties:

(1) $P \cap R = P_o$,
(2) $G \subseteq P$,
(3) P is "compatible" with v in the sense that $1 + \mathfrak{m} \subset P$.

Then show that P is uniquely determined by these three properties.

Solution. (1) and (2) are both obvious. For (3), look at any $\alpha = \sum \alpha_g g \in \mathfrak{m}$. By definition, this means that

$$\alpha_g \neq 0 \Longrightarrow g > 1 \quad \text{in} \quad G.$$

Thus, $\min(\operatorname{supp}(1 + \alpha))$ is the group element $1 \in G$, which occurs in $1 + \alpha$ with coefficient 1. Therefore, $1 + \alpha \in P$, which shows that $1 + \mathfrak{m} \subseteq P$.

Now let P' be *any* ordering on A such that

$$P' \cap R = P_o, \quad G \subseteq P',$$

and $1 + \mathfrak{m} \subseteq P'$. *We claim that* $P \subseteq P'$. Once this is proved, it is easy to see that $P = P'$ since P, P' are both orderings. Let $\alpha \in P$, with $\min(\operatorname{supp}(\alpha)) = g_o$. Then

$$\alpha = \alpha_{g_o} g_o + \cdots + \alpha_h h + \cdots = \alpha_{g_o} g_o (1 + \cdots + \alpha_{g_o}^{-1} \alpha_h g_o^{-1} h + \cdots).$$

The term in parentheses is in $1 + \mathfrak{m} \subseteq P'$. Since we also have $\alpha_{g_o} \in P_o \subseteq P'$ and $g_o \in P'$, it follows that $\alpha \in P'$, so $P \subseteq P'$, as desired.

Ex. 18.3. In the exercise above, show that the valuation ring V of v is given by the convex hull of R in A with respect to the ordering P, i.e.

$$(*) \qquad V = \{a \in A : -r \leq a \leq r \text{ for some } r \in P_0 \subseteq R\}.$$

Moreover, show that the residue division ring $V/\mathfrak{m}$ is isomorphic to R.

Solution. Let V' be the RHS of the equation $(*)$. First consider any $\alpha = \sum \alpha_g g \in V \cap P$. Then

$$g_o := \min(\operatorname{supp}(\alpha)) \geq 1, \quad \text{and} \quad \alpha_{g_o} > 0.$$

If $g_o > 1$, then $1 - \alpha \geq 0$ so $\alpha \in V'$. If $g_o = 1$, then

$$(\alpha_{g_o} + 1) - \alpha \in 1 + \mathfrak{m} \subseteq P$$

by Exercise 2. This shows that $\alpha \leq \alpha_{g_o} + 1 \in R$, so again $\alpha \in V'$. We have therefore $V \subseteq V'$. Now, changing notations, consider any positive $\alpha \notin V$. Then $g_o < 1$ and $\alpha_{g_o} > 0$. For any $r \in R$, $\alpha - r$ still has "lowest term" $\alpha_{g_o} g_o$, so $\alpha - r \in P$. This means that $\alpha > R$ in the ordering P, so $\alpha \notin V'$. This completes the proof that $V = V'$. Since $V = R \oplus \mathfrak{m}$, it follows further that the residue division ring of the valuation ring V is $V/\mathfrak{m} \cong R$.

Ex. 18.4. Keep the notations in Exercise 2. Define two elements $a, b \in A^*$ to be *in the same archimedean class* relative to R if $|a| \leq r|b|$, $|b| \leq r'|a|$ for suitable $r, r' \in P_o \subseteq R$. (Here, the absolute values are with respect to P, and are defined in the usual way.) Write $[a]$ for the archimedean class of a (relative to R). We multiply these classes by the rule $[a] \cdot [b] = [ab]$, and order them by the rule: $[a] < [b]$ *iff* $r|a| < |b|$ for all $r \in P_o$. Show that all archimedean classes relative to R form an ordered group which is anti-order-isomorphic to $(G, <)$ itself.

Solution. Let us write $a \sim b$ if $a, b \in A^*$ are in the same archimedean class (always relative to R). We claim that

(*) $$a \sim b \quad \text{iff} \quad v(a) = v(b) \in G,$$

where $v : A^* \to G$ is the Krull valuation on A in the two previous exercises. First assume $a \sim b$, with $|a| \le r|b|$ and $|b| \le r'|a|$ where $r, r' \in P_o$. Changing the signs of a, b if necessary, we may assume $a, b > 0$. Let

$$g = v(a) = \min (\mathrm{supp}(a)), \quad \text{and}$$
$$h = v(b) = \min (\mathrm{supp}(b)).$$

If $g < h$, then the "lowest term" of $a - rb$ has positive coefficient, so $a - rb > 0$, a contradiction. Similarly, $h < g$ also leads to a contradiction. Therefore, we must have

$$v(a) = g = h = v(b).$$

Conversely, if $v(a) = v(b) = g$ and

$$a = a_o g + \cdots, \quad b = b_o g + \cdots,$$

with (say) $a_o, b_o \in P_o$, we can choose $r, r' \in P_o$ such that $a_o < rb_o$ and $b_o < r'a_o$. Then $a < rb$ and $b < r'a$, so we have $a \sim b$.

Let $U = U(V)$ where V is the valuation ring of v. Then $U = \ker(v)$ is a normal subgroup of A^* with $A^*/U \cong G$. The work in the paragraph above shows that the archimedean classes are just the cosets of A^* with respect to U; in fact, $[a] = aU = Ua$ for every $a \in A^*$. This accounts for the group structure on the set of all archimedean classes. By a slight modification of the argument in the first paragraph, we can also show that

$$r|a| < |b| \text{ for all } r \in P_o \iff v(a) > v(b) \text{ in } G.$$

Therefore, the isomorphism $A^*/U \to G$ induced by φ is *order-reversing* from the ordered group of archimedean classes to the given ordered group G.

Ex. 18.5. Keep the notations in Exercise 2, and define the valuation topology on A by taking as a system of neighborhoods at 0 the sets

$$\{\alpha \in A : \alpha = 0 \text{ or } v(\alpha) > g\} \quad (g \in G).$$

Show that the valuation topology, as a uniform structure, is complete, i.e. every Cauchy net in A has a limit.

Solution. The idea here is the same as that used in the proof of the completeness of the field of p-adic numbers. To simplify the notations, let us just show that any Cauchy sequence in A is convergent. (The case of

Cauchy nets can be done in essentially the same way.) For convenience, we shall take $v(0) = \infty$.

Let $\{\alpha^{(n)} : n \in \mathbb{N}\}$ be a Cauchy sequence in A. It is enough to find a convergent subsequence $\{\alpha^{(n_i)}\}$. Indeed, assume such a subsequence exists, with $\lim \alpha^{(n_i)} = \alpha \in A$. For any $g \in G$, take an $N \in \mathbb{N}$ such that

$$v(\alpha^{(m)} - \alpha^{(n)}) > g \quad \text{whenever} \quad m, n > N.$$

For any $n > N$, take $n_i > n$ to be large enough so that $v(\alpha - \alpha^{(n_i)}) > g$. Then

$$v(\alpha - \alpha^{(n)}) = v(\alpha - \alpha^{(n_i)} + \alpha^{(n_i)} - \alpha^{(n)})$$
$$\geq \min\{v(\alpha - \alpha^{(n_i)}),\; v(\alpha^{(n_i)} - \alpha^{(n)})\} \geq g,$$

and hence $\lim \alpha^{(n)} = \alpha$.

To find a convergent subsequence of $\{\alpha^{(n)}\}$, we may assume that $\{\alpha^{(n)}\}$ contains no constant subsequence. After passing to a suitable subsequence, we may assume that, for $g_n = v(\alpha^{(n+1)} - \alpha^{(n)})$, we have

$$1 < g_1 < g_2 < \cdots.$$

We shall now show that $\{\alpha^{(n)}\}$ is convergent by constructing its limit $\alpha = \sum \alpha_g g \in A$. For any $g \in G$, take any n such that $g < g_n$. (Such n exists since $\{\alpha^{(n)}\}$ is Cauchy.) Then define $\alpha_g = \alpha_g^{(n)}$ (the coefficient of g in $\alpha^{(n)}$). Note that, since $v(\alpha^{(n+1)} - \alpha^{(n)}) = g_n$, we have

$$\alpha^{(n+1)} - \alpha^{(n)} = rg_n + \text{``higher terms''} \quad (r \in R),$$

so the coefficients of g in $\alpha^{(n)}$, $\alpha^{(n+1)}$ (and $\alpha^{(n+2)}$ etc.) are the same. Assuming for the moment that the formal sum $\alpha = \sum \alpha_g g$ is an element of A, we see immediately that $\lim \alpha^{(n)} = \alpha$, since, for any $g \in G$, we have $g_n > g$ for large n, and the latter implies $v(\alpha - \alpha^{(n)}) \geq g_n$ since α and $\alpha^{(n)}$ have the same coefficients for every $g < g_n$.

To see that $\alpha \in A$, we must show that every nonempty $T \subseteq \text{supp}(\alpha)$ has a least element. Fix any $g \in T$, and choose an n such that $g_n > g$. Then for any $g' \leq g$ in T, we have $0 \neq \alpha_{g'} = \alpha_{g'}^{(n)}$. Therefore,

$$\{g' \in T : g' \leq g\} \subseteq \text{supp}\,(\alpha^{(n)}).$$

Since $\text{supp}(\alpha^{(n)})$ is well-ordered, the LHS has a least element, which is also the least element of α. Therefore, $\alpha \in A$ as claimed.

Comment. In case A is a field, the same kind of arguments above can be used to show that any "pseudo-convergent" sequence in A has a "pseudo-limit." Thus, A is in fact a "maximally valued field" in the sense of Krull and Kaplansky.

Ex. 18.6. (K.H. Leung) An ordered division ring $(D, <)$ is said to be *archimedean* over a division subring D_0 if, for every $d \in D$, we have $d \leq d_0$ for some $d_0 \in D_0$. True or False: "*If $(D, <)$ is archimedean over $Z(D)$, then D is commutative*"?

Solution. Let us first explain why this is an interesting question. Assume that the answer to the question is "yes." Then we can deduce from it two well-known theorems on ordered division rings. First, if $(D, <)$ is archimedean (i.e., archimedean over $\mathbb{Q}$), then D is commutative (Hilbert's Theorem: Ex. 17.10). Secondly, if $(D, <)$ is algebraic over $Z(D)$, then D is commutative (Albert's Theorem: FC-(18.10)).[3] Thus, if the answer to the question is in the affirmative, we would have a common generalization to the two theorems cited above.

Unfortunately, the answer to the question is "no" in general. To construct a counterexample, start with any noncommutative ordered division ring $(K, <)$ with center F. Then the division ring $D = K((t))$ of Laurent series over K has center $F((t))$. We extend "$<$" to D as in FC-(18.5): a Laurent series is positive *iff* its lowest degree coefficient is positive. Denoting the extended ordering again by "$<$," *we claim that $(D, <)$ is archimedean over $F((t))$.* Indeed, for any

$$d = a_n t^n + a_{n+1} t^{n+1} + \cdots \in D,$$

we have $d < t^{n-1} \in F((t))$ since

$$t^{n-1} - d = t^{n-1} - a_n t^n - \cdots > 0.$$

However, since $F((t)) \neq D$, D is *not* commutative!

Ex. 18.7. A theorem of K.H. Leung states that, *if $(R, <)$ is an ordered ring such that $a^2 + b^2 \geq 2ab$ for all $a, b \in R$, then R must be commutative.* Prove this theorem in the special case when R is an (ordered) division ring.

Solution. Replacing b by $a + rb$ (where r is any element in R), we get the inequality

$$a^2 + a^2 + (rb)^2 + arb + rba \geq 2a(a + rb) = 2a^2 + 2arb,$$

so $(rb)^2 \geq arb - rba$. Since $a \in R$ is arbitrary, we also have

$$(rb)^2 \geq (-a)rb - rb(-a) = -(arb - rba),$$

and so

(1) $$|arb - rba| \leq |rb|^2 \quad (\forall a, b, r \in R).$$

By the triangular inequality,

$$|r(ba - ab)| \leq |rba - arb| + |(ar - ra)b| \leq |rb|^2 + |r|^2 |b|,$$

---

[3] Note that if $(D, <)$ is algebraic over $Z(D)$, then $(D, <)$ is archimedean over D: see the "root estimate" argument in the solution to Exercise 17.12.

where the second inequality follows by applying (1) twice (first in general and secondly for $b = 1$). Therefore,

(2) $$|ba - ab| \leq (|b| + 1) \cdot |r| \cdot |b| \quad \text{whenever } r \neq 0.$$

Let c be an arbitrary positive element in R, and assume $b \neq 0$. If R is a division ring, there certainly exists $r \in R \backslash \{0\}$ such that

$$(|b| + 1) \cdot |r| \cdot |b| \leq c.$$

Therefore, (2) yields $|ba - ab| \leq c$. Since this holds for all $c > 0$, we must have $ba = ab$, as desired.

Comment. Leung has proved various other forms of commutativity theorems over ordered rings, all in generalization of the above exercise. For instance, if $(R, <)$ is any ordered ring such that, for any $a, b \in R$, there exists a positive integer n such that $a^{2n} + b^{2n} \geq 2a^n b^n$, then R must be commutative. For more details, see Leung's paper "Positive semidefinite forms over ordered skew fields," in Proc. Amer. Math. Soc. *106* (1989), 933–942.

Chapter 7
Local Rings, Semilocal Rings, and Idempotents

§19. Local Rings

In commutative algebra, a local ring is defined to be a (nonzero) ring that has a unique maximal ideal. This definition generalizes readily to arbitrary rings: a (nonzero) ring A is said to be *local* if A has a unique maximal left ideal. This definition turns out to be left-right symmetric, and is equivalent to the condition that $A/\mathrm{rad}\,A$ be a division ring.

In commutative algebra, one of the most powerful tools is the association of a family of local rings to any commutative ring A, namely, the family of the localizations $A_{\mathfrak{p}}$ of A at the various prime ideals $\mathfrak{p} \subset A$. This theory does not generalize well to the noncommutative setting. Therefore, the role of local rings in noncommutative algebra is not as prominent as in the commutative case. Nevertheless, noncommutative local rings do arise naturally, and form an important class for study. For instance, they arise frequently as endomorphism rings of modules, as valuation rings associated with Krull valuations of division rings, and as group algebras of finite p-groups over fields of characteristic $p > 0$.

A nonzero module M_R over a ring R is said to be *strongly indecomposable* if $\mathrm{End}(M_R)$ is a local ring. Such M is always indecomposable. Conversely, if M_R is indecomposable and has a composition series, then M is strongly indecomposable (cf. *FC*-(19.17)). The significance of strong indecomposability can be seen in part from the Krull-Schmidt-Azumaya Theorem, which states the following: *Let M_R be an R-module with the decompositions*

$$M = M_1 \oplus \cdots \oplus M_r = N_1 \oplus \cdots \oplus N_s,$$

where the N_i's are indecomposable, and the M_i's are strongly indecomposable. Then $r = s$, and after a reindexing, we have $M_i \cong N_i$ for all i. The assumption that the M_i's be *strongly* indecomposable is essential for this theorem. In fact, if the M_i's are only indecomposable (as are the N_i's), the conclusions of the above theorem may fail rather miserably, even over a commutative noetherian ring R.

For finitely generated right modules M over a right artinian ring R, the theorem cited above is classically known as the Krull-Schmidt Theorem. In this case, it is enough to assume that the M_i's and N_i's are indecomposable. This result, along with other closely related results such as the Noether-Deuring Theorem (FC-(19.25)), are indispensable tools in the study of representations of finite-dimensional algebras over fields.

The exercises in this section explore different examples and properties of noncommutative local rings. If R is local, so is its center $Z(R)$ (Ex. 6), and every finitely generated projective R-module is free (Ex. 11). However, if R is right noetherian, it may not be left noetherian, and the intersection $\bigcap_{i=1}^{\infty} (\mathrm{rad}\ R)^i$ may not be zero (Ex. 12), etc.

Exercises for §19

Ex. 19.1. For any local ring R, show that (1) the opposite ring R^{op} is local, and (2) any nonzero factor ring of R is local.

Solution. (1) Since R is local, R has a unique maximal left ideal $\mathfrak{m}$ by FC-(19.1). Clearly,

$$\mathfrak{m}^{\mathrm{op}} = \{a^{\mathrm{op}} : a \in \mathfrak{m}\}$$

is the unique maximal right ideal of R. By FC-(19.1) again, R^{op} is a local ring.

(2) Consider any factor ring $\overline{R} = R/I$, where I is any proper ideal of R. Since I cannot contain any units, we have $I \subseteq \mathfrak{m}$ (where $\mathfrak{m}$ is as in (1)). Clearly, $\mathfrak{m}/I$ is the unique maximal left ideal of $\overline{R}$, so $\overline{R}$ is local, again by FC-(19.1).

Ex. 19.2. (a) For any field k, show that

$$(1) \qquad R = \left\{ \begin{pmatrix} a & 0 & b \\ 0 & a & c \\ 0 & 0 & a \end{pmatrix} : a, b, c \in k \right\}$$

is a commutative local ring whose unique maximal ideal $\mathfrak{m}$ has square zero.
(b) Assume that char $k = p > 0$, and let G be an elementary p-group of order p^2 generated by x, y. Let V be the 3-dimensional right kG-module $k^3 = e_1 k \oplus e_2 k \oplus e_3 k$ with

$$(2) \qquad e_1 x = e_1, \quad e_2 x = e_2, \quad e_3 x = e_1 + e_3 \quad \text{and}$$

$$(3) \qquad e_1 y = e_1, \quad e_2 y = e_2, \quad e_3 y = e_2 + e_3.$$

Show that $\text{End}(V_{kG}) \cong R$ and deduce that V is a strongly indecomposable kG-module.

Solution. (a) Let $\mathfrak{m}$ be the set of matrices in (1) with $a = 0$. It is straightforward to check that $\mathfrak{m}$ is an ideal of R with $\mathfrak{m}^2 = 0$, and that $U(R) = R \backslash \mathfrak{m}$. Therefore, by FC-(19.1), $(R, \mathfrak{m})$ is a local ring.

(b) Since G is abelian, we may also think of V as a left kG-module, on which x and y act by left multiplication of the matrices

$$X = \begin{pmatrix} 1 & 0 & 1 \\ 0 & 1 & 0 \\ 0 & 0 & 1 \end{pmatrix} \quad \text{and} \quad Y = \begin{pmatrix} 1 & 0 & 0 \\ 0 & 1 & 1 \\ 0 & 0 & 1 \end{pmatrix}$$

with respect to the basis $\{e_1, e_2, e_3\}$. A matrix $A \in M_3(k)$ gives a kG-endomorphism of V iff A commutes with X and Y. For $A = (a_{ij})$, these conditions amount to:

$$a_{12} = a_{21} = a_{31} = a_{32} = 0 \quad \text{and} \quad a_{11} = a_{22} = a_{33}.$$

Therefore, $\text{End}(V_{kG})$ corresponds exactly to $R \subseteq M_3(k)$ with respect to the basis $\{e_1, e_2, e_3\}$. Since R is a local ring, V_{kG} is by definition strongly indecomposable.

Ex. 19.3. What can you say about a local ring $(R, \mathfrak{m})$ that is von Neumann regular?

Solution. Since a von Neumann regular ring is always J-semisimple (cf. FC-(4.24)), we have $\mathfrak{m} = \text{rad } R = 0$. Therefore, $R \cong R/\mathfrak{m}$ is a division ring.

Ex. 19.4. Let $R = kG$ where k is a field and G is a nontrivial finite group. Show that the following statements are equivalent:

(1) R is a local ring.
(2) R_R is an indecomposable R-module.
(3) $R/\text{rad } R$ is a simple ring.
(4) k has characteristic $p > 0$ and G is a p-group.

Solution. (1) $\Longrightarrow$ (2) follows from the fact that a local ring has only trivial idempotents (cf. FC-(19.2)).

(2) $\Longrightarrow$ (1). Since R_R is a module with a composition series, the fact that it is indecomposable implies that $\text{End}(R_R)$ is a local ring, by FC-(19.17). But $\text{End}(R_R) \cong R$, so R is a local ring.

(1) $\Longrightarrow$ (3). If R is local, $R/\text{rad } R$ is in fact a division ring.

(3) $\Longrightarrow$ (4). Assume instead char $k = 0$. Then rad $R = 0$ by Maschke's Theorem, so (3) implies that R is a simple ring. This is impossible since

the augmentation ideal I is an ideal $\neq 0$, R in R (thanks to $|G| \neq 1$). Thus, $p = \mathrm{char}\ k > 0$. Under the augmentation map $R \to k$, rad R maps into rad $k = 0$, so we have rad $R \subseteq I$. From (3), we see that rad $R = I$. Since R is artinian, I is nilpotent by FC-(4.12). Therefore, for any $g \in G$,

$$0 = (g-1)^{p^n} = g^{p^n} - 1 \in R$$

for a sufficiently large integer n. This gives $g^{p^n} = 1$, so G is a (finite) p-group.

(4) $\Longrightarrow$ (1) follows from FC-(8.8) (cf. FC-(19.10)). In fact, under the hypotheses in (4), rad R is just the augmentation ideal of R, and $R/\mathrm{rad}\ R \cong k$.

Ex. 19.5. Let $\mathfrak{A}$ be an ideal in a ring R such that $\mathfrak{A}$ is maximal as a left ideal. Show that $\overline{R} = R/\mathfrak{A}^n$ is a local ring for every integer $n \geq 1$.

Solution. Since the image $\overline{\mathfrak{A}}$ of $\mathfrak{A}$ in $\overline{R}$ is nilpotent, we have $\overline{\mathfrak{A}} \subseteq \mathrm{rad}\ \overline{R}$. Clearly, this implies that $\overline{\mathfrak{A}}$ is the *unique* maximal left ideal of $\overline{R}$. Therefore, $(\overline{R}, \overline{\mathfrak{A}})$ is a local ring.

Ex. 19.6. Show that if a ring R has a unique maximal ideal $\mathfrak{m}$, then the center Z of R is a local ring. (In particular, the center of a local ring is a local ring.)

Solution. First observe that $\mathrm{U}(Z) = Z \cap \mathrm{U}(R)$. Consider any element $a \in Z \backslash \mathrm{U}(Z)$. Then $1 \notin Ra$, (for otherwise $a \in \mathrm{U}(R)$), so $Ra = aR$ is contained in a maximal ideal of R. By the hypothesis on $\mathfrak{m}$, we must have $Ra \subseteq \mathfrak{m}$. Therefore, for $a, b \in Z \backslash \mathrm{U}(Z)$, we have $a + b \in \mathfrak{m}$; in particular,

$$a + b \in Z \backslash \mathrm{U}(Z).$$

This shows that $Z \backslash \mathrm{U}(Z)$ is closed under addition, so Z is a (commutative) local ring, by FC-(19.1).

Comment. A ring R with a unique maximal ideal $\mathfrak{m}$ need not be a local ring. For instance, consider $V = e_1 k \oplus e_2 k \oplus \cdots$ where k is a division ring, and let $R = \mathrm{End}(V_k)$. By Exercise 3.15, R has a unique maximal ideal

$$\mathfrak{m} = \{f \in R : \ \dim_k f(V) < \infty\}.$$

However, R is not a local ring. The center Z of R can be shown to be $\cong Z(k)$, which is a field.

Ex. 19.7. A domain R is called a *right discrete valuation ring* if there is a nonunit $\pi \in R$ such that every nonzero element $a \in R$ can be written in the form $\pi^n u$ where $n \geq 0$ and u is a unit. Show that

(1) R is a local domain;
(2) every nonzero right ideal in R has the form $\pi^i R$ for some $i \geq 0$;
(3) each $\pi^i R$ is an ideal of R; and
(4) $\bigcap_{i>1}(\pi^i R) = 0$.

Give an example of a noncommutative right discrete valuation ring by using the twisted power series construction.

Solution. First we show that π has neither a right inverse nor a left inverse. Indeed, assume $\pi a = 1$ where $a \in R$. Write $a = \pi^n u$ where $n \geq 0$ and $u \in \mathrm{U}(R)$. Then $\pi^{n+1} u = 1$, which implies $\pi \in \mathrm{U}(R)$, a contradiction. Next assume $b\pi = 1$ and write $b = \pi^m v$ where $m \geq 0$ and $v \in \mathrm{U}(R)$. Then $\pi^m v \pi = 1$. From the case we have just treated, m must be zero, so again $\pi = v^{-1} \in \mathrm{U}(R)$, a contradiction.[1]

Now let $\mathfrak{m} = \pi R \neq R$. If $\mathfrak{A} \neq 0$, R is any right ideal, let $0 \neq a \in \mathfrak{A}$ and write $a = \pi^n u$, where $n \geq 0$ and $u \in \mathrm{U}(R)$. Clearly $n \neq 0$, so $a \in \pi R$. This shows that $\mathfrak{A} \subseteq \mathfrak{m}$, so $\mathfrak{m}$ is the unique maximal right ideal of R. Therefore, R is a local domain with rad $R = \mathfrak{m}$. In the above notation, we also have $\pi^n \in \mathfrak{A}$. Now choose i minimal such that $\pi^i \in \mathfrak{A}$. For any nonzero $a \in \mathfrak{A}$ above, we must have $n \geq i$, so

$$a = \pi^i \pi^{n-i} u \in \pi^i R.$$

This shows that $\mathfrak{A} = \pi^i R$, as required in (2). (Note that this makes R a *principal right ideal domain*.)

The right ideals of R form a chain

$$(0) \subsetneq \cdots \subsetneq \pi^{i+1} R \subsetneq \pi^i R \subsetneq \cdots \subsetneq \pi R \subsetneq R,$$

where the strict inclusions follow from the fact that π has no right inverse. For any $u \in \mathrm{U}(R)$, conjugation by u takes the above chain to another chain with strict inclusions. Therefore, we must have $u(\pi^i R)u^{-1} = \pi^i R$ for every $i \geq 0$. This shows that $u(\pi^i R) = \pi^i Ru = \pi^i R$, from which we see that

$$(\pi^n u)(\pi^i R) = \pi^{n+i} R \subseteq \pi^i R \quad (\forall n \geq 0),$$

so $\pi^i R$ is an ideal.

Finally, if

$$a = \pi^n u \in \bigcap_{i \geq 1} (\pi^i R) \quad (n \geq 0,\ u \in \mathrm{U}(R)),$$

then $\pi^n u = \pi^{n+1} r$ for some $r \in R$. But then $u = \pi r$ implies that π has a right inverse, which is not the case. This proves (4).

To construct an example of a noncommutative right discrete valuation ring, we take a field k with an endomorphism σ, and let $R = k[[x; \sigma]]$ be the ring of formal power series of the form $\sum_{i \geq 0} x^i a_i$ ($a_i \in k$), with multiplication induced by the twist $ax = x\sigma(a)$ for all $a \in k$. (Here, we have reversed our usual notational conventions for the twist equation.

---

[1] Actually, the fact that π has no left inverse will not be needed in the rest of the proof.

But of course this does not matter.) Taking $\pi = x$, we have for any $\alpha = x^n a_n + x^{n+1} a_{n+1} + \cdots$ with $a_n \neq 0$:

$$\alpha = x^n(a_n + x a_{n+1} + \cdots) \in \pi^n U(R) \quad (\text{cf. } FC\text{-p. 9}).$$

Therefore, R is a right discrete valuation ring. (Note that if σ is not surjective, then $R\pi^i \subsetneq \pi^i R$ for all $i \geq 1$, and the $R\pi^i$'s are not ideals in R.)

Ex. 19.8. (Brungs) Let R be a nonzero ring such that any nonempty collection of right ideals of R has a largest member (i.e. one that contains all the others). Show that (1) R is a local ring, (2) every right ideal of R is principal, and is an ideal.

Solution. Let $\mathfrak{m}$ be the largest right ideal among all right ideals $\neq R$. Since $\mathfrak{m}$ contains all right ideals $\neq R$, $\mathfrak{m}$ is the unique maximal right ideal of R. Therefore, R is a local ring with $\mathrm{rad}\, R = \mathfrak{m}$. Let $\mathfrak{A}$ be any right ideal of R. Among all right ideals $\{aR : a \in \mathfrak{A}\}$, let $a_0 R$ be the largest. Then $a_0 R \supseteq aR$ for all $a \in \mathfrak{A}$, so $\mathfrak{A} = a_0 R$. To see that all right ideals are ideals, let us assume the contrary. Among all right ideals which are not ideals, let aR be maximal. Then, for some $b \in R$, we have $baR \not\subseteq aR$. Comparing the two right ideals baR and aR, we must have $aR \subsetneq baR$. Write $a = bar$, where $r \in R$. Necessarily, $r \notin U(R)$, for otherwise $ba = ar^{-1} \in aR$. Therefore, $r \in \mathfrak{m}$. By the maximal choice of aR, baR is an ideal; in particular, $b \cdot ba = bas$ for some $s \in R$. Then

$$ba = b(bar) = (b^2 a)r = (bas)r$$

gives $ba(1 - sr) = 0$. Here

$$1 - sr \in 1 + \mathfrak{m} \subseteq U(R),$$

so we get $ba = 0$, which contradicts $baR \not\subseteq aR$.

Ex. 19.9. For a division ring D and a (not necessarily abelian) ordered group $(G, <)$, a function $v : D^* \to G$ is called a (Krull) *valuation* if $v(ab) = v(a)v(b)$ for all $a, b \in D^*$ and $v(a + b) \geq \min\{v(a), v(b)\}$ for all $a, b \in D^*$ such that $a + b \neq 0$. Given such a valuation, let

$$R = \{0\} \cup \{r \in D^* : v(r) \geq 1\}.$$

(1) Show that R is a local ring.
(2) Show that $aRa^{-1} = R$ for any $a \in D^*$.
(3) Show that for any $a \in D^*$, either $a \in R$ or $a^{-1} \in R$.
(4) Show that any right (resp. left) ideal in R is an ideal.
(5) Show that the (right) ideals in R form a chain with respect to inclusion.
(6) Show that any finitely generated right ideal in R is principal.

Solution. The two valuation axioms clearly imply that R is a subring of D.

(1) Let $\mathfrak{m} = \{0\} \cup \{r \in D^* : v(r) > 1\}$, which is easily checked to be an ideal of R. For any $r \in R\backslash\mathfrak{m}$, we have $v(r) = 1$, so $v(r^{-1}) = 1$ as well, and we have $r^{-1} \in R\backslash\mathfrak{m}$. This shows that $R\backslash\mathfrak{m} = U(R)$, so R is a local ring with rad $R = \mathfrak{m}$.

(2) For $r \in R\backslash\{0\}$, we have $v(r) \geq 1$, so

$$v(ara^{-1}) = v(a)v(r)v(a)^{-1} \geq 1,$$

since the ordering cone in $(G, <)$ is closed under conjugation (see *FC*-p. 95). Therefore, $ara^{-1} \in R$, from which we see easily $aRa^{-1} = R$.

(3) Assume $a \notin R$. Then $v(a) < 1$ implies $v(a^{-1}) > 1$, so by definition $a^{-1} \in \mathfrak{m} \subseteq R$.

(4) follows from (2) since (2) implies $aR = Ra$ ($\forall a \in R$).

(5) For given (right) ideals $A, B \subseteq R$, consider the following two cases (where A^*, B^* denote $A\backslash\{0\}, B\backslash\{0\}$ respectively):

Case (i). $v(A^*) \subseteq v(B^*)$. For any $a \in A^*$, we have $v(a) = v(b)$ for some $b \in B^*$. Then, $v(b^{-1}a) = 1$ implies $b^{-1}a \in U(R)$, so $a = b \cdot b^{-1}a \in B$. Thus, in this case, $A \subseteq B$.

Case (ii). $v(A^*) \not\subseteq v(B^*)$; say $v(a) \notin v(B^*)$, where $a \in A^*$. For any $b \in B^*$, we cannot have $b^{-1}a \in R$ (for otherwise $v(a) = v(b \cdot b^{-1}a) \in v(B^*)$). By (3), we must have $a^{-1}b \in R$. Therefore, $b = a \cdot a^{-1}b \in A$, so we are in this case $B \subseteq A$.

(6) Let $A = a_1 R + \cdots + a_r R$ be a finitely generated right ideal in R. Since $a_1 R, \ldots, a_r R$ form a chain, $A = a_i R$ for the largest member $a_i R$ in the chain.

Comment. In the theory of Dedekind cuts, a set S in the ordered group $(G, <)$ is called an *upper class* if $s \leq s'$ and $s \in S$ imply that $s' \in S$. For any ideal $A \subseteq R$, $v(A^*)$ is an upper class in

$$G^+ = \{g \in G : g \geq 1\}.$$

The ideas in the proof of (5) above can be used to show that $A \mapsto v(A^*)$ *gives a one-one correspondence between the ideals of R and the upper classes in G^+* (with (0) corresponding to the empty upper class, and R corresponding to G^+). Moreover, under this correspondence, multiplication of ideals corresponds to multiplication of upper classes in G^+.

Any ordered group $(G, <)$ (abelian or otherwise) is the recipient group of a surjective valuation from a suitable division ring D. For an explicit construction of D and v, see Exercise 14.10.

Ex. 19.10. Let R be a subring of a division ring D which satisfies the two properties (2), (3) in Exercise 9. Show that there exists an ordered group $(G, <)$ and a valuation $v : D^* \to G$ such that

$$R = \{0\} \cup \{d \in D^* : v(d) \geq 1\}.$$

(Such a subring R of a division ring D is called an *invariant valuation ring* of D. If D is a field, the property (2) is automatic; in this case, we get back the usual (commutative) valuation rings in D).

Solution. Let $U = \mathrm{U}(R)$. From $aRa^{-1} = R$, we see that U is a normal subgroup of D^*. Therefore, we can form the quotient group $G = D^*/U$, whose elements are written as $\bar{d} = dU$ (for $d \in D^*$). Let

$$P = \{\bar{d} \in G : d \in R\backslash \mathrm{U}(R)\}.$$

We claim that P satisfies the following three axioms for a "positive cone" on the multiplicative group G:

(a) $P \cdot P \subseteq P$;
(b) $gPg^{-1} \subseteq P$ for any $g \in G$;
(c) $G\backslash\{1\}$ is the disjoint union of P and P^{-1}.

Clearly, (b) follows from (2), and (c) follows from (3). (a) follows from the fact that the nonunits in R are closed under multiplication. Therefore, P defines an ordering "$<$" on G (cf. FC-p.100). We now define $v : D^* \to G$ by $v(d) = \bar{d}$ (for any $d \in D^*$), which is just the projection map from D^* to its quotient group D^*/U. Clearly, the set

$$\{0\} \cup \{d \in D^* : v(d) \geq 1\}$$

is just our given ring R. To show that v is a (Krull) valuation, we need only check that:

$$v(a + b) \geq \min\{v(a), v(b)\} \quad \text{whenever} \quad a, b, a + b \in D^*.$$

We may assume that $v(a) \leq v(b)$. Then $\bar{a} \leq \bar{b}$ implies that $\bar{b}\,\bar{a}^{-1} \geq 1$, so $ba^{-1} \in R$. Therefore, $1 + ba^{-1} \in R$, and we have

$$v(a + b) = v((1 + ba^{-1})a) = v(1 + ba^{-1})v(a) \geq v(a),$$

as desired.

Comment. For a general division ring D, a subring $R \subseteq D$ may satisfy the property (3) in Exercise 9 without satisfying the "invariance" property (2). Such subrings are local rings and are called *valuation rings* in the literature (cf. *FC*-(19.3)(b)), though they do not give rise to Krull valuations as we saw in the two exercises above.

Ex. 19.11. Deduce the fact that a finitely generated projective right module P over a local ring $(R, \mathfrak{m})$ is free (*FC*-(19.29)) from the Krull-Schmidt-Azumaya Theorem (*FC*-(19.21)).

Solution. Since $P/P\mathfrak{m}$ is a finite-dimensional vector space over $R/\mathfrak{m}$, P has a decomposition into a finite direct sum of indecomposable R-modules. (If $P = P_1 \oplus \cdots \oplus P_n$ where each $P_i \neq 0$, then $P_i/P_i\mathfrak{m} \neq 0$ by Nakayama's Lemma, and hence $n \leq \dim_{R/\mathfrak{m}} P/P\mathfrak{m}$.) Say $P \oplus Q \cong R^r$. Since P, Q are finite direct sums of indecomposables and R_R is strongly indecomposable[2], the Krull-Schmidt-Azumaya Theorem implies that $P \cong R^s$ for some $s \leq r$.

Ex. 19.12. Let k be a field with the property that $k(t)$ (the rational function field in one variable over k) is isomorphic to k. (For instance, $\mathbb{Q}(x_1, x_2, \ldots)$ is such a field.) Let $\theta : k(t) \to k$ be a fixed isomorphism. Let $p(t)$ be a fixed irreducible polynomial in $k[t]$, and let A be the discrete valuation ring obtained by localizing $k[t]$ at the prime ideal $(p(t))$. On $R = A \oplus A$, define a multiplication by

$$(a, b)(a', b') = (aa', ba' + \theta(a)b').$$

Then R is a ring with identity (1,0).

(1) Show that R is a local ring with rad $R = A \cdot p(t) \oplus A$.
(2) Show that $\bigcap_{i=1}^{\infty} (\text{rad } R)^i = (0) \oplus A$, and that this is the prime radical of R.
(3) Show that every right ideal of R is an ideal.
(4) Show that R is right noetherian but not left noetherian.

Solution. The distributive laws on R result from $\theta(a + a') = \theta(a) + \theta(a')$, and the associative law on R results from $\theta(aa') = \theta(a)\theta(a')$. The other ring axioms are easily verified.

(1) From the multiplication formula on $R = A \oplus A$, it is clear that

$$\mathfrak{m} := A \cdot p(t) \oplus A$$

is an ideal. On the other hand, if $(a, b) \notin \mathfrak{m}$, then $a \in U(A)$, and it is easy to verify that (a, b) has inverse $(a^{-1}, -b\theta(a)^{-1}a^{-1})$ in R. It follows that R is a local ring with rad $R = \mathfrak{m}$ (cf. *FC*-(19.1)).

---

[2] Recall that $\text{End}(R_R) \cong R$ (which is a local ring).

(2) By (1) and induction on i, we see easily that $\mathfrak{m}^i = A \cdot p(t)^i \oplus A$. Therefore,

$$\bigcap_{i=1}^{\infty} \mathfrak{m}^i = \left(\bigcap_{i=1}^{\infty} A \cdot p(t)^i \right) \oplus A = (0) \oplus A.$$

To see that this is the prime radical of R, note first that $(0, b)(0, b') = (0, 0)$ implies that $(0) \oplus A$ has square zero. This gives $(0) \oplus A \subseteq \mathrm{Nil}_* R$. Secondly, $R/((0) \oplus A) \cong A$ is an integral domain, so $\mathrm{Nil}_* R \subseteq (0) \oplus A$, and hence equality holds. (Incidentally, we also have $\mathrm{Nil}^* R = \mathrm{Nil}_* R$.)

(3) It suffices to check that any principal *right* ideal $(a, b) \cdot R$ is an ideal in R. To see that

$$(a', b')(a, b) \in (a, b) \cdot R,$$

we need only find $x, y \in A$ such that

$$(a'a, b'a + \theta(a')b) = (a, b)(x, y) = (ax, bx + \theta(a)y).$$

If $a = 0$, we are done by choosing $x = \theta(a')$ and $y = 0$. Now assume $a \neq 0$. We choose $x = a'$ and find y from the equation

$$b'a + \theta(a')b = ba' + \theta(a)y.$$

Since $a, b, a', b' \in A$ and $\theta(a) \in k^*$, the unique solution y for this equation is in A, as desired.

(4) To see that R_R is a noetherian module, it suffices to check that $(0) \oplus A$ and $R/((0) \oplus A)$ are noetherian right R-modules. For $R/((0) \oplus A) \cong A$, this is clear since A is a (commutative) noetherian ring. To analyze $(0) \oplus A$, note that

$$(0, b)(x, y) = (0, bx),$$

so R acts on the right of $(0) \oplus A$ via the natural (first coordinate) projection $R \to A$. Since A_A is noetherian, it follows that $((0) \oplus A)_R$ is noetherian. The *left* R-structure on $(0) \oplus A$ is, however, quite different. Since

$$(x, y)(0, b) = (0, \theta(x)b),$$

we have $R \cdot (0, b) \subseteq (0, k \cdot b)$. From this (and the fact that $\dim_k A = \infty$), it is clear that ACC fails for $_R((0) \oplus A)$. In particular, R is not left noetherian.

Comment. Note that both $\mathfrak{m}$ and $(0) \oplus A$ are *principal* right ideals. In fact,

$$\mathfrak{m} = (p(t), 0) \cdot R, \quad \text{and} \quad (0) \oplus A = (0, 1)R.$$

However, they are not principal left ideals. The fact that

$$\bigcap_{i=1}^{\infty} \mathfrak{m}^i = (0) \oplus A \neq 0$$

for this example is to be contrasted with Krull's Intersection Theorem in commutative algebra: for more commentary on this, see Exercise 4.23.

Ex. 19.13. Show that any finitely generated projective right module P over a right artinian ring R is isomorphic to a finite direct sum of principal indecomposable right modules of R.

Solution. Let $R_R \cong n_1 P_1 \oplus \cdots \oplus n_r P_r$ where the P_i's are the distinct principal indecomposable right R-modules. Since P is finitely generated and projective, P is a direct summand of some

$$R_R^n \cong nn_1 P_1 \oplus \cdots \oplus nn_r P_r.$$

By the Krull-Schmidt Theorem, it follows that

$$P \cong m_1 P_1 \oplus \cdots \oplus m_r P_r$$

for suitable integers $m_i \geq 0$.

Ex. 19.14. Give an example of a local ring whose unique maximal ideal is nil but not nilpotent.

Solution. Let R be the commutative ring $\mathbb{Q}[x_1, x_2, \ldots]$ with the relations

$$x_1^2 = x_2^2 = \cdots = 0.$$

Then R is a local ring with unique maximal ideal $\mathfrak{m} = \sum R x_i$. Since each x_i is nilpotent, $\mathfrak{m}$ is a nil ideal. However, $\mathfrak{m}^n \neq 0$ for any integer n, since $\mathfrak{m}^n$ contains $x_1 x_2 \cdots x_n \neq 0$. Another example is given by $R = kG$ where k is any field of characteristic $p > 0$, and G is the infinite abelian p-group $\mathbb{Z}_p \oplus \mathbb{Z}_p \oplus \cdots$.

Ex. 19.15. Let (R, J) be a local ring, and M be a finitely generated left R-module. If $\text{Hom}_R(M, R/J) = 0$, show that $M = 0$.

Solution. Assume that $M \neq 0$. Since M is finitely generated, Nakayama's Lemma $(FC\text{-}(4.22))$ implies that $M/JM \neq 0$ as a left vector space over the division ring R/J. Let

$$f : M/JM \to R/J$$

be a nonzero R/J-homomorphism. Composing this with the projection map $M \to M/JM$, we get a nonzero R-homomorphism $\tilde{f} : M \to R/J$, so we have $\text{Hom}_R(M, R/J) \neq 0$.

Ex. 19.16. Let (R, J) be a left noetherian local ring, and M be a finitely generated left R-module. Show that M is a free R-module iff, for any exact sequence $0 \to A \to B \to M \to 0$ of left R-modules, the induced sequence

$$0 \to A/JA \to B/JB \to M/JM \to 0$$

remains exact.

Solution. The "only if" part is clear since

$$0 \to A \to B \to M \to 0$$

splits in case M is a free module. For the "if" part, note first that M/JM is a finite-dimensional vector space over R/J. Choose $m_1, \ldots, m_r \in M$ such that $\overline{m}_1, \ldots, \overline{m}_r$ form an R/J-basis of M/JM. By Nakayama's Lemma, $\sum R m_i = M$. Let $B = R^r$, with the usual basis $e_1, \ldots, e_r$, and let $\varphi : B \to M$ be the R-epimorphism defined by $\varphi(e_i) = m_i$. Then, for $A = ker(\varphi)$, the hypothesis gives an exact sequence

$$0 \to A/JA \to B/JB \xrightarrow{\overline{\varphi}} M/JM \to 0.$$

Here $\overline{\varphi}$ takes the basis $\overline{e}_1, \ldots, \overline{e}_r$ of B/JB to the basis $\overline{m}_1, \ldots, \overline{m}_r$ of M/JM, so $\overline{\varphi}$ is an isomorphism. It follows that $A/JA = 0$. Since R is left noetherian, $A \subseteq R^r$ implies that A is finitely generated over R. Thus, Nakayama's Lemma yields $A = 0$, and we have the desired conclusion $M \cong R^r$.

Ex. 19.17. Let $M \in \mathbb{M}_n(R)$ where $(R, \mathfrak{m})$ is a local ring. If each diagonal entry of M is a unit and each off-diagonal entry of M is a nonunit, show that $M \in \mathrm{GL}_n(R)$.

Solution. From $\mathrm{rad}(R) = \mathfrak{m}$, we know that $\mathrm{rad}\, \mathbb{M}_n(R) = \mathbb{M}_n(\mathfrak{m})$ (see *FC*-p. 57). Thus,

$$\mathbb{M}_n(R)/\mathrm{rad}\, \mathbb{M}_n(R) = \mathbb{M}_n(R)/\mathbb{M}_n(\mathfrak{m}) \cong \mathbb{M}_n(k),$$

where k is the division ring $R/\mathfrak{m}$. The given conditions on the matrix M means that, under the natural surjection $\mathbb{M}_n(R) \to \mathbb{M}_n(k)$, M goes to a diagonal matrix with unit diagonal entries over k. Since such a matrix is in $\mathrm{GL}_n(k)$, it follows from *FC*-(4.8) that $M \in \mathrm{GL}_n(R)$.

Comment. This exercise can easily be strengthened as follows. For an arbitrary ring R, *let $M \in \mathbb{M}_n(R)$ be such that each diagonal entry of M is a unit and each off-diagonal entry is in* rad R. *Then $M \in \mathrm{GL}_n(R)$.* Essentially the same proof works in this more general case (except that the new formulation has no longer anything to do with local rings!).

§20. Semilocal Rings

In commutative algebra, a commutative ring is called semilocal if it has only finitely many maximal ideals. Such rings arise upon localizing a commutative ring A at a multiplicative set $S = A \backslash \bigcup_{i=1}^m \mathfrak{p}_i$ where $\mathfrak{p}_1, \ldots, \mathfrak{p}_m$ are prime ideals in A.

For arbitrary rings, we take a different approach. By definition, a ring R is *semilocal* if $R/\mathrm{rad}\ R$ is artinian (and hence semisimple). In the commutative case, this turns out to be equivalent to the earlier definition. In general, if a ring R has finitely many maximal left ideals, then R is semilocal. However, a semilocal ring (e.g. a simple artinian ring) may have infinitely many maximal left ideals.

While there is no obvious process of "semilocalization" for noncommutative rings, many semilocal rings arise naturally. For instance, all local rings and 1-sided artinian rings are semilocal rings. Moreover, if k is any commutative semilocal ring (e.g. a field), and R is a k-algebra that is finitely generated as a k-module, then R is always a semilocal ring (see *FC*-(20.6)).

Semilocal rings R have many nice properties, most of which are proved by "lifting" corresponding properties from the semisimple ring $R/\mathrm{rad}\ R$. One of the most important properties of a semilocal ring R is that it has "left stable range 1." This property has played such an interesting role in modern ring theory and algebraic K-theory that it certainly deserves a mention here.

A ring E is said to have *left stable range* 1 if, whenever $Ea + Eb = E$ $(a, b \in E)$, there exists $e \in E$ such that $a + eb \in \mathrm{U}(E)$. "Right stable range 1" can be defined similarly by looking at principal right ideals. Amazingly, "left stable range 1" and "right stable range 1" turn out to be equivalent conditions. This result is due to Vaserstein; an elementary proof of it is contained in the solution to Exercise 1.25. As we have mentioned above, semilocal rings have stable range 1. However, rings of stable range 1 need not be semilocal (see Ex. 10C).

Stable range conditions were first introduced by Bass in his investigation of the stability properties of the "K_1" of a ring. Independently of this, the stable range 1 condition turned out to be of significance in the study of the cancellation problem of modules. For instance, E.G. Evans has proved that, *if A, B, C are right modules over any ring R such that $\mathrm{End}(A_R)$ has (left) stable range 1, then $A \oplus B \cong A \oplus C$ implies that $B \cong C$* (cf. *FC*-(20.11)). Note that this is a very strong cancellation theorem, since no conditions were imposed upon the modules B and C! In particular, if R itself has stable range 1, then finitely generated projective R-modules can be "cancelled" from direct sum decompositions (cf. *FC*-(20.13)).

The exercises in this section offer more characterizations of semilocal rings (Exercises 1, 13, and *Comment* on Exercise 11) and some of their additional properties (Exercises 3, 6, 7 and 12). The class of semilocal rings with a nilpotent Jacobson radical is especially susceptible to analysis: these are called the *semiprimary rings*. Over such rings, for instance, any nonzero module has a maximal submodule and a minimal submodule (Exercise 5*). Finally, the stable range 1 condition is related to a matrix-theoretic "stable finiteness" condition in (the *Comment* on) Exercise 8.

Exercises for §20

Ex. 20.1. Show that a ring R is a semilocal iff, for every left R-module M,

$$(*) \qquad \operatorname{soc}(M) = \{m \in M : (\operatorname{rad} R) \cdot m = 0\}.$$

Solution. The "only if" part should look familiar: we have solved it before in Exercise 4.18! For the converse, assume that $(*)$ holds for all left R-modules M (or just for all *cyclic* left R-modules). Applying $(*)$ to the cyclic left R-module

$$M := R/\operatorname{rad}(R),$$

we deduce that $\operatorname{soc}(M) = M$. This means that M is a semisimple module over R, and hence over $R/\operatorname{rad}(R)$. Thus, $R/\operatorname{rad}(R)$ is a semisimple ring, so R is semilocal, as desired.

Comment. Some further characterizations of semilocal rings in terms of the behavior of their simple and semisimple modules can be found in Exercise 13. The verification for these characterizations is again based on the useful principle that semisimple R-modules are essentially the same as the semisimple modules over the canonical factor ring $R/\operatorname{rad}(R)$.

If R is a semiprimary ring, more can be said about the socles of modules. For any left R-module M, it is easy to deduce from Ex. 5* below that $\operatorname{soc}(M)$ *is always essential in* M; that is $M' \cap \operatorname{soc}(M) \neq 0$ for any nonzero submodule M' of M. In fact, the same property holds for the socles of arbitrary left modules *over any right perfect ring* R, if we apply Ex. 24.7 instead of Ex. 5*.

Ex. 20.2. Let A, B, C be right modules over a ring R. If A has a composition series, show that $A \oplus B \cong A \oplus C$ implies $B \cong C$.

Solution. We can decompose A into $A_1 \oplus \cdots \oplus A_n$ where each A_i is indecomposable and has a composition series. Therefore, it suffices for us to handle the case $A = A_i$. In this case, $E = \operatorname{End}(A_R)$ is a local ring by FC-(19.17); in particular, E has left stable range 1. By the Cancellation Theorem in FC-(20.11), we conclude that

$$A \oplus B \cong A \oplus C \Longrightarrow B \cong C.$$

Comment. For a stronger conclusion, see the *Comment* on Exercise 21.24.

Ex. 20.3. Show that, over any commutative semilocal domain R, any finitely generated projective R-module P is free.

Solution. Let K be the quotient field of R. Then $P \otimes_R K \cong K^n$ for some integer n. For any prime ideal $\mathfrak{p}$ of R, the localized module $P_{\mathfrak{p}}$ over the local ring $R_{\mathfrak{p}}$ is also free, so $P_{\mathfrak{p}} \cong R_{\mathfrak{p}}^m$ for some integer m (by FC-(19.29)). Localizing further to the quotient field K, we see that $m = n$. Now consider the quotient map $i : R \to S$, where $S = R/\mathrm{rad}\, R$ is semisimple (and hence a finite direct product of fields). For *any* prime ideal $\mathfrak{p}'$ of S, $\mathfrak{p} := i^{-1}(\mathfrak{p}')$ is a prime ideal of R, and we have a ring homomorphism $R_{\mathfrak{p}} \to S_{\mathfrak{p}'}$. From

$$(P \otimes_R S) \otimes_S S_{\mathfrak{p}'} \cong (P \otimes_R R_{\mathfrak{p}}) \otimes_{R_{\mathfrak{p}}} S_{\mathfrak{p}'} \cong R_{\mathfrak{p}}^n \otimes_{R_{\mathfrak{p}}} S_{\mathfrak{p}'} \cong S_{\mathfrak{p}'}^n \quad (\forall \mathfrak{p}'),$$

it is easy to see that $P \otimes_R S \cong S^n$. (Use Exercise 19.13 for the ring S.) It now follows from FC-(19.27) that $P \cong R^n$ as R-modules.

Ex. 20.4. (Jensen-Jøndrup) Can every commutative semiprimary ring C be embedded in a right (resp. left) noetherian ring?

Solution. The answer is "no" in general. To give a counterexample, let C be the commutative $\mathbb{Q}$-algebra generated by $x_1, x_2, \ldots$ with the relations $x_i x_j = 0$ for $i \neq j$, and $x_i^3 = 0$ for all i. It is easy to see that C is a local ring with

$$J := \mathrm{rad}\, C = \bigoplus_i (\mathbb{Q} x_i \oplus \mathbb{Q} x_i^2).$$

Since $x_i x_j x_k = 0$ for all i, j, k, we have $J^3 = 0$, so C is a semiprimary ring. Assume, for the moment, that C can be embedded as a subring of a right noetherian ring R. Then $JR = \sum_i x_i R$ is a finitely generated right ideal in R, so we have $JR = \sum_{i=1}^n x_i R$ for some n. But then

$$x_{n+1} = x_1 r_1 + \cdots + x_n r_n$$

for suitable $r_i \in R$, and left multiplication by x_{n+1} (in R) gives $x_{n+1}^2 = 0$, a contradiction.

Comment. In contrast, if C is any commutative semiprimary ring with $(\mathrm{rad}\, C)^2 = 0$, Jensen and Jøndrup have shown that C can then be embedded in an artinian ring! See their paper in Math. Zeit. *130* (1973), 189–197.

Ex. 20.5. The ring C in the above exercise is a commutative non-noetherian semiprimary ring. Show that $R = \begin{pmatrix} \mathbb{Q} & \mathbb{R} \\ 0 & \mathbb{Q} \end{pmatrix}$ is a noncommutative semiprimary ring that is neither right noetherian nor left noetherian.

Solution. Clearly, $J := \begin{pmatrix} 0 & \mathbb{R} \\ 0 & 0 \end{pmatrix}$ is an ideal of square zero in R. Therefore, $J \subseteq \mathrm{rad}\, R$. On the other hand, $R/J \cong \mathbb{Q} \times \mathbb{Q}$ is semisimple, so by Exercise 4.11, $\mathrm{rad}\, R \subseteq J$. This shows that $\mathrm{rad}\, R = J$, so

$$R/\mathrm{rad}\, R \cong \mathbb{Q} \times \mathbb{Q}.$$

Hence R is semilocal. Since $(\mathrm{rad}\, R)^2 = J^2 = 0$, R is a semiprimary ring. According to FC-(1.22), R is neither right noetherian nor left noetherian.

Ex. 20.5*. Let R be any semiprimary ring, and M be any nonzero left R-module. Show that $(\mathrm{rad}\, R)M \neq M$ and $\mathrm{soc}(M) \neq 0$.

Solution. Let $J = \operatorname{rad} R$. The hypothesis on R means that R/J is semisimple and $J^n = 0$ for some n. If $JM = M$, successive multiplication by J yields $M = J^n M = 0$, a contradiction. Therefore, $JM \neq M$. Let m be the smallest integer such that $J^m M = 0$. Then $m \geq 1$, and $N := J^{m-1}M$ is a nonzero R/J-module. Since R/J is a semisimple ring, $_R N$ is a semisimple R-module, so $\operatorname{soc}(M) \supseteq N \neq 0$.

Comment. For a much more general result, see Exercise 24.7 below.

Ex. 20.6. Let R be a left noetherian semilocal ring such that $\operatorname{rad} R$ is a nil ideal. Show that R is left artinian.

Solution. Since R is left noetherian, the nil ideal $\operatorname{rad} R$ must be nilpotent by Levitzki's Theorem (FC-(10.30)). Therefore, R is a semiprimary ring. By the Hopkins-Levitzki Theorem (FC-(14.15)), the fact that R is left noetherian implies that R is left artinian.

Ex. 20.7. (cf. Exercise 4.21) Let R be any semilocal ring. For any ideal $I \subseteq R$, show that the natural map $\varphi : \operatorname{GL}_n(R) \to \operatorname{GL}_n(R/I)$ is onto.

Solution. We first prove the desired conclusion for $n = 1$. In this case, we have to prove the surjectivity of $\operatorname{U}(R) \to \operatorname{U}(R/I)$. Let $a \in R$ be such that $\bar{a} \in \operatorname{U}(R/I)$. Then $Ra + I = R$. By Bass' Theorem (FC-(20.9)), $a + I$ contains a unit $u \in \operatorname{U}(R)$. Clearly, this unit lifts the given unit $\bar{a} \in \operatorname{U}(R/I)$. For a general n, let

$$R' = \operatorname{M}_n(R) \quad \text{and} \quad I' = \operatorname{M}_n(I).$$

By FC-(20.4), R' is also semilocal. By the $n = 1$ case settled above, $\operatorname{U}(R') \to \operatorname{U}(R'/I')$ is onto. Since $\operatorname{U}(R') = \operatorname{GL}_n(R)$ and

$$\operatorname{U}(R'/I') = \operatorname{U}(\operatorname{M}_n(R/I)) = \operatorname{GL}_n(R/I),$$

we conclude that $\varphi : \operatorname{GL}_n(R) \to \operatorname{GL}_n(R/I)$ is onto.

Ex. 20.8. For any ring R and any integer $n \geq 1$, show that the following statements are equivalent:

(1) $\operatorname{M}_n(R)$ is Dedekind-finite;
(2) for any right R-module M, $R^n \cong R^n \oplus M$ implies that $M = (0)$;
(3) R_R^n is hopfian, i.e. any right module epimorphism $\alpha : R^n \to R^n$ is an isomorphism.

Solution. (1) $\Rightarrow$ (3). Let $\beta : R^n \to R^n$ be a splitting of the epimorphism α, so that $\alpha\beta = I$. Identifying $\operatorname{End}(R_R^n)$ with $\operatorname{M}_n(R)$ in the usual way, we have $\alpha\beta = I \Longrightarrow \beta\alpha = I$, so α is also one-one, hence an isomorphism.

(3) $\Rightarrow$ (2). Fix an isomorphism $\varphi : R^n \to R^n \oplus M$ and let $\pi : R^n \oplus M \to R^n$ be the first-component projection onto R^n. Then $\pi\varphi$ is an epimorphism

and hence an isomorphism. This implies that π is also an isomorphism, so $M = \ker(\pi) = 0$.

$(2) \Rightarrow (1)$. Say $\alpha\beta = I \in \mathbb{M}_n(R)$. Then α defines an epimorphism $R^n \to R^n$ with a splitting β. Letting $M = \ker(\alpha)$, we have $R^n \cong R^n \oplus M$. Now (2) implies that $M = 0$, and we conclude easily that $\beta\alpha = I \in \mathbb{M}_n(R)$.

Comment. A ring R satisfying any of the conditions (1), (2), (3) above is called *stably n-finite* (or, according to P.M. Cohn, *weakly n-finite*). Note that if R is stably n-finite, then R is stably m-finite for any $m \leq n$. For instance, *if R has stable range 1, then R is stably n-finite for all n*, by FC-(20.13).

Ex. 20.9. According to Ex. 1.12, if a ring R is right noetherian, then any finitely generated right R-module is hopfian. Show that the same conclusion holds also over any S-algebra R over a commutative ring S such that R is finitely generated as a module over S. (In particular, right noetherian rings and S-algebras of the above type are stably n-finite for any n.)

Solution. In view of Ex. 4.16*, it is enough to handle the case $R = S$; that is, we may assume that R is a commutative ring.

Given a surjective $f : M \to M$ where M_R is finitely generated, we may view M as a right $R[t]$-module by letting t act via f. Let $\mathfrak{A}$ be the ideal $tR[t]$ in $R[t]$. Then $M\mathfrak{A} = M$ (since f is onto). Pick $m_1, \ldots, m_n \in M$ such that $M = \sum_{i=1}^n m_i R$. We can write $m_i = \sum_j m_j \alpha_{ij}$ where $\alpha_{ij} \in \mathfrak{A}$. By a familiar argument in linear algebra, we see that

$$M \cdot \det(I_n - (\alpha_{ij})) = 0.$$

Now $\det(I_n - (\alpha_{ij}))$ has the form $1 + \alpha$ for some $\alpha \in \mathfrak{A}$, say $\alpha = r_1 t + \cdots + r_k t^k$. For any $m \in \ker(f)$, we have then

$$0 = m(1 + \alpha) = m + f(m)r_1 + \cdots + f^k(m)r_k = m,$$

which shows that f is surjective.

Comment. The result in this exercise first appeared in print in Vasconcelos's paper in Án. Acad. Brasil Ci. *37* (1965), 389–393. Our proof above follows that of Prop. 1.2 in a later paper of Vasconcelos in Trans. Amer. Math. Soc. *138* (1969), 505–512. Essentially the same result was independently proved in §4 of Strooker's paper in Nagoya J. Math. *27* (1966), 747–751. (Strooker assumed only that R is a direct limit of a directed system of right noetherian rings, which is certainly the case when R is commutative, by the Hilbert Basis Theorem.) The commutative case of Strooker's result (for finitely presented modules) also appeared on p. 35 in Grothendieck's "Éléments de Géométrie Algébrique", IV. 3, Publ. I.H.E.S., no. 28, 1966, and was proved by a familiar reduction to the case of noetherian rings.

By standard localization techniques, it follows immediately from this exercise that, for R commutative, the hopfian property already holds for *locally* finitely generated R-modules.

The most general result on the hopfian property was that obtained by K. Goodearl. In his paper in Comm. Algebra 15 (1987), 589–609, Goodearl defined a (possibly noncommutative) ring R to be *right repetitive* if, for any $a, b \in R$, $a^k b = \sum_{i=0}^{k-1} a^i b$ for some k. With this definition, Goodearl showed that *all finitely generated right modules over a ring R are hopfian iff all matrix rings* $\mathbb{M}_n(R)$ *are right repetitive*. Goodearl also pointed out that these conditions are *not* left-right symmetric. From Goodearl's results, it is easy to deduce, for instance, that commutative rings and right noetherian rings both have the hopfian property discussed in this exercise. An example due to Armendariz, Fisher, and Snider showed, however, that P.I. rings (rings satisfying a polynomial identity) need not have this hopfian property; see their paper in Comm. Algebra 6 (1978), 659–672.

The cohopfian analogue of Goodearl's result, due to Dischinger, was actually obtained some ten years earlier; see Ex. 23.9 below.

Ex. 20.9*. (Partial generalization of Ex. 9.) A ring R is called *right duo* if every right ideal of R is an ideal. Show that such a ring R is stably n-finite for all n.

Solution. The following succinct solution was suggested to us by K. Goodearl, who also pointed out that it is sufficient to assume R is *right quasi-duo*; that is, every maximal right ideal of R is an ideal. For such a ring R, let $\{\mathfrak{m}_i\}$ be the set of maximal right ideals, with rad $R = \bigcap \mathfrak{m}_i$. Since $\mathfrak{m}_i$ is an ideal, $R/\mathfrak{m}_i$ is a ring; this is a division ring since its only right ideals are (0) and itself. Thus, for any n,

$$\mathbb{M}_n(R)/\mathbb{M}_n(\mathfrak{m}_i) \cong \mathbb{M}_n(R/\mathfrak{m}_i)$$

is a simple artinian ring, which we know is Dedekind-finite (by linear algebra, or by one of Ex. 3.10, Ex. 4.17). The natural ring homomorphism $\mathbb{M}_n(R) \to \prod_i \mathbb{M}_n(R/\mathfrak{m}_i)$ has kernel $\mathbb{M}_n(\text{rad } R) = \text{rad } \mathbb{M}_n(R)$ (see *FC*-p. 57). Thus, $\mathbb{M}_n(R)/\text{rad } \mathbb{M}_n(R)$ is Dedekind-finite, and *FC*-(4.8) implies the same for $\mathbb{M}_n(R)$.

Comment. This exercise, brought to my attention by J. Sjogren, was a remark from p. 479 of Kaplansky's classic paper "Elementary divisors and modules," Trans. Amer. Math. Soc. *66* (1949), 464–491. No proof was given for this remark.

In the terminology of *FC*-§12, the proof above showed that, for a right quasi-duo ring R, $\mathbb{M}_n(R)/\text{rad } \mathbb{M}_n(R)$ is a subdirect product of simple artinian rings. The reader might try to investigate what other properties of $\mathbb{M}_n(R)$ can be deduced from this fact.

Deciding whether a particular class of rings is stably finite (i.e. stably n-finite for all n) is often not easy. For an interesting example, consider the class of group algebras kG where k is a field and G is a group. If char $k = 0$, Kaplansky has shown that kG is stably finite, but the same conclusion is not known if char $k = p > 0$. For more detailed discussions on this (including references to the work of I. Kaplansky, S. Montgomery and D. Passman in this direction), see I. Herstein's lovely "Notes from a Ring Theory Conference," CBMS Monograph Series No. 9, Amer. Math. Soc., 1971. Some cases in which the characteristic p question has an affirmative answer are: (1) G is finite or abelian, (2) G is an ordered group, and (3) G is an amenable group. Here, (1) is easy, (2) follows from FC-(14.24), and (3) is a recent result of Ara, O'Meara and Perera from their paper "Stable finiteness of group rings in arbitrary characteristic," Adv. Math. *170* (2002), 224–238.

Ex. 20.10A. (Kaplansky) Show that a ring R has left stable range 1 iff $Ra + Rb = R$ implies that $R \cdot (a + xb) = R$ for some $x \in R$.

Solution. ("If" part) Assuming that $Ra + Rb = R$ implies $R \cdot (a + xb) = R$ for some $x \in R$, we are done if we can show that R is Dedekind-finite. Suppose $au = 1$. From $Ra + R(1 - ua) = R$, our assumption gives a left-invertible element $v = a + x(1 - ua)$ for some $x \in R$. Then $vu = au + xu - xu(au) = 1$ shows that $v \in U(R)$, and hence $u, a \in U(R)$ also.

Comment. The result in this exercise appeared in an unpublished (and mimeographed!) note of I. Kaplansky, entitled "Bass's first stable range condition", written in 1971. The general (left) stable range n condition, first introduced by Bass in his work on algebraic K-theory, requires that, whenever $Ra_1 + \cdots + Ra_{n+1} = R$, there exist $x_1, \ldots, x_n \in R$ such that

$$R(a_1 + x_1 b_{n+1}) + \cdots + R(a_n + x_n b_{n+1}) = R.$$

In the case $n = 1$, we used a slightly different definition in FC-p. 300. The point of this exercise is to show that the two definitions are equivalent (for $n = 1$). For general n, L. Vaserstein proved the rather amazing result that the stable range n condition is left-right symmetric. We have alluded to this in the case $n = 1$ in the Introduction to this section.

Ex. 20.10B. For any ring R with left stable range 1, show that rad (R) is given by $\{r \in R : r + U(R) \subseteq U(R)\}$.

Solution. It suffices to show that $r + U(R) \subseteq U(R)$ implies $r \in \text{rad}(R)$. By FC-(4.1), the latter will follow if we can show that, for any $s \in R$, $1 - sr$ has a left inverse. From $Rr + R(1 - sr) = R$, the left stable range 1 condition gives a unit

$$r + x(1 - sr) \in U(R)$$

for some $x \in R$. But then $x(1 - sr) \in r + U(R) \subseteq U(R)$, which clearly implies that $1 - sr$ has a left inverse.

Comment. It was Sam Perlis who first proved that the (Wedderburn) radical of a finite-dimensional algebra R over a field is given by the set $\{r \in R : r + U(R) \subseteq U(R)\}$. (See his paper "A characterization of the radical of an algebra", Bull. Amer. Math. Soc. *48* (1942), 128–132.) This is a special case of the present exercise, since any such algebra is certainly semilocal, and semilocal rings have left stable 1 according to Bass's Theorem (*FC*-(20.9)). The following is, for instance, a consequence of this exercise: *if R is a semisimple ring, then the only element $r \in R$ with the property $r + U(R) \subseteq U(R)$ is zero.* Note, however, that the "stable range 1" condition on R in this exercise *cannot* be dropped, as we have seen earlier in Ex. 4.24.

Ex. 20.10C. Without assuming the results in the next two exercises, show that a ring of stable range 1 need not be semilocal.

Solution. It is easy to see that if R_i $(i \in I)$ are rings of stable range 1, then so is their direct product $R = \prod_i R_i$. Therefore, if we take $\{R_i\}$ to be an infinite family of fields, then $R = \prod_i R_i$ has stable range 1. But R is clearly not semilocal.

Another construction can be given if we notice that the direct limit of a family of rings of stable range 1 is also of stable range 1. Thus, if we fix any field k and embed $\mathbb{M}_{2^i}(k)$ into $\mathbb{M}_{2^{i+1}}(k)$ by

$$M \mapsto \begin{pmatrix} M & 0 \\ 0 & M \end{pmatrix},$$

then $R = \bigcup_i \mathbb{M}_{2^i}(k)$ is of stable range 1. But R is a nonartinian simple ring (see *FC*-p. 40), so it is again not semilocal.

Comment. In case the above examples look a bit "pathological" to some readers, we point out that there are many commutative integral domains which are of stable range 1 but are not semilocal. Some examples are: the ring of all algebraic integers, and the ring of all holomorphic functions on $\mathbb{C}$. (Both of these are Bézout domains.) For other examples, see the papers "Stable range in commutative rings" by D. Estes and J. Ohm, J. Algebra *7* (1967), 343–362, and "Stable range one for rings with many units" by K. Goodearl and P. Menal, J. Pure and Applied Alg. *54* (1988), 261–287.

The class of rings of stable range 1 turns out to be veritably larger than what one might initially expect. For more information on this, see the next two exercises.

Ex. 20.10D. For any von Neumann regular ring R, show that the following conditions are equivalent:

(1) R has (left) stable range 1;
(2) R is unit-regular (in the sense defined in Ex. 4.14B);
(3) finitely generated projective (right) R-modules satisfy the cancellation law.

Solution. (1) $\Longrightarrow$ (3) follows from FC-(20.13) (even without the regularity assumption on R). (3) $\Longrightarrow$ (2) follows from Ex. 4.14C (applied to the module $M = R_R$). Thus, we need only prove (2) $\Longrightarrow$ (1).

Assume R is unit-regular. We can show that R has left stable range 1 by repeating Swan's proof of the same conclusion for semisimple rings, as given in FC-(20.9). Starting with $Ra + Rb = R$, there are exactly three steps in Swan's proof that depended on semisimplicity, namely:

(A) We can write $Rb = (Ra \cap Rb) \oplus \mathfrak{B}'$ for some left ideal $\mathfrak{B}'$.

(B) If $K = \{r : ra = 0\}$, the sequence $0 \to K \to R \xrightarrow{f} Ra \to 0$ (defined by $f(1) = a$) splits.

(C) From $Ra \oplus K \cong R = Ra \oplus \mathfrak{B}'$, we can deduce $K \cong \mathfrak{B}'$.

If we can justify each of these steps over the unit-regular ring R, Swan's proof will go through. Now (B) is clear, since Ra (as a direct summand of $_R R$) is projective. (A) follows from the natural exact sequence

$$0 \to Ra \cap Rb \to R \longrightarrow (R/Ra) \oplus (R/Rb) \to 0.$$

(Since $(R/Ra) \oplus (R/Rb)$ is projective, this sequence splits, so $Ra \cap Rb$ is a direct summand of R, hence of Rb.) So far, we have used only the von Neumann regularity of R. Finally, (C) follows from Ex. 4.14C in light of the *unit-regularity* of R.

Comment. A direct proof for (1) $\Longrightarrow$ (2) is also possible; see, for instance, Prop. 4.12 in K. Goodearl's book "von Neumann Regular Rings", 2nd edition, Krieger Publ. Co., Malabar, Florida, 1991.

Ex. 20.10E. For any ring R, show that the following conditions are equivalent:

(1) R has left stable range 1;
(2) $R/\mathrm{rad}\, R$ has left stable range 1;
(3) $R[[x]]$ has left stable range 1.

From this, deduce that, if R is a commutative ring of (Krull) dimension 0, then $R[[x_1, \dots, x_n]]$ has stable range 1 (for any $n \geq 0$).

Solution. The equivalence (1) $\Longleftrightarrow$ (2) follows quickly from the following two observations on the factor ring $\overline{R} = R/J$, where J is any ideal of R contained in $\mathrm{rad}\, R$:

(A) For $r \in R$, $r \in \mathrm{U}(R)$ iff $\overline{r} \in \mathrm{U}(\overline{R})$;
(B) For $a, b \in R$, $Ra + Rb = R$ iff $\overline{R}\overline{a} + \overline{R}\overline{b} = \overline{R}$.

In fact, these observations show that (1) is equivalent to R/J having left stable range 1. Applying this to the ideal $(x) \subseteq \mathrm{rad}\, R[[x]]$, we see also that (1) $\Longleftrightarrow$ (3).

Now assume R is commutative, with dim $R = 0$. By the *Comment* on Ex. 4.15, $R/\text{rad } R$ is unit-regular. Applying Ex. 10D, we see that $R/\text{rad } R$ has left stable range 1, so by (2) $\Longrightarrow$ (1) above, R also has left stable range 1. Applying (1) $\Longrightarrow$ (3) n times, we see that $R[[x_1, \ldots, x_n]]$ has left stable range 1.

Comment. The implication (1) $\Longrightarrow$ (3) was observed by Brewer, Katz, and Ullery in their paper in J. Algebra *106* (1987), 265–286. The fact that 0-dimensional commutative rings have stable range 1 has a powerful generalization: P. Ara has proved that *any strongly π-regular ring has stable range* 1; see his paper in Proc. Amer. Math. Soc. *124* (1996), 3293–3298.

Ex. 20.11. (Camps-Menal) Recall that a subring $A \subseteq R$ is *full* if $A \cap U(R) \subseteq U(A)$ (see Exercise 5.0). If $R = D_1 \times \cdots \times D_n$ where each D_i is a division ring, show that any full subring $A \subseteq R$ is a semilocal ring.

Solution. We proceed by induction on n. If $n = 1$, a full subring $A \subseteq R = D_1$ is a division ring and hence semilocal. Now assume $n > 1$. If A contains a nontrivial idempotent e, then e must be central in R and we have

$$(1) \qquad eA \subseteq eR, \quad (1-e)A \subseteq (1-e)R,$$

where eR, $(1-e)R$ are direct products of fewer than n division rings. It is easy to check that, in (1), eA and $(1-e)A$ are *full* subrings of eR and $(1-e)R$. (Use the fact that if $x \in eA \cap U(eR)$, then $x + (1-e) \in A \cap U(R)$.) Therefore, by the inductive hypothesis, eA and $(1-e)A$ are semilocal, which implies that $A = eA \times (1-e)A$ is semilocal.

In the following, we assume that A contains no trivial idempotents. We may also assume that A is not a division ring. Pick an element $a \in A\backslash\{0\}$ with the largest number of zero coordinates, say

$$a = (0, \ldots, 0, a_{r+1}, \ldots, a_n),$$

where all $a_i \neq 0$. Then $0 < r < n$. Let

$$\pi : A \to D_1 \times \cdots \times D_r$$

be the projection map into the first r components. We claim that

$$(2) \qquad b \in \ker(\pi), \ x \in A \Longrightarrow 1 - xb \in U(R).$$

In fact, let $b = (0, \ldots, 0, b_{r+1}, \ldots, b_n)$ and $x = (x_1, \ldots, x_n)$ in A. Then

$$(3) \qquad 1 - xb = (1, \ldots, 1, 1 - x_{r+1}b_{r+1}, \ldots) \in A,$$

$$(4) \qquad xb(1 - xb) = (0, \ldots, 0, x_{r+1}b_{r+1}(1 - x_{r+1}b_{r+1}), \ldots) \in A.$$

Assume, for the moment, that $1 - xb \notin U(R)$. Then, from (3), we have, say, $1 - x_{r+1}b_{r+1} = 0$. Now the element in (4) has at least $r+1$ zero coordinates, so $xb(1 - xb) = 0$. This means that xb is an idempotent in A; hence $xb = 0$ or 1. Both are contradictions, so we have proved (2).

Since $A \subseteq R$ is full, we have in fact $1 - xb \in U(A)$ in (2). Therefore, $\ker(\pi) \subseteq \operatorname{rad} A$. Next we claim that

(5)   *The subring $A/\ker(\pi) \subseteq D_1 \times \cdots \times D_r$ is full.*

If so, then, by the inductive hypothesis, $A/\ker(\pi)$ is semilocal. Since $\ker(\pi) \subseteq \operatorname{rad} A$, this implies that A is semilocal, as desired.

To prove (5), consider $x = (x_1, \ldots, x_n) \in A$ such that

$$\pi(x) \in U(D_1 \times \cdots \times D_r),$$

i.e. $x_1, \ldots, x_r \neq 0$. We want to prove that *all* $x_i \neq 0$, for then $x \in A \cap U(R) \subseteq U(A)$. Assume, for the moment, that, say $x_{r+1} = 0$. Then (for the element a constructed earlier) xa has at least $r + 1$ zero coordinates, so $xa = 0$. This means

$$x_{r+1} = \cdots = x_n = 0,$$

and hence

$$x + a = (x_1, \ldots, x_r, a_{r+1}, \ldots, a_n) \in A \cap U(R) \subseteq U(A).$$

Since $a \in \ker(\pi) \subseteq \operatorname{rad} A$, this gives $x \in U(A)$, which is clearly a contradiction.

Comment. The above result of Camps and Menal appeared in Comm. Algebra *19* (1991), 2081–2095. It was subsequently further generalized by Camps and Dicks (in Israel J. Math. *81* (1993), 203–211) as follows. Define a ring homomorphism $f : S \to R$ to be *local* if $f^{-1}(U(R)) \subseteq U(S)$. (Note that a subring $A \subseteq R$ is full iff the inclusion map $A \to R$ is local.) In their joint paper, Camps and Dicks proved that *if $f : S \to R$ is local and R is semilocal, then S is also semilocal.*

For any ring S, the projection map $S \to S/\operatorname{rad} S$ is always a local homomorphism (by *FC*-(4.8)). Therefore, the result of Camps and Dicks can also be stated in the following equivalent form: *A ring S is semilocal iff there exists a local homomorphism from S to some semisimple ring.*

Ex. 20.12. (Cf. Exercise 19.6) For any semilocal ring R, show that $C := Z(R)$ is semilocal.

Solution. By the solution to Exercise 5.0, C is full in R, and

$$J := C \cap \operatorname{rad} R \subseteq \operatorname{rad} C.$$

We claim that

(1) $C/J \hookrightarrow R/\operatorname{rad} R$ *is a full subring.*

In fact, if $c \in C$ is such that $\bar{c} \in U(R/\operatorname{rad} R)$, then by *FC*-(4.8),

$$c \in C \cap U(R) \subseteq U(C).$$

From (1), it follows that

(2) $C/J \hookrightarrow Z(R/\mathrm{rad}\ R)$ is also a full subring.

By Exercise 3.4,
$$Z(R/\mathrm{rad}\ R) \cong D_1 \times \cdots \times D_n,$$

where the D_i's are fields. Therefore, by the last exercise, C/J is semilocal. Since $J \subseteq \mathrm{rad}\ C$, this implies that C is semilocal.

Comment. Of course, if we assume the more general result of Camps and Dicks mentioned in the *Comment* on Exercise 11, then the fact that $C \hookrightarrow R$ is a local homomorphism implies directly that C is a semilocal ring.

The following two exercises were worked out in collaboration with H.W. Lenstra, Jr.

Ex. 20.13. For any ring R, show that the following are equivalent:

(1) R is semilocal;
(2) Any direct product of semisimple right R-modules is semisimple;
(3) Any direct product of copies of a semisimple right R-module is semisimple;
(4) Any direct product of simple right R-modules is semisimple.

Solution. Let $\overline{R} = R/J$, where $J = \mathrm{rad}\ (R)$.

(1) $\Longrightarrow$ (2). If R is semilocal, $\overline{R}$ is a semisimple ring. Let $\{M_i\}$ be any family of semisimple right R-modules. Since $M_i J = 0$ for all i, each M_i is an $\overline{R}$-module, and therefore so is $\prod_i M_i$. But then $\prod_i M_i$ is semisimple over $\overline{R}$, and hence also over R.

(2) $\Longrightarrow$ (3) and (2) $\Longrightarrow$ (4) are tautologies.

(3) $\Longrightarrow$ (1). Let $\{S_j\}$ be a representative set of simple right R-modules. Then $M = \bigoplus_i S_j$ is a faithful semisimple right module over $\overline{R}$ (as in *FC*-(11.1)). For each $i \in M$, let $\varphi_i : \overline{R} \to M$ be the $\overline{R}$-module homomorphism sending $\overline{r}$ to $i\overline{r}$. Since $M_{\overline{R}}$ is faithful, these φ_i's define an embedding of $\overline{R}_{\overline{R}}$ into $\prod_{i \in M} M$. By (3), the latter is semisimple, so $\overline{R}_{\overline{R}}$ is also semisimple. This means $\overline{R}$ is a semisimple ring, so we have (1).

(4) $\Longrightarrow$ (1). For $\{S_j\}$ as above, let $N = \prod_j S_j$. By (4), N_R is semisimple (and still faithful over $\overline{R}$). Repeating the above argument with N replacing M (and noting that $\prod_{i \in N} N$ is semisimple), we see that $\overline{R}_{\overline{R}}$ is semisimple as before.

Comment. It should not go unnoticed, perhaps, that the choices of our implications were designed in such a way that we stay clear from any slippery considerations of arbitrary direct products of direct sums of modules!

Ex. 20.14. For any ring R, show that the following are equivalent:

(5) Any right primitive factor ring of R is artinian;
(6) Any direct product of copies of a simple right R-module is semisimple.

Solution. (5) $\Longrightarrow$ (6). Given any simple module M_R, let $\overline{R} := R/\mathrm{ann}(M)$. This ring is right primitive, since it has a faithful simple right module M. Thus, by (5), $\overline{R}$ is simple artinian. Any direct product of copies of M is semisimple over $\overline{R}$, and hence also over R.

(6) $\Longrightarrow$ (5). Any right primitive factor ring $\overline{R}$ of R has a faithful simple right module M. Arguing as in (3) $\Longrightarrow$ (1) of the previous exercise, we see that $\overline{R}_{\overline{R}}$ can be embedded into $\prod_{i \in M} M$. Thus, (6) implies that $\overline{R}$ is a semisimple (and hence artinian) ring.

Comment. Of course, the conditions in Ex. 13 imply those in Ex. 14. The converse is not true. For instance, any *commutative* ring R satisfies (5), but not necessarily (1)!

Since the conditions in Ex. 13 are left-right symmetric (specifically, (1) is), it is natural to ask if the same is true for the conditions (5), (6) in Ex. 14. In pp. 60–61 of K. Goodearl's book on von Neumann regular rings (Krieger Publ. Co., 1991), it is proved that, if R is von Neumann regular, the condition (5) is equivalent to prime factor rings of R being artinian. This implies, in particular, that (5) (and hence (6)) is left-right symmetric for von Neumann regular rings. In general, however, (5) *is not equivalent to its left analogue*, according to my colleague G. Bergman. In his 1964 paper in Proc. A.M.S. *15*, 473–475, Bergman constructed the first example of a ring (in fact an infinite-dimensional $\mathbb{Q}$-algebra) that is right primitive but not left primitive. Clearly, R cannot satisfy (5). In an email to me on 3/28/03, Bergman proved further that all simple *left* R-modules are finite-dimensional over $\mathbb{Q}$. In particular, every left primitive factor ring of R is finite-dimensional over $\mathbb{Q}$, and therefore artinian. This sharpens the earlier conclusion that R is not left primitive into the statement that R actually satisfies the left analogue of (5)—almost forty years after the publication of the original Proc. A.M.S. paper!

§21. The Theory of Idempotents

If e is a central idempotent in a ring R (that is, $e = e^2 \in Z(R)$), then, for the complementary idempotent $f = 1 - e$, we have a decomposition of R into a direct product of the two rings eR and fR. Conversely, any decomposition of R into a direct product of two rings arises in this fashion.

If $e = e^2$ but e is not necessarily central, then for $f = 1 - e$ we have $R = eR \oplus fR$ where eR, fR are now right ideals, and any decomposition of R into a direct sum of two right ideals arises in this fashion. (A similar result holds for left ideals.) In general, even if a ring R has only the trivial central idempotents (0 and 1), it may have lots of nontrivial idempotents. The example of a matrix ring $R = \mathbb{M}_n(k)$ ($n > 1$, k any domain) is an important case in point.

In view of what we said above, it is perhaps not surprising that idempotents play a much larger role in the structure theory of noncommutative rings. Given $e = e^2 \in R$, we can associate with it the ring eRe, which has identity e, and can be identified with the endomorphism ring of $(eR)_R$ (cf. FC-(21.7)). In a manner of speaking, the ring eRe contains the "local information" about the idempotent e. The close relationship between the 1-sided (resp. 2-sided) ideals of R and eRe is described in detail in FC-(21.11).

For any nonzero idempotent $e \in R$, consider the following properties for the right module eR:

(1) eR is a simple module.
(2) eR is a strongly indecomposable module.
(3) eR is an indecomposable module.

These properties define, respectively, the notions of *right irreducible, local,* and *primitive* idempotents, in decreasing strength. The latter two turn out to be left-right symmetric conditions: e local means eRe is a local ring, while e primitive means eRe has only trivial idempotents. In general, e right irreducible is different from e left irreducible, although, in the case of a *semiprime* ring R, these two notions do coincide, and they both amount to eRe being a division ring (FC-(21.16)).

For two given idempotents $e, e' \in R$, consider the following possible relationships between them:

(1) $eR = e'R$.
(2) $eR \cong e'R$ and $(1 - e)R \cong (1 - e')R$ (as right R-modules).
(3) $eR \cong e'R$ (as right R-modules).

Here again, we have (1) $\Rightarrow$ (2) $\Rightarrow$ (3) (the first implication is Exercise 4 below), and (2), (3) turn out to be left-right symmetric relations. The relation (3) amounts to $e = ab$ and $e' = ba$ for suitable $a, b \in R$: in this case, we write $e \cong e'$, and say that e, e' are *isomorphic* idempotents (FC-(21.20)). The relation (2) amounts to e, e' being *conjugate* in R: see Exercise 16.

For an ideal $I \subseteq R$, it is of interest to study the relationship between the idempotents of R and those of $\overline{R} = R/I$. For instance, when can an idempotent in $\overline{R}$ be lifted to an idempotent in R? This turns out to be possible in at least two important cases, first when I is nil, and secondly when R is I-adically complete: see FC-(21.28) and FC-(21.31). The liftability of idempotents in these cases is a fundamental result in ring theory with a wide range of applications to number theory, integral and modular representation theory, and algebraic K-theory.

The many exercises in this section expand the themes discussed above, and provide examples and counterexamples to illustrate the general theory of idempotents. Much of this belongs to the folklore of noncommutative ring theory. A few exercises (e.g. Ex's 16*, 23, 28, 29, 30) are taken directly from original papers in the literature.

Exercises for §21

Ex. 21.1. Let e be an idempotent in a ring R. For any right R-module V, we can view Ve as a right eRe-module.

(1) Show that if $0 \to V' \to V \to V'' \to 0$ is an exact sequence of right R-modules, then $0 \to V'e \to Ve \to V''e \to 0$ is an exact sequence of right eRe-modules.

(2) If V_R is irreducible, show that Ve is either zero or is irreducible as an eRe-module.

(3) Show that for any irreducible right eRe-module W, there exists an irreducible right R-module V, unique up to isomorphism, such that $W \cong Ve$.

Solution. (1) It suffices to show that, whenever $V' \overset{\varphi}{\to} V \overset{\psi}{\to} V''$ is exact, so is $V'e \overset{\alpha}{\to} Ve \overset{\beta}{\to} V''e$, where α, β are induced by φ, ψ. Clearly, $\mathrm{im}(\alpha) \subseteq \ker(\beta)$. Suppose $\beta(ve) = 0$, where $v \in V$. Then $\psi(ve) = 0$ too, so $ve = \varphi(v')$ for some $v' \in V'$. But then

$$ve = vee = \varphi(v')e = \varphi(v'e) = \alpha(v'e).$$

(2) Assume $Ve \neq 0$. To prove the simplicity of $(Ve)_{eRe}$, we'll show that any $ve \neq 0$ generates this module. Noting that $veR = V$, we see that

$$ve \cdot eRe = (veR)e = Ve.$$

(3) Since eR is a left eRe-module, we can form the tensor product

$$M = W \otimes_{eRe} eR,$$

and this is a right R-module, with

$$Me = W \otimes_{eRe} eRe \cong W.$$

It is convenient to think of this isomorphism as an identification, so that we can view W as Me. In general, however, M_R may not be irreducible. Let N be the sum of all R-submodules of M which are killed by e. Since $N \cap W = 0$, $N \neq M$. We claim that N contains all proper R-submodules X

of M. (*In particular, N is the unique maximal R-submodule of M.*) To see this, consider the exact sequence

$$0 \to X \to M \to M/X \to 0,$$

and the associated exact sequence

$$0 \to Xe \to Me \to (M/X)e \to 0.$$

If $Xe \neq 0$, then $Xe = Me$ (since $Me = W$ is simple), and hence $(M/X)e = 0$. This implies that $X \supseteq Me = W$ and hence $X \supseteq W \cdot R = M$, a contradiction. Thus, $Xe = 0$, so $X \subseteq N$ as claimed. Now, for the *simple* R-module $V := M/N$, we have

$$Ve = (Me + N)/N = (W + N)/N \cong W.$$

$(3)'$ To prove the *uniqueness* of V, let V'_R be another simple R-module with $V'e \cong W$. Using the notations in (3), we can construct an R-epimorphism $\varphi : M \to V'$ which "extends" the embeddings of W into M and V'. Since $Ne = 0$ but $V'e \neq 0$, the R-submodule $\varphi(N) \subseteq V'$ must be zero. Therefore, φ induces $V = M/N \to V'$, which must be an isomorphism by the irreducibility of V.

Ex. 21.2. Define a partial ordering on the set of all idempotents in R by: $e' \leq e$ iff $ee' = e'e = e'$. Call a nonzero idempotent e *minimal* if there is no idempotent strictly between 0 and e. Show that the minimal idempotents in this sense are precisely the primitive idempotents of R.

Solution. The condition $ee' = e'e = e'$ is easily seen to be equivalent to $e' \in eRe$. Therefore, the minimality of $e (\neq 0)$ amounts to the fact that eRe has no nontrivial idempotents. By *FC*-(21.8), this is equivalent to e being a primitive idempotent of R.

Ex. 21.2*. Describe the primitive idempotents in R if R is (1) the Boolean ring of all subsets of a nonempty set S, or (2) the ring $\operatorname{End}(V_k)$ where V is a nonzero vector space over a division ring k.

Solution. (1) Here, every $e \in R$ is an idempotent. If $e, e' \in R$, then e, e' are subsets of S, and $e' \leq e$ (in the sense of Exercise 2) means that $e' \subseteq e \subseteq S$. Therefore, the primitive idempotents of R are given by the singleton subsets of S.

(2) For $R = \operatorname{End}(V_k)$, every idempotent $e \in R$ gives rise to a direct sum decomposition $V = \ker(e) \oplus \operatorname{im}(e)$. Conversely, for any direct sum decomposition $V = U \oplus W$, the projection of V onto W with kernel U is an idempotent in R. If e, e' are idempotents in R, then $e' \leq e$ (in the sense of Exercise 2) is easily seen to be equivalent to

$$\operatorname{im}(e') \subseteq \operatorname{im}(e) \quad \text{and} \quad \ker(e') \supseteq \ker(e).$$

From this, it follows that $e = e^2$ is a primitive idempotent in R iff $\dim_k e(V) = 1$, i.e. iff e is a projection of V onto a line.

Ex. 21.3. Let $e \in R$ be an idempotent, and $f = 1 - e$. Show that for any $r \in R$, $e' = e + erf$ is an idempotent. Writing $f' = 1 - e'$, show that $e = e' + e'sf'$ for some $s \in R$.

Solution. First, we have

$$e'^2 = (e + erf)e(1 + rf) = e(1 + rf) = e'.$$

Noting that $ee' = e'$ and $e'e = (e + erf)e = e$, we have further

$$ff' = (1 - e)(1 - e') = 1 - e - e' + ee' = f, \quad \text{so}$$
$$e = e' - erf = e' - e'erff' = e' + e'sf'$$

for $s = -erf$.

Comment. The apparent symmetry of the situation studied above will be elucidated by the next exercise.

Ex. 21.4. For idempotents $e, e' \in R$, show that the following statements are equivalent:

(1) $eR = e'R$;
(2) $ee' = e'$ and $e'e = e$;
(3) $e' = e + er(1 - e)$ for some $r \in R$;
(4) $e' = eu$ where $u \in U(R)$;
(5) $R(1 - e) = R(1 - e')$.

If these conditions hold, show that $e' = u^{-1}eu$ for some $u \in U(R)$. Does the converse hold?

Solution. $(1) \Longrightarrow (2)$. Write $e = e'r$ and $e' = es$, where $r, s \in R$. Left multiplying by e' and e respectively, we get (2).

$(2) \Longrightarrow (3)$. Given (2), $e + ee'(1 - e) = e + e'(1 - e) = e + e' - e'e = e'$.

$(3) \Longrightarrow (4)$. From (3), $e' = eu$ where $u = 1 + er(1 - e)$. Here $u \in U(R)$ since it has inverse $1 - er(1 - e)$.

$(4) \Longrightarrow (1)$. Given (4), we have $e'R = euR = eR$, since $u \in U(R)$.

$(1) \Longrightarrow (5)$. From (1), we have (2), whence

$$1 - e = 1 - e - e' + ee' = (1 - e)(1 - e') \in R(1 - e').$$

Similarly, we have $1 - e' \in R(1 - e)$, so $R(1 - e) = R(1 - e')$.

$(5) \Longrightarrow (1)$ follows from the above, by left-right symmetry.

Now suppose the above conditions hold. For the unit u constructed explicitly in $(3) \Longrightarrow (4)$, we have

$$u^{-1}e = (1 - er(1 - e))e = e,$$

so $u^{-1}eu = eu = e'$. Alternatively, note that

$$(*) \qquad (1-e)R \cong R/eR = R/e'R \cong (1-e')R$$

as right R-modules. From this and (1), we can also show directly that e' is conjugate to e: see Exercise 16(1) below.

In general, the relation $e' = u^{-1}eu$ (for some unit u) *does not* imply the conditions (1)–(5). For instance, for $e = \begin{pmatrix} 1 & 0 \\ 0 & 0 \end{pmatrix}$ and $u = \begin{pmatrix} 0 & 1 \\ 1 & 0 \end{pmatrix} = u^{-1}$

in $R = \mathbb{M}_2(\mathbb{Q})$, we have $e' := u^{-1}eu = \begin{pmatrix} 0 & 0 \\ 0 & 1 \end{pmatrix}$. But

$$eR = \begin{pmatrix} \mathbb{Q} & \mathbb{Q} \\ 0 & 0 \end{pmatrix} \neq e'R = \begin{pmatrix} 0 & 0 \\ \mathbb{Q} & \mathbb{Q} \end{pmatrix}.$$

Comment. We should also note that the conditions (1)–(5) above *do not* imply $(1-e)R = (1-e')R$ (although they do imply the isomorphism $(*)$). For instance, for the idempotents $e = \begin{pmatrix} 1 & 0 \\ 0 & 0 \end{pmatrix}$ and $e' = \begin{pmatrix} 1 & 1 \\ 0 & 0 \end{pmatrix}$ in $R = \mathbb{M}_2(\mathbb{Q})$, we have

$$eR = e'R = \begin{pmatrix} \mathbb{Q} & \mathbb{Q} \\ 0 & 0 \end{pmatrix}, \quad R(1-e) = R(1-e') = \begin{pmatrix} 0 & \mathbb{Q} \\ 0 & \mathbb{Q} \end{pmatrix}, \text{ but}$$

$$(1-e)R = \left\{ \begin{pmatrix} 0 & 0 \\ x & y \end{pmatrix} : x, y \in \mathbb{Q} \right\}$$

$$\neq (1-e')R = \left\{ \begin{pmatrix} -x & -y \\ x & y \end{pmatrix} : x, y \in \mathbb{Q} \right\}.$$

However, we do have $e' = eu = u^{-1}eu$ for the unit $u = \begin{pmatrix} 1 & 1 \\ 0 & 1 \end{pmatrix}$, as predicted by the exercise.

In the next exercise, we shall show that the situation is quite a bit simpler with a class of idempotents called "projections" in rings with involutions.

Ex. 21.4*. Let $(R, *)$ be a ring with an involution $*$. (This means R is equipped with an additive endomorphism $*$ such that $a^{**} = a$ and $(ab)^* = b^*a^*$ for all $a, b \in R$.) An idempotent $e \in R$ with $e = e^*$ is called a *projection*. For projections e, f in $(R, *)$, show that

(1) $e \leq f$ in the sense of Exercise 2 iff $eR \subseteq fR$, iff $Re \subseteq Rf$.
(2) $e = f$ iff $eR = fR$.
(3) $eR = fR$ iff $Re = Rf$, iff $(1-e)R = (1-f)R$, iff $R(1-e) = R(1-f)$.

Solution. (1) If $e \leq f$, then by definition $ef = fe = e$. In particular, $eR = feR \subseteq fR$. Conversely, assume that $eR \subseteq fR$. Then $e = fr$ for some $r \in R$, so $fe = f^2 r = fr = e$. But then

$$ef = e^* f^* = (fe)^* = e^* = e$$

also, so we have $e \leq f$. This establishes the first "iff"; the second "iff" follows from left-right symmetry.

(2) ("If" part) Assume that $eR = fR$. From the last paragraph, we have $e = ef = fe$. By symmetry, we also have $f = ef = fe$, so $e = f$.

(3) is now clear, since each of the conditions in fact amounts to $e = f$.

Ex. 21.5. Let e be an idempotent in a semiprime ring R, and let $S = eRe$. Show that the following are equivalent:

(1) $(eR)_R$ is semisimple;
(2) $_S(eR)$ is semisimple;
(3) S is a semisimple ring;
(1)′ $_R(Re)$ is semisimple;
(2)′ $(Re)_S$ is semisimple.

Solution. Clearly, it suffices to prove the equivalence of (1), (2) and (3).

(1) $\Longrightarrow$ (3). Say $(eR)_R \cong n_1 S_1 \oplus \cdots \oplus n_r S_r$ where $S_1, \ldots, S_r$ are distinct simple right R-modules. By FC-(21.7):

$$eRe \cong \operatorname{End}(eR)_R \cong \prod_i \mathbb{M}_{n_i}(\operatorname{End}(S_i)_R),$$

which is a semisimple ring.

(3) $\Longrightarrow$ (2). All (left) modules over a semisimple ring are semisimple.

(2) $\Longrightarrow$ (1) (cf. solution to Exercise 3.6A). Every nonzero element $m \in eR$ can be written as $m_1 + \cdots + m_n$ where each Sm_i is a simple S-submodule of eR. *We claim that each $m_i R$ is a simple R-module.* Once this is proved, then m is contained in the semisimple R-module $\sum m_i R$, and we are done. To show the simplicity of $m_i R$, it suffices to check that, for any $r \in R$ such that $m_i r \neq 0$, $m_i r R$ contains m_i. Since R is semiprime, $m_i r a m_i r \neq 0$ for some $a \in R$. Note that

$$(Sm_i)(ram_i) = eRem_i raem_i \subseteq Sm_i.$$

Therefore, right multiplication by ram_i defines an S-endomorphism φ of Sm_i. Since $\varphi(m_i) = m_i ram_i \neq 0$, we have $\varphi \neq 0$, so φ is an isomorphism (by the simplicity of Sm_i). Let ψ be the inverse of φ. Then, in view of $m_i = em_i$ again:

$$m_i = \psi\varphi(m_i) = \psi(m_i ram_i) = \psi\left((em_irae)m_i\right)$$
$$= (em_irae)\psi(m_i) \in m_i rR,$$

as desired.

Comment. The proof above shows that

$$(1) \Longrightarrow (3) \Longrightarrow (2) \quad \text{and} \quad (1)' \Longrightarrow (3) \Longrightarrow (2)'$$

are true over *any* ring R. However, the *equivalence* of the five conditions depends on the semiprimeness of R. For instance, for any field k and the idempotent $e = \begin{pmatrix} 0 & 0 \\ 0 & 1 \end{pmatrix}$ in the nonsemiprime ring $R = \begin{pmatrix} k & k \\ 0 & k \end{pmatrix}$, we have

$$eR = \begin{pmatrix} 0 & 0 \\ 0 & k \end{pmatrix} \quad \text{and} \quad Re = \begin{pmatrix} 0 & k \\ 0 & k \end{pmatrix}.$$

Here, (1), (2), (3), (2)' are satisfied, but (1)' is not. (For more details, see the solution to Exercise 3.6B.) If we choose $e = \begin{pmatrix} 1 & 0 \\ 0 & 0 \end{pmatrix}$ instead, then (1)', (2)', (3), (2) are satisfied, but (1) is not.

Ex. 21.6. Show that in a von Neumann regular ring R, the intersection of any two principal left ideals $A, B \subseteq R$ is a principal left ideal.

Solution. We know that A, B and $C := A + B$ are all direct summands of $_RR$ (cf. *FC*-(4.23)). Thus, R/B is a projective R-module, and so is its direct summand C/B. Therefore, the exact sequence

$$0 \to A \cap B \to A \to C/B \to 0$$

splits, so $A \cap B$ is a direct summand of A. It follows that $A \cap B$ is a direct summand of $_RR$, so $A \cap B = Re$ for some idempotent $e \in R$.

Comment. In general, the set of all principal left ideals in a von Neumann regular ring R forms a "complemented modular lattice," with "sup" given by the sum, and "inf" given by the intersection. In fact, it follows easily from Ex. 6.19 that, *in any projective module $_RP$, the intersection of a finite number of finitely generated submodules of P is always finitely generated, and is a direct summand of P.*

Ex. 21.7. Let R be a von Neumann regular ring.

(1) Show that the center $Z(R)$ is also von Neumann regular.
(2) If R is indecomposable as a ring, then $Z(R)$ is a field.

Solution. (1) Let $c \in R$ and write $c = cxc$. Note that we can always "replace" x by $y = xcx$, since

$$cyc = cxcxc = cxc = c.$$

Thus, we are done if we can show that, whenever c is central, so is y. Assume $c \in Z(R)$ and let $r \in R$. Then

$$yr = xcxr = x^2rc = x^2rcxc = x^2c^2rx = xcrx,$$

and similarly $ry = xrcx$, so $yr = ry$ as desired.

(2) Assume R is indecomposable. Then $R \neq 0$ and the only central idempotents are $0, 1$. For $0 \neq c \in Z(R)$, consider the equation $c = cyc$ obtained in (1). Since $cy \neq 0$ is a central idempotent, we have $cy = 1$. This shows that $Z(R)$ is a field.

Ex. 21.8. Show that an idempotent e in a von Neumann regular ring R is primitive iff e is right (resp. left) irreducible, iff eRe is a division ring.

Solution. If $(eR)_R$ is irreducible, it is certainly indecomposable. Conversely, assume $(eR)_R$ is indecomposable. If $(eR)_R$ was not irreducible, it would contain a proper submodule $aR \neq 0$. But aR is a direct summand in R_R, and hence also in $(eR)_R$, a contradiction. This proves the first "iff" statement, and the second follows from *FC*-(21.16), since R is a semiprime ring.

Ex. 21.9. Let $e = e^2 \in R$. Show that if R is semilocal (resp. von Neumann regular, unit-regular, strongly regular), so is $S : = eRe$.

Solution. (1) First assume R is *semilocal*, and let $J = \operatorname{rad} R$. From *FC*-(21.10), we have $S/\operatorname{rad} S \cong \bar{e}\bar{R}\bar{e}$, where $\bar{R} = R/J$. By assumption, $\bar{R}$ is a semisimple ring, so by *FC*-(21.13), $\bar{e}\bar{R}\bar{e}$ is also semisimple. Therefore, S is a semilocal ring.

(2) Assume now R is *von Neumann regular*. Let $a \in S$ and write $a = axa$ where $x \in R$. Since $ae = a = ea$, we have

$$a = (ae)x(ea) = aya$$

where $y = exe \in S$. This verifies that S is also von Neumann regular.

(3) Assume R is *strongly regular*. Let $a \in S$ and write $a = a^2x$ where $x \in R$. Since $ae = a = ea$, we have

$$a = (a^2x)e = a^2(exe) \in a^2S,$$

so S is strongly regular.

(4) Assume, finally, that R is *unit-regular*. Identifying S with $\mathrm{End}\,(eR)_R$, we shall show the unit-regularity of S by checking that $eR = N \oplus K = N' \oplus K'$ and $N \cong N'$ imply $K \cong K'$. (We use here Exercise 4.14C.) From

$$R = (fR \oplus N) \oplus K = (fR \oplus N') \oplus K' \quad (f := 1 - e)$$

and $fR \oplus N \cong fR \oplus N'$, we conclude that $K \cong K'$, since $R \cong \mathrm{End}\,(R_R)$ is unit-regular. (Here, we use again Exercise 4.14C.)

Comment. It is natural to ask if one can prove (4) directly from the definition of a unit-regular ring, rather than from a characterization thereof. For such a proof, see the note by T.Y. Lam and W. Murray: "Unit regular elements in corner rings," Bull. Hong Kong Math. Soc. *1* (1997), 61–65. On the other hand, the proof of (4) using Exercise 4.14C is remarkably easy.

Ex. 21.10A. (McCoy's Lemma) An element $a \in R$ is said to be (von Neumann) *regular* if $a = ara$ for some $r \in R$. Show that a is regular iff there exists $x \in R$ such that $axa - a$ is regular.

Solution. The "only if" part is trivial, since we can choose x to be zero. For the "if" part, assume $axa - a$ is regular (for some x), and choose $y \in R$ such that

$$axa - a = (axa - a)y(axa - a).$$

Since the RHS lies in aRa, transposition yields $a \in aRa$.

Ex. 21.10B. Using Exercise 9 and direct matrix computations (but not using Ex. (6.19) or Ex. (6.20)), show that, for any $n \geq 1$, R is von Neumann regular iff $S_n = \mathbb{M}_n(R)$ is.

Solution. First suppose S_n is von Neumann regular. For $e =$ (the matrix unit) E_{11}, we have $eS_n e \cong R$, so by Exercise 9, R is also von Neumann regular. (Alternatively, for $a \in R \subseteq S_n$, $a \in aS_n a$ implies that $a \in aRa$.) Conversely, assume R is von Neumann regular. We consider first the case $n = 2$, and show that any $\begin{pmatrix} a & b \\ c & d \end{pmatrix}$ is regular (i.e. S_2 is a von Neumann regular ring). Writing $c = cxc$, we have

$$\begin{pmatrix} a & b \\ c & d \end{pmatrix} \begin{pmatrix} 0 & x \\ 0 & 0 \end{pmatrix} \begin{pmatrix} a & b \\ c & d \end{pmatrix} - \begin{pmatrix} a & b \\ c & d \end{pmatrix} = \begin{pmatrix} * & * \\ 0 & * \end{pmatrix}.$$

By McCoy's Lemma, it suffices to show that any $\begin{pmatrix} a & b \\ 0 & d \end{pmatrix}$ is regular. Writing $a = aya$ and $d = dzd$, we have

$$\begin{pmatrix} a & b \\ 0 & d \end{pmatrix} \begin{pmatrix} y & 0 \\ 0 & z \end{pmatrix} \begin{pmatrix} a & b \\ 0 & d \end{pmatrix} - \begin{pmatrix} a & b \\ 0 & d \end{pmatrix} = \begin{pmatrix} 0 & * \\ 0 & 0 \end{pmatrix},$$

so we are now reduced to showing that any $\begin{pmatrix} 0 & b \\ 0 & 0 \end{pmatrix}$ is regular. This follows by noting that, if $b = bwb$, then

$$\begin{pmatrix} 0 & b \\ 0 & 0 \end{pmatrix} \begin{pmatrix} 0 & 0 \\ w & 0 \end{pmatrix} \begin{pmatrix} 0 & b \\ 0 & 0 \end{pmatrix} = \begin{pmatrix} 0 & b \\ 0 & 0 \end{pmatrix}.$$

Since $S_{2^{i+1}} \cong M_2(S_{2^i})$, it follows by induction that S_{2^i} is von Neumann regular for any $i \geq 1$. To handle the general case of S_n, choose i such that $2^i \geq n$. To show that (any) $M \in S_n$ is regular in S_n, consider the block matrix $\begin{pmatrix} M & 0 \\ 0 & 0 \end{pmatrix} \in S_{2^i}$. Writing

$$\begin{pmatrix} M & 0 \\ 0 & 0 \end{pmatrix} = \begin{pmatrix} M & 0 \\ 0 & 0 \end{pmatrix} \begin{pmatrix} A & B \\ C & D \end{pmatrix} \begin{pmatrix} M & 0 \\ 0 & 0 \end{pmatrix}$$

in S_{2^i}, block multiplication of the RHS leads to $M = MAM$, so M is regular in S_n, as desired.

Comment. We have given a proof for the "only if" part of this exercise before in the solution to Ex. 6.20, using projective modules. The proof above has the virtue of being direct and very elementary. There is at least one more proof available, which is based on the following two facts: (1) R and S_n have equivalent right module categories; (2) a ring is von Neumann regular iff all of its right modules are flat. We cannot advocate such a proof here since the notions used in the above two facts are not "officially" at our disposal.

Ex. 21.10C. Let P be any finitely generated projective right module over a von Neumann regular ring A. Show that $\mathrm{End}(P_A)$ is a von Neumann regular ring.

Solution. Say $P \oplus Q = A_A^n$, and let e be the projection of A_A^n onto P with respect to this decomposition. As usual, we identify $R = \mathrm{End}(A_A^n)$ with $M_n(A)$, which is von Neumann regular by Exercise 10B. It is easy to see that $\mathrm{End}(P_A) \cong eRe$, and this is von Neumann regular by Exercise 9.

Ex. 21.11. For any idempotent $e \in R$, show that $\mathrm{End}_R(eR/eJ) \cong eRe/eJe$, where $J = \mathrm{rad}\, R$.

Solution. Viewing eR/eJ as a module over $\overline{R} := R/J$, we have a natural $\overline{R}$-epimorphism $\varphi : eR/eJ \to \overline{e}\overline{R}$. This is in fact an isomorphism. Indeed, if $0 = \varphi(er + eJ) = \overline{e}\overline{r}$, then $er \in J$ and hence $er = e \cdot er \in eJ$. Therefore,

$$\mathrm{End}_R(eR/eJ) \cong \mathrm{End}_{\overline{R}}(\overline{e}\overline{R}) \cong \overline{e}\overline{R}\overline{e} \cong eRe/eJe$$

by *FC*-(21.7) and *FC*-(21.10).

Ex. 21.12. Give an example of a nonzero ring in which 1 is not a sum of primitive idempotents. More generally, give an example of a nonzero ring which has *no* primitive idempotents.

Solution. Let S be an infinite set, and R be the Boolean ring of all subsets of S. By Exercise 2*, the primitive idempotents of R are given by the singleton subsets of S. Since the identity $1 \in R$ is given by the (infinite) set S (as a subset of itself), 1 is clearly not a sum of primitive idempotents in R. To construct the second example, note that

$$\mathfrak{F} = \{T \subseteq S : |T| < \infty\}$$

is an ideal of R, so we can form the quotient ring $\overline{R} = R/\mathfrak{F}$. We claim that $\overline{R}$ has no primitive idempotents. Indeed, consider any nonzero element of $\overline{R}$, say $A + \mathfrak{F}$, where $A \subseteq S$ is an infinite set. We can write $A = B \cup C$ where B, C are disjoint infinite sets. Then A is also the symmetric difference of B and C, so $A + \mathfrak{F}$ is the sum of the two nonzero orthogonal idempotents $B + \mathfrak{F}$ and $C + \mathfrak{F}$ in $\overline{R}$. Therefore, $A + \mathfrak{F}$ cannot be a primitive idempotent in $\overline{R}$.

Comment. Since $\overline{R}$ has no primitive idempotents, it follows from Exercise 2*(1) that the Boolean ring $\overline{R}$ is *not* isomorphic to the Boolean ring of all subsets of any set. Nevertheless, according to *Stone's Representation Theorem*, any Boolean ring B (with or without identity) is isomorphic to the Boolean ring of all compact open subsets of the "Stone space" associated to B.

Ex. 21.13. Recall that two idempotents $e, f \in R$ are *isomorphic* (written $e \cong f$, or, if necessary, $e \cong_R f$) if $eR \cong fR$ as right R-modules, or equivalently, if $Re \cong Rf$ as left R-modules (cf. FC-(21.20)). Let S be a subring of R and let e, f be idempotents in S. Does $e \cong_S f$ imply $e \cong_R f$? How about the converse? What happens in the case when $S = gRg$ where $g = g^2$?

Solution. First assume $e \cong_S f$. By FC-(21.20), there exist $a, b \in S$ such that $e = ab$ and $f = ba$. Since a, b are also elements of R, the same result FC-(21.20) yields $e \cong_R f$. (Notice that this argument does not require S, R to have the same identity.) Next, assume that $S = gRg$ where $g = g^2$. This is a ring with identity g. Suppose $e, f \in S$ and $e \cong_R f$. By FC-(21.20) again, there exist $a \in eRf$ and $b \in fRe$ such that $e = ab$ and $f = ba$. But $e = ge = eg$ and $f = gf = fg$ imply that

$$a \in geRfg \subseteq S \quad \text{and} \quad b \in gfReg \subseteq S,$$

so we have $e \cong_S f$. If S is an *arbitrary* subring of R and $e, f \in S$, then $e \cong_R f$ need not imply $e \cong_S f$. For instance, let $R = \mathbb{M}_2(\mathbb{Q})$ and S be the subring of diagonal matrices. Then, for

$$e = \begin{pmatrix} 1 & 0 \\ 0 & 0 \end{pmatrix} \quad \text{and} \quad f = \begin{pmatrix} 0 & 0 \\ 0 & 1 \end{pmatrix} \quad \text{in } S,$$

we have $e \cong_R f$, but $e \not\cong_S f$ since S is a commutative ring.

Ex. 21.14. Let e, e' be idempotents in R.

(1) If $e \cong e'$ and e is primitive, local, or right irreducible, show that so is e'.
(2) If $e \cong e'$, with $e = ab$, $e' = ba$ where $a, b \in R$, construct an explicit ring isomorphism from eRe to $e'Re'$ using a, b.
(3) Conversely, does $eRe \cong e'Re'$ imply $e \cong e'$?
(4) For any $u \in U(R)$, show that $e \cong u^{-1}eu$.

Solution. (1) is clear since the notions of e being primitive, local, or right irreducible are defined via properties of the right module $(eR)_R$.

(2) If $e \cong e'$, then $eR \cong e'R$ as R-modules. Taking their endomorphism rings and using FC-(21.7), we see that eRe and $e'Re'$ are isomorphic as rings. An *explicit* isomorphism $\alpha : eRe \to e'Re'$ can be defined by:

$$\alpha(ere) = b(ere)a = (ba)(bra)(ba)$$
$$= e'(bra)e' \in e'Re'.$$

Indeed, we have $\alpha(e) = b(ab)a = e'^2 = e'$, and

$$\alpha(ere)\,\alpha(ese) = b(ere)a \cdot b(ese)a = b(ere^3se)a$$
$$= b(ere \cdot ese)a = \alpha(ere \cdot ese),$$

so α is a ring homomorphism. Similarly, $\beta : e'Re' \to eRe$ defined by $\beta(e're') = a(e're')b$ is a ring homomorphism. A straightforward calculation shows that $\alpha\beta = I$ and $\beta\alpha = I$, so α, β are mutually inverse ring isomorphisms.

(3) In general, $eRe \cong e'Re'$ (as rings) does not imply $e \cong e'$ (as idempotents). An example for this is given by

$$e = \begin{pmatrix} 1 & 0 \\ 0 & 0 \end{pmatrix} \quad \text{and} \quad e' = \begin{pmatrix} 0 & 0 \\ 0 & 1 \end{pmatrix}$$

in the ring $R = \begin{pmatrix} k & k \\ 0 & k \end{pmatrix}$, where k is any field. For these idempotents, we have

$$eRe = \begin{pmatrix} k & 0 \\ 0 & 0 \end{pmatrix}, \quad e'Re' = \begin{pmatrix} 0 & 0 \\ 0 & k \end{pmatrix},$$

both of which are isomorphic to k. However,

$$eR = \begin{pmatrix} k & k \\ 0 & 0 \end{pmatrix} \quad \text{and} \quad e'R = \begin{pmatrix} 0 & 0 \\ 0 & k \end{pmatrix}$$

are not isomorphic as right R-modules; in fact, they are not even isomorphic as k-modules!

(4) Let $e' = u^{-1}eu$. Then $ue' = eu$ so $ue'R = euR = eR$. Therefore, left multiplication by u defines a right module isomorphism $e'R \to eR$, showing that $e \cong e'$.

Ex. 21.15. Let $1 = e_1 + \cdots + e_r = e_1' + \cdots + e_r'$ be two decompositions of 1 into sums of orthogonal idempotents. If $e_i \cong e_i'$ for all i, show that there exists $u \in U(R)$ such that $e_i' = u^{-1}e_iu$ for all i.

Solution. We have

$$R_R = e_1R \oplus \cdots \oplus e_rR = e_1'R \oplus \cdots \oplus e_r'R.$$

For each i, fix an R-isomorphism $\varphi_i : e_i'R \to e_iR$. Then $\varphi_1 \oplus \cdots \oplus \varphi_r$ is an automorphism of R_R, given by left multiplication by some unit $u \in U(R)$. Since $ue_i'u^{-1} \in e_iRu^{-1} = e_iR$ and

$$\sum ue_i'u^{-1} = u\left(\sum e_i'\right)u^{-1} = 1,$$

the direct sum decomposition $R = e_1R \oplus \cdots \oplus e_rR$ shows that $ue_i'u^{-1} = e_i$ for all i.

Ex. 21.16. Let e, e' be idempotents in R, and $f = 1 - e$, $f' = 1 - e'$ be their complementary idempotents.

(1) Show that e and e' are conjugate in R iff $e \cong e'$ and $f \cong f'$.
(2) If eRe is a semilocal ring, show that e and e' are conjugate in R iff $e \cong e'$.
(3) Is (2) still true if eRe is not assumed to be semilocal?

Solution. (1) The "if" part is a special case of Exercise 15. For the "only if" part, assume $e' = u^{-1}eu$ where $u \in U(R)$. Then $e \cong e'$ by Exercise 14, and since

$$u^{-1}fu = u^{-1}(1 - e)u = 1 - e' = f',$$

we have similarly $f \cong f'$.

(2) Now assume eRe is a semilocal ring, and that $e \cong e'$. Then $\text{End}(eR)_R \cong eRe$ by FC-(21.7), and this ring has left stable range 1 by Bass' Theorem FC-(20.9). From

$$R_R = eR \oplus fR = e'R \oplus f'R \quad \text{and} \quad eR \cong e'R,$$

the Cancellation Theorem FC-(20.11) implies that $fR \cong f'R$. Therefore, e and e' are conjugate in R by (1).

(3) If no assumption is imposed upon e, the conclusion in (2) is no longer true in general. To see this, consider the $\mathbb{Q}$-algebra R generated by x, y with a single relation $xy = 1$. Then $e' = yx \in R$ is an idempotent, and $e' \cong e = 1$ by FC-(21.20). But since $e' \neq e$, e' and e are certainly not conjugate in R! (Alternatively, $0 \neq f'R \ncong fR = 0$.)

Ex. 21.16*. (Ehrlich, Handelman) Let R be a von Neumann regular ring. Show that R is unit-regular (in the sense of Exercise 4.14B) iff, for any two idempotents $e, e' \in R$, $e \cong e'$ implies $1 - e \cong 1 - e'$.

Solution. By assumption, $R \cong \operatorname{End}(R_R)$ is von Neumann regular. By Exercise 4.14C, this endomorphism ring is unit regular iff whenever

$$R_R = K \oplus N = K' \oplus N'$$

(in the category of right R-modules), $N \cong N'$ implies $K \cong K'$. Since a decomposition $R_R = K \oplus N$ arises precisely by taking $N = eR$ and $K = (1 - e)R$ where e is an idempotent, the desired condition amounts to

$$e \cong e' \Longrightarrow 1 - e \cong 1 - e'$$

for all idempotents $e, e' \in R$.

Ex. 21.17. Let $1 = e_1 + \cdots + e_r = e'_1 + \cdots + e'_s$ be two decompositions of 1 into sums of orthogonal local idempotents. Show that $r = s$ and that there exists $u \in \operatorname{U}(R)$ such that $e'_{\pi(i)} = u^{-1} e_i u$ for all i, where π is a suitable permutation of $\{1, 2, \ldots, r\}$.

Solution. We have the decompositions

$$R_R = e_1 R \oplus \cdots \oplus e_r R = e'_1 R \oplus \cdots \oplus e'_s R$$

where the $e_i R, e'_j R$ are strongly indecomposable right R-modules (cf. FC-(21.9)). By the Krull-Schmidt-Azumaya Theorem (FC-(19.21)), we have $r = s$, and $e_i R \cong e'_{\pi(i)} R$ for some permutation π of $\{1, 2, \ldots, r\}$. The desired conclusion now follows from Exercise 15.

Ex. 21.18. Let $e_1, \ldots, e_r$ be idempotents in R which are pairwise orthogonal and isomorphic. For $e = e_1 + \cdots + e_r$, show that $eRe \cong \mathbb{M}_r(e_i R e_i)$ for any i.

Solution. Clearly, e is an idempotent. Let $P = e_i R$ for a fixed i. Then $eR \cong r \cdot P$ (direct sum of r copies of P), so by FC-(21.7):

$$eRe \cong \operatorname{End}(eR) \cong \operatorname{End}(r \cdot P) \cong \mathbb{M}_r(\operatorname{End}(P)) \cong \mathbb{M}_r(e_i R e_i).$$

Ex. 21.19. Let A be a ring which has no nontrivial idempotents. Let $\{E_{ij}\}$ be the matrix units in $R = \mathbb{M}_n(A)$. True or False: Every idempotent in R is conjugate to $E_{11} + E_{22} + \cdots + E_{ii}$ for some $i \leq n$?

Solution. Any idempotent $e \in R$ defines an A-endomorphism

$$e : A_A^n \longrightarrow A_A^n$$

with $A_A^n = \ker(e) \oplus \operatorname{im}(e)$. Therefore, e gives rise to a finitely generated projective right A-module $\operatorname{im}(e)$ (and conversely, every finitely generated projective right A-module arises in this way). If two idempotents e, e' are conjugate in R, it is easy to see that $\mathrm{im}(e) \cong \mathrm{im}(e')$ as right A-modules. In particular, if e is conjugate to $e' = E_{11} + \cdots + E_{ii}$, then

$$\mathrm{im}(e) \cong \mathrm{im}(e') \cong A_A^i.$$

Therefore, if there exist finitely generated projective right A-modules that are not free, then there are idempotents in some $\mathbb{M}_n(A)$ which are not conjugate to any $E_{11} + \cdots + E_{ii}$.

An explicit example for A is the ring of algebraic integers $\mathbb{Z}[\theta]$ where $\theta^2 = -5$. The class number of A is 2, and the class group of A is generated by the projective ideal

$$P = 2R + (1 - \theta)R.$$

It is easy to verify that

$$e = \begin{pmatrix} -2 & \theta + 1 \\ \theta - 1 & 3 \end{pmatrix}$$

is an idempotent in $\mathbb{M}_2(A)$ whose kernel and image are both isomorphic to P. Since P is not free, e is not conjugate[3] to the idempotent $\begin{pmatrix} 1 & 0 \\ 0 & 0 \end{pmatrix}$ in $\mathbb{M}_2(A)$. (This can also be verified directly by a matrix computation, using the fact that $U(A) = \{\pm 1\}$.)

Ex. 21.20. Let J be an ideal in R which contains no nonzero idempotents (e.g. $J \subseteq \mathrm{rad}\, R$). Let e, f be commuting idempotents in R.

(1) If $\bar{e} = \bar{f}$ in R/J, show that $e = f$ in R.
(2) If $\bar{e}, \bar{f}$ are orthogonal in R/J, show that e, f are orthogonal in R.

Is any of these results true if e, f do not commute?

Solution. (1) Since $ef = fe$, we have

$$(e - ef)^2 = e^2(1 - f)^2 = e(1 - f),$$

so $e - ef$ is an idempotent. On the other hand, $e - f \in J$ implies that

$$e - ef = e(e - f) \in J,$$

so $e - ef = 0$. Similarly, $f - ef = 0$, so $f = ef = e$. For (2), assume that $\bar{e}\bar{f} = 0 \in R/J$. Then $ef \in J$. Since

$$(ef)^2 = e^2 f^2 = ef,$$

we have $ef = 0$.

---

[3] Of course, e cannot be conjugate to 0 or I.

If e and f do not commute, neither (1) *nor* (2) *is true in general.* To construct counterexamples, consider the ring $R = M_2(\mathbb{Z})$ and the ideal $J = M_2(2\mathbb{Z})$. For the two (noncommuting) idempotents $e = \begin{pmatrix} 1 & 0 \\ 0 & 0 \end{pmatrix}$ and $f = \begin{pmatrix} 1 & 2 \\ 0 & 0 \end{pmatrix}$, we have

$$f - e = \begin{pmatrix} 0 & 2 \\ 0 & 0 \end{pmatrix} \in J,$$

so $\bar{e} = \bar{f} \in R/J$, but $e \neq f$. Similarly, for f as above and

$$e' = \begin{pmatrix} 0 & 0 \\ 0 & 1 \end{pmatrix} = e'^2,$$

we have $e'f = 0$, $fe' = \begin{pmatrix} 0 & 2 \\ 0 & 0 \end{pmatrix} \in J$, so e', f are orthogonal in R/J, but not in R.

We finish by showing that, for any integer $n > 1$, the ideal $J = M_2(n\mathbb{Z})$ contains no nonzero idempotents. Assume, for the moment, that $nM \in J$ is a nonzero idempotent, where $M \in M_2(\mathbb{Z})$. Then $n^2M^2 = nM$ implies $nM^2 = M$. If $\det M \neq 0$, this yields $nM = I$, a contradiction. Therefore, $\det M = 0$, and the Cayley-Hamilton Theorem gives $M^2 = tM$, where $t = \operatorname{tr}(M) \in \mathbb{Z}$. But then

$$M = nM^2 = ntM$$

implies $nt = 1$, a final contradiction.

Ex. 21.21. Let R be a semilocal ring whose radical is nil. Let I be a right ideal of R. Show that I is indecomposable (as a right R-module) and nonnil iff $I = eR$ where e is a primitive idempotent.

Solution. The "if" part follows directly from FC-(21.8). Conversely, assume I_R is indecomposable and nonnil. Then I contains a nonzero idempotent e by FC-(21.29). From

$$I = eR \oplus (I \cap (1 - e)R),$$

it follows that $I = eR$, and that e is a primitive idempotent.

Ex. 21.22. Let R be as in Exercise 21. Show that a nonzero idempotent $e \in R$ is primitive iff every right ideal properly contained in eR is nil.

Solution. For the "if" part, assume $eR = A \oplus B$, where A, B are nonzero right ideals. Then A, B are nil and so

$$e \in A + B \subseteq \operatorname{rad} R,$$

which is impossible. Therefore, e is primitive. For this part of the proof, we do not need the given hypotheses on R.

Now assume R is semilocal with rad R nil, and let $e \in R$ be a primitive idempotent. Let $A \subsetneq eR$ be any right ideal. If A is nonnil, FC-(21.29) implies that it contains a nonzero idempotent e'. But then $e'R$ is a nonzero direct summand $\subsetneq eR$, a contradiction.

Ex. 21.23. (Asano) Let R be a ring for which $J = \text{rad } R$ is nil and $\overline{R} = R/J$ is unit-regular (in the sense of Exercise 4.14B). Show that any nonunit $a \in R$ is a left (resp. right) 0-divisor in R. (In particular, "left 0-divisor" and "right 0-divisor" are both synonymous with "nonunit" in R. In the terminology of Exercise 4.16, $_RR$ and R_R are both cohopfian (and hence also hopfian).)

Solution. Since $\overline{R}$ is unit-regular, $\overline{a} = \overline{u}\,\overline{e}$, where

$$\overline{u} \in U(\overline{R}) \quad \text{and} \quad \overline{e}^2 = \overline{e} \in \overline{R}.$$

Therefore, $u \in U(R)$ (by FC-(4.8)), and since J is nil, we may assume $e \in R$ is chosen such that $e^2 = e$ (by FC-(21.28)). Note that $f = 1 - e \neq 0$ (for otherwise $\overline{a} = \overline{u} \in U(\overline{R})$ implies that $a \in U(R)$). Letting $x = a - ue \in J$ and $y = u^{-1}x \in J$, we have

$$a = u(e + y) = u[e + (e + f)y] = u[e(1 + y) + fy].$$

Let $v = (1 + y)^{-1} \in U(R)$. Then $av = u(e + fyv)$, where $z :\, = fyv \in J$. If $z = 0$, then $avf = uef = 0$ with $vf \neq 0$, so we are done. Now assume $z \neq 0$. Then $z^r \neq 0 = z^{r+1}$ for some integer $r \geq 1$. Since $ez = 0$, we have

$$avz^r = u(ez^r + z^{r+1}) = 0 \quad \text{with} \quad vz^r \neq 0,$$

so we are also done.

Comment. Since any semisimple ring is unit-regular (by Exercise 3.10(3)), the conclusions of the Exercise apply to all semilocal rings R for which rad R is nil. These include, for instance, the 1-sided perfect rings to be studied in §23. In particular, the conclusions of this exercise apply to all semiprimary rings and all 1-sided artinian rings.

Ex. 21.24. Let M_k be a module of finite length n over a ring k, and let $R = \text{End}(M_k)$, $J = \text{rad } R$. Show that R is a semilocal ring with $J^n = 0$. (In particular, R is a semiprimary ring.)

Solution. (Following Rowen's book "Ring Theory I," p. 240.) Certainly there exists a decomposition

$$M = M_1 \oplus \cdots \oplus M_r$$

where the M_i's are indecomposable k-modules. Let $e_i \in R$ be the projection of M onto M_i with respect to this decomposition. Then the e_i's are orthogonal idempotents in R with sum 1, and it is easy to see that $e_i R e_i \cong \mathrm{End}(M_i)_k$. By $FC\text{-}(19.17)$, the latter is a local ring, so each $e_i \in R$ is a local idempotent, and by $FC\text{-}(21.18)$, $\bar{e}_i \bar{R}$ is a minimal right ideal in $\bar{R} := R/J$. Since

$$\bar{R} = \bar{e}_1 \bar{R} \oplus \cdots \oplus \bar{e}_r \bar{R},$$

$\bar{R}$ is a semisimple ring, so R is semilocal.

Next we'll show that $f \in J \Rightarrow f^n = 0$. Since M has length n, the chains

$$M \supseteq \mathrm{im}(f) \supseteq \mathrm{im}(f^2) \supseteq \cdots \quad \text{and}$$

$$(0) \subseteq \ker(f) \subseteq \ker(f^2) \subseteq \cdots$$

show that $\mathrm{im}(f^n) = \mathrm{im}(f^{n+1}) = \cdots$ and $\ker(f^n) = \ker(f^{n+1}) = \cdots$. As in the proof of $FC\text{-}(19.16)$, we see that

$$M = \ker(f^n) \oplus \mathrm{im}(f^n).$$

The restriction of f to $\mathrm{im}(f^n)$ is then an automorphism, say with inverse g. Now view g as an element of R by defining $g(\ker(f^n)) = 0$. Then $1 - gf$ is zero on $\mathrm{im}(f^n)$. Since $1 - gf \in \mathrm{U}(R)$, it follows that $\mathrm{im}(f^n) = 0$, as desired.

Finally, we claim that *if $S \subseteq R$ is any set of nilpotent endomorphisms closed under multiplication, then $S^m = 0$ for some m.* (If so, then J itself is nilpotent. Since

$$M \supseteq JM \supseteq J^2 M \supseteq \cdots$$

is a chain of k-modules, the fact that $\mathrm{length}(M) = n$ implies that $J^n M = 0$, so $J^n = 0$ as desired.) To prove our claim, we distinguish the two cases: $J = 0$, and $J \neq 0$. If $J = 0$, R is semisimple and we are done by Exercise 3.24(2). Now assume $J \neq 0$. Since $f^n = 0$ for every $f \in J$, J contains a nonzero nilpotent ideal I by the solution of Exercise 10.13. Then $0 \neq IM \neq M$, so IM and M/IM are k-modules of lengths $< n$. Invoking an inductive hypothesis at this point, we may assume that $S^a(M/IM) = 0$ and $S^b(IM) = 0$ for suitable a, b. Therefore

$$S^{b+a} M \subseteq S^b(IM) = 0,$$

so $S^{a+b} = 0$. (Of course, S may not act faithfully on IM and M/IM. But we may replace S by its images in $\mathrm{End}(IM)$ and $\mathrm{End}(M/IM)$ in order to apply the inductive hypothesis.)

Comment. One might wonder what happens if the assumption that M_k has finite length is weakened to M_k being artinian. In this case, Camps and Menal have shown that $\mathrm{End}(M_k)$ may not be semiprimary. However, Camps and Dicks have shown that $\mathrm{End}(M_k)$ is still *semilocal*: see their paper referenced in the solution to Exercise 20.11. Since semilocal rings have stable range 1, this result of Camps and Dicks, together with $FC\text{-}(20.11)$, imply the following strong cancellation theorem (improving Exercise 20.2): *if M_k is artinian and B, C are arbitrary k-modules, then*

$$M \oplus B \cong M \oplus C \Longrightarrow B \cong C.$$

Ex. 21.25. Let R be any right artinian ring, and let $C = Z(R)$.

(1) Show that C is a semiprimary ring.
(2) Deduce from (1) that $C \cap \operatorname{rad} R = \operatorname{rad} C$.

Solution. (1) Let $S = R^{\mathrm{op}} \otimes_{\mathbb{Z}} R$ and view R as a right S-module via the action

$$r(a^{\mathrm{op}} \otimes b) = arb.$$

The S-endomorphisms of R commute with both left and right multiplications by elements of R, so they are given by multiplications by *central* elements of R, and consequently $\operatorname{End}(R_S) \cong C$. Now, by the Hopkins-Levitzki Theorem (FC-(4.15)), R_R satisfies *both* ACC and DCC. Since S-submodules of R are necessarily right ideals of the ring, R_S also satisfies ACC and DCC, so R_S is a module of finite length. From Exercise 24, we conclude that $C \cong \operatorname{End}(R_S)$ is a semiprimary ring.

(2) Since $\operatorname{rad} C$ is a nilpotent ideal of C, $R \cdot \operatorname{rad} C$ is a nilpotent ideal of R. Therefore, $R \cdot \operatorname{rad} C \subseteq \operatorname{rad} R$, so we have $\operatorname{rad} C \subseteq C \cap \operatorname{rad} R$. The reverse inclusion holds for any ring R by Exercise 5.0, so we have $\operatorname{rad} C = C \cap \operatorname{rad} R$.

Comment. In the solution to Exercise 20.4, we have constructed a commutative semiprimary ring C with

$$(\operatorname{rad} C)^3 = 0 \neq (\operatorname{rad} C)^2$$

such that C does not embed in any right noetherian ring. Since any right artinian ring is right noetherian, it follows in particular that C cannot be the center of any right artinian ring. In contrast, Jensen and Jøndrup have shown that, if C is a commutative semiprimary ring such that $(\operatorname{rad} C)^2 = 0$ *and $C/\operatorname{rad} C$ is countable*, then C can be realized as the center of a suitable artinian ring. (See their paper cited in the *Comment* on Exercise 20.4.) Since C need not be noetherian, it follows that *the center of an artinian (or noetherian) ring need not be noetherian (let alone artinian).*

Ex. 21.26. Let a be a nonsquare element in a field F of characteristic not 2, and let A be a commutative F-algebra with basis $\{1, x, y, xy\}$ such that $x^2 = y^2 = a$. Find the primitive idempotents in A, and show that $A \cong F(\sqrt{a}) \times F(\sqrt{a})$.

Solution. Let $e = \frac{1}{2}\left(1 + \frac{xy}{a}\right)$ and $e' = \frac{1}{2}\left(1 - \frac{xy}{a}\right)$ in A. Since

$$ee' = \frac{1}{4}\left(1 - \frac{x^2 y^2}{a^2}\right) = 0,$$

the equation $e + e' = 1$ implies that $e^2 = e$ and $e'^2 = e'$. We have therefore a direct product decomposition $A = eA \times e'A$ (cf. Exercise 1.7). To determine eA, note that

$$eA = F\text{-span}\{e, ex, ey, exy\}$$
$$= F\text{-span}\{e, \tfrac{1}{2}(x + y), \tfrac{1}{2}(x + y), ae\}$$
$$= F\text{-span}\{e, \tfrac{1}{2}(x + y)\}.$$

Since $\left[\frac{1}{2}(x + y)\right]^2 = \frac{1}{4}(2a + 2xy) = ae$, we see that $eA \cong F(\sqrt{a})$. A similar calculation shows that $e'A \cong F(\sqrt{a})$. Therefore, e, e' are primitive idempotents, and they are the *only* primitive idempotents in the algebra A.

Ex. 21.27. Let A be a "look-alike" quaternion algebra over a field F of characteristic 2, i.e. $A = F1 \oplus Fi \oplus Fj \oplus Fk$, where $i^2 = j^2 = -1$ and $k = ij = -ji$. (Of course, $-1 = 1 \in A$, since char $F = 2$.) What are the primitive idempotents in A, and what kind of ring is A?

Solution. The first is a trick question. Since

$$(a + bi + cj + dk)^2 = a^2 + b^2 + c^2 + d^2 \in F \quad (\text{for } a, b, c, d \in F),$$

the only idempotents in A are 0 and 1. Therefore, the only primitive idempotent is 1. In particular, by FC-(19.19), A is a (commutative) local ring. If we don't want to assume FC-(19.19), the structure of A can also be described explicitly as follows. Since

$$1 + i, \ 1 + j \ \text{ and } \ 1 + k$$

are linearly independent nilpotent elements in the commutative F-algebra A of dimension 4, they must span $\mathrm{Nil}(A) = \mathrm{rad}(A)$. Thus, $A/\mathrm{rad}\,A \cong F$, so A is local with $(\mathrm{rad}\,A)^3 = 0$. Note that

$$F1 \oplus Fi \cong F1 \oplus Fj \cong F[x]/(x^2 - 1) \cong F[t]/(t^2).$$

This is known as the algebra $\mathfrak{D}$ of dual numbers over F. It is easy to see that $A \cong \mathfrak{D} \otimes_F \mathfrak{D}$ as F-algebras, with

$$\mathrm{Nil}(A) \leftrightarrow (\mathrm{Nil}\,\mathfrak{D}) \otimes \mathfrak{D} + \mathfrak{D} \otimes (\mathrm{Nil}\,\mathfrak{D}).$$

Ex. 21.28. (Bass) Let G be an abelian group and H be its torsion subgroup. For any commutative ring k, show that any idempotent e of kG belongs to kH.

Solution. We may assume that G is finitely generated. Write

$$G = F \times H, \quad \text{where} \quad F \cong \mathbb{Z}^r \quad \text{and} \quad |H| < \infty.$$

Then $kG \cong (kH)[F]$. If we know the conclusion in the case $F \cong \mathbb{Z}^r$, then the given idempotent e must lie in kH and we are done. Therefore, we are reduced to the case $G = F \cong \mathbb{Z}^r$. By induction, it suffices to handle the case $r = 1$. In this case, $kG = k[t, t^{-1}]$ for an indeterminate t. Let

$$e = e^2 = \sum a_i t^i \in k[t, t^{-1}].$$

For any prime ideal $\mathfrak{p} \subset k$, $(k/\mathfrak{p})[t, t^{-1}]$ is an integral domain, in which the only idempotents are 0 and 1. Therefore, $a_i \in \mathfrak{p}$ for all $i \neq 0$. Writing $N = \text{Nil}(k)$, we have then $a_i \in N$ for all $i \neq 0$. Since now

$$\bar{a}_0 = \bar{e} \in (k/N)[t, t^{-1}],$$

we see that $\bar{a}_0$ is an idempotent in k/N. By FC-(21.28), there exists an idempotent $a \in k$ which lifts $\bar{a}_0$. In particular, $e - ea$ and $a - ea$ are nilpotent (since they lie in $N[t, t^{-1}]$). But

$$e - ea = e(1 - a) \quad \text{and} \quad a - ea = a(1 - e)$$

are also idempotent. Therefore,

$$e - ea = 0 = a - ea,$$

so $e = ea = a \in k$, as desired.

Comment. Bass' result above appeared as Lemma 6.7 in his paper "Euler characteristics and characters of discrete groups," Invent. Math. *35* (1976), 155–196. This paper contains many other interesting results on idempotents in group rings.

Ex. 21.29. (Bergman) Let A be the real coordinate ring of the 2-sphere S^2, i.e. $A = \mathbb{R}[x, y, z]$ with the relation $x^2 + y^2 + z^2 = 1$. Let σ be the $\mathbb{R}$-automorphism of A defined by

$$\sigma(x) = -x, \quad \sigma(y) = -y, \quad \text{and} \quad \sigma(z) = z.$$

Let $R = A \oplus Ar$, where $r^2 = 1$ and $rh = \sigma(h)r$ for every $h \in A$. Show that for the idempotent $e_0 = (1 - r)/2$ in the ring R, $R/Re_0R \cong \mathbb{R} \times \mathbb{R}$, but the two nontrivial idempotents of $\mathbb{R} \times \mathbb{R}$ cannot be lifted to R.

Solution. In the $\mathbb{R}$-algebra R/Re_0R we have the relation $r = 1$ in addition to the other relations on x, y, z, r in R. The two relations $rx = -xr$ and $ry = -yr$ now simplify to $x = 0$ and $y = 0$. Therefore,

(1) $$R/Re_0R \cong \mathbb{R}[z]/(z^2 - 1) \cong \mathbb{R} \times \mathbb{R},$$

with the isomorphism induced by

(2) $\pi : f + gr \mapsto (f(N) + g(N), f(S) + g(S)) \in \mathbb{R} \times \mathbb{R},$

where $N = (0,0,1)$ and $S = (0,0,-1)$ are the north and south poles respectively. If e is any idempotent in R, we would like to prove that $\pi(e)$ is either $(0,0)$ or $(1,1)$ (i.e. not $(1,0)$ or $(0,1)$).

In the following, we'll give a sketch of Bergman's argument. (More information can be found in his paper "Some examples in PI ring theory" in Israel J. Math. *18* (1974), 257–277.) Defining the map

$$P = (x,y,z) \mapsto P' = (-x,-y,z)$$

on S^2, we have clearly $\sigma(h)(P) = h(P')$. Note that A is a $\mathbb{Z}_2$-graded $\mathbb{R}$-algebra with

(3) $A_0 = \{h \in A : \sigma(h) = h\}$ and $A_1 = \{h \in A : \sigma(h) = -h\}.$

We claim that R is isomorphic to the algebra of matrices

(4) $$T = \begin{pmatrix} A_0 & A_1 \\ A_1 & A_0 \end{pmatrix} \subseteq M_2(A).$$

The isomorphism is established here by identifying

$$A = \{\alpha + \beta : \alpha \in A_0, \quad \beta \in A_1\}$$

with $\left\{\begin{pmatrix} \alpha & \beta \\ \beta & \alpha \end{pmatrix}\right\} \subseteq T$, and then identifying $r \in R$ with $\begin{pmatrix} 1 & 0 \\ 0 & -1 \end{pmatrix} \in T$. The relation

$$r(\alpha + \beta) = (\alpha - \beta)r \in R$$

is checked by the matrix equation

$$\begin{pmatrix} 1 & 0 \\ 0 & -1 \end{pmatrix} \begin{pmatrix} \alpha & \beta \\ \beta & \alpha \end{pmatrix} = \begin{pmatrix} \alpha & -\beta \\ -\beta & \alpha \end{pmatrix} \begin{pmatrix} 1 & 0 \\ 0 & -1 \end{pmatrix}.$$

Note also that the image of $A \oplus Ar$ consists of matrices

$$\begin{pmatrix} \alpha & \beta \\ \beta & \alpha \end{pmatrix} + \begin{pmatrix} \gamma & \delta \\ \delta & \gamma \end{pmatrix} \begin{pmatrix} 1 & 0 \\ 0 & -1 \end{pmatrix} = \begin{pmatrix} \alpha + \gamma & \beta - \delta \\ \beta + \delta & \alpha - \gamma \end{pmatrix},$$

which, of course, constitute the subalgebra T in (4). Finally, note that under the above isomorphism, we have

(5) $x \mapsto \begin{pmatrix} 0 & x \\ x & 0 \end{pmatrix}, \; y \mapsto \begin{pmatrix} 0 & y \\ y & 0 \end{pmatrix}, \; z \mapsto \begin{pmatrix} z & 0 \\ 0 & z \end{pmatrix}, \; e_0 \mapsto \begin{pmatrix} 0 & 0 \\ 0 & 1 \end{pmatrix}.$

In the following, we shall "identify" R with T, and thus think of elements of R as (continuous) matrix-valued functions from S^2 to $M_2(\mathbb{R})$. The map π in (2) now simply "becomes":

$$
(6) \qquad \begin{pmatrix} \alpha & \beta \\ \delta & \gamma \end{pmatrix} \mapsto (\alpha(N), \alpha(S)),
$$

since this rule produces the right values on the generators $x, y, z, r \in R$, namely

$$
x \mapsto (0,0), \quad y \mapsto (0,0), \quad z \mapsto (1,-1), \quad r \mapsto (1,1).
$$

Now consider any *nontrivial* idempotent $e \in R$. As a matrix function $S^2 \to M_2(\mathbb{R})$, e assigns continuously to every point $P \in S^2$ an idempotent matrix, that is, a projection of $\mathbb{R}^2$ to a subspace. By continuity (and the connectedness of S^2), the ranks of these subspaces must be all 1. [Note that the only idempotent matrix "very close" to I_2 (resp. $0 \cdot I_2$) is I_2 (resp. $0 \cdot I_2$) itself.] Thus, *to each $P \in S^2$, e assigns continuously the projection of $\mathbb{R}^2$ to a line $L(P)$ through the origin.*

Note that if e corresponds to $\begin{pmatrix} \alpha & \beta \\ \delta & \gamma \end{pmatrix}$, then

$$
e(P') = \begin{pmatrix} \alpha(P') & \beta(P') \\ \delta(P') & \gamma(P') \end{pmatrix} = \begin{pmatrix} \alpha(P) & -\beta(P) \\ -\delta(P) & \gamma(P) \end{pmatrix},
$$

so $L(P')$ is just $L(P)'$, the reflection of the line $L(P)$ with respect to the x-axis. In particular, since $N' = N$, $L(N)$ must be either the x-axis or the y-axis. As a matter of fact, the rank 1 idempotent matrix

$$
e(N) = \begin{pmatrix} \alpha(N) & 0 \\ 0 & \gamma(N) \end{pmatrix}
$$

can only be $\begin{pmatrix} 1 & 0 \\ 0 & 0 \end{pmatrix}$ or $\begin{pmatrix} 0 & 0 \\ 0 & 1 \end{pmatrix}$. (A similar remark holds for the south pole S.) Since $\pi(e) = (\alpha(N), \alpha(S))$ by (6), our goal is to show that $\alpha(N) = \alpha(S)$, which simply means that $L(N) = L(S)$.

For a point P moving along a path C from N to S, let w denote the total winding number through which the line $L(P)$ turns. At the same time, the point P' will travel the path C' from N to S, and the associated line $L(P') = L(P)'$ will clearly turn through a winding number of $-w$. But the paths C and C' are homotopic, since S^2 is simply connected. Hence we must have $w = -w$, so $w = 0$. This proves that $L(N) = L(S)$, as desired.

Ex. 21.30. (Stanley) Let e, f, e', f' be idempotents in a ring R such that $e - e', f - f' \in J$, where J is an ideal in R such that $\bigcap_{n=1}^{\infty} J^n = 0$. Show that $eRf = 0$ iff $e'Rf' = 0$.

Solution. It suffices to show that $eRf = 0 \Longrightarrow e'Rf = 0$. Write $e' = e + a$, where $a \in J$. Then $e'^2 = e'$ amounts to $a = ae + ea + a^2$. Assuming $eRf = 0$, *we claim that* $e'Rf \subseteq a^n Rf$ *for all* $n \geq 1$. This is true for $n = 1$, since

$$e'Rf \subseteq eRf + aRf = aRf.$$

Inductively, if $e'Rf \subseteq a^n Rf$ for some $n \geq 1$, then

$$e'Rf = e'(e'Rf) \subseteq (e + a)a^n Rf \subseteq a^{n+1} Rf,$$

which proves our claim. Since $a \in J$ and J is an ideal, it follows that $e'Rf \subseteq \bigcap_{n=1}^{\infty} J^n = 0$.

Comment. The fact about idempotents in this exercise was discovered by R. Stanley in his investigation on the incidence algebras of locally finite posets. Using this fact, Stanley proved that a locally finite poset is uniquely determined by its incidence algebra (over a fixed field). See his paper "Structure of incidence algebras and their automorphism groups," in Bull. Amer. Math. Soc. *76* (1970), 1236–1239.

§22. Central Idempotents and Block Decompositions

A central idempotent e in a ring R is said to be *centrally primitive* if $e \neq 0$, and e cannot be written as a sum of two orthogonal nonzero central idempotents. If a ring R has nontrivial central idempotents, a first impulse might be to try to decompose 1 into a (finite) sum of mutually orthogonal centrally primitive idempotents. This is not always possible (as is shown by the example of $\mathbb{Z} \times \mathbb{Z} \times \cdots$), but whenever it is possible, the decomposition is essentially unique $(FC\text{-}(22.1))$, and R decomposes into a finite direct product of indecomposable rings (called the "blocks" of R). This is always the case, for instance, when R satisfies certain reasonable chain conditions on its ideals $(FC\text{-}(22.2))$.

If the identity element $1 \in R$ can be written as a (finite) sum of mutually orthogonal primitive idempotents, say $1 = e_1 + \cdots + e_n$, then by suitably grouping the e_i's and forming their sums in groups, we obtain a decomposition of 1 into a sum of mutually orthogonal centrally primitive idempotents (and therefore obtain a block decomposition R). This result is proved in detail in $FC\text{-}(22.5)$. The proof makes interesting use of a certain equivalence relation among the primitive idempotents, known as "linkage." In our case, retrospectively, two primitive idempotents are "linked" iff they belong to the same block.

For certain ideals $I \subseteq R$, there are interesting (and sometimes surprising) results relating the central idempotents of R and those of R/I. In some cases, the natural map $e \mapsto e + I$ may define a one-one correspondence between these central idempotents: see, for instance, $FC\text{-}(22.9)$ and

FC-(22.11). Again, results of this nature are of significance in the integral and modular representation theory of finite groups, as well as in other areas of algebra.

The exercises in this section present a few more results on central idempotents. Exercise 4A introduces the notion of a right duo ring, and Exercise 4B gives various new characterizations of strongly regular (a.k.a. abelian regular) rings. Exercise 5 shows that any ring R gives rise to a Boolean ring $B(R)$, whose elements are the central idempotents in R.

Exercises for §22

Ex. 22.1. Show that the ring R of $n \times n$ upper triangular matrices over any indecomposable ring k is indecomposable.

Solution. A routine calculation shows that $Z(R)$ consists of matrices of the form aI_n where $a \in Z(k)$. The matrix aI_n is idempotent in R iff a is idempotent in k. If k is indecomposable, then k has no nontrivial central idempotents. It follows that R has no nontrivial central idempotents, so R is indecomposable as well.

Ex. 22.2. For two central idempotents e, f in a ring R, show that $e \cong f$ iff $e = f$. Using this fact and Exercise 21.16*, show that a strongly regular ring must be unit-regular (a fact proved earlier in Exercise 12.6C).

Solution. ("Only if") One characterization of $e \cong f$ is that $e = ab$, $f = ba$ for some $a, b \in R$ (cf. FC-(21.20)). Since e is central,

$$f = f^2 = (ba)(ba) = bea = ef.$$

Similarly, we have $e = fe$, and so $f = e$.

Now let R be strongly regular. Let e, f be idempotents in R. By Exercise 12.6A, e, f are central. Therefore,

$$e \cong f \Longrightarrow e = f \Longrightarrow 1 - e \cong 1 - f.$$

By Exercise 21.16*, this guarantees that R is unit-regular.

Ex. 22.3A. For $e = e^2 \in R$, show that the following are equivalent:

(1) $e \in Z(R)$,
(2) $eR = Re$,
(3) e commutes with all the idempotents of R which are isomorphic to e.

Solution. (1) clearly implies (2) and (3). Let $f = 1 - e$.

(2) $\Longrightarrow$ (1). For any $r \in R$, $erf \in Ref = 0$ implies that $er = ere$. Similarly, $fre \in feR = 0$ gives $re = ere$. Therefore, $er = re$ for all $r \in R$.

(3) $\implies$ (1). Again, consider $r \in R$. By Exercise 21.4, $e + erf$ and similarly $e + fre$ are idempotents $\cong e$. By the given hypothesis, we have

$$e(e + erf) = (e + erf)e \quad \text{and} \quad e(e + fre) = (e + fre)e.$$

Since $fe = ef = 0$, these simplify to $erf = 0$ and $fre = 0$. It follows as in (2) $\implies$ (1) that $er = ere = re$.

Ex. 22.3B. Let $S = eR$ where $e = e^2 \in R$. Suppose S is an ideal of R not containing any nonzero nilpotent ideal. Show that $e \in Z(R)$, and conclude that R is a direct product of the semiprime ring S and the ring $(1 - e)R$.

Solution. Let $f = 1 - e$ and consider any $r \in R$. Since $Re \subseteq eR$, we have $fRe = 0$ so $ere = re$ as before. However, some work is needed to show that $eRf = 0$. For $s := erf \in S$, note that

$$sRs = erfRerf \subseteq erfeRf = 0,$$

so $(RsR)^2 = 0$. Since $RsR \subseteq S$, the hypothesis yields $s = 0$. Therefore $eRf = 0$ after all, and we get

$$er = ere = re,$$

so $e \in Z(R)$. From this, it follows that $R = eR \times fR$ as rings. Since $eR = S$ contains no nonzero nilpotent ideals, it is a semiprime ring.

Comment. The hypothesis that S contains no nonzero nilpotent ideals cannot be dropped from the above exercise. In fact, if $R = \begin{pmatrix} k & k \\ 0 & k \end{pmatrix}$ where $k \neq 0$ is any ring, then for $e = \begin{pmatrix} 1 & 0 \\ 0 & 0 \end{pmatrix}$, $S = eR = \begin{pmatrix} k & k \\ 0 & 0 \end{pmatrix}$ is an ideal, but $e \notin Z(R)$. Here,

$$S \supseteq J = \begin{pmatrix} 0 & k \\ 0 & 0 \end{pmatrix}$$

with $J^2 = 0$. Note also that

$$eR \supsetneq Re = \begin{pmatrix} k & 0 \\ 0 & 0 \end{pmatrix}.$$

Ex. 22.3C. Let S be an ideal in a ring R such that S_R is an artinian R-module and S contains no nonzero nilpotent ideals of R. Show that S is a semisimple ring with an identity e, and that R is the direct product of the ring S with the ring $(1 - e)R$.

Solution. By Exercise 10.9, S_R is a semisimple R-module, and $S = eR$ for an idempotent $e \in S$. Since S is an ideal, Exercise 3B implies that $e \in Z(R)$. Therefore,

$$R = S \times (1 - e)R.$$

It follows that S_S is a semisimple S-module, so S is a semisimple ring (with identity e).

Ex. 22.4A. In Ex. 12.20, a ring R is defined to be *right duo* (resp. *left duo*) if every right (resp. left) ideal in R is an ideal. Show that a right duo ring R is always *abelian*; that is all idempotents in R are central. Is every right duo ring also left duo?

Solution. The first part is just a slight modification of $(2) \Rightarrow (1)$ in Exercise 3A. Let $e = e^2 \in R$ and $f = 1 - e$. The fact that R is right duo amounts to $Ra \subseteq aR$ for every $a \in R$. In particular, $Re \subseteq eR$ and $Rf \subseteq fR$. Therefore, $fRe = 0 = eRf$ and we are done as before.

The answer to the question in the exercise has to be "no," for otherwise we would not have bothered about the "left," "right" terminology! A right duo ring that is not left duo has, in fact, been constructed in the solution to Exercise 19.7 (cf. also Exercise 19.12). By using essentially the same method, we can construct an artinian example. Let $R = K \oplus uK$ where $K = \mathbb{Q}(x)$. We make R into a ring by using the multiplication rules $u^2 = 0$ and

$$f(x) \cdot u = uf(x^2) \quad \text{for} \quad f(x) \in K.$$

Then R is an artinian local ring with $J := \operatorname{rad} R = uK$ having square zero. Since R_K is 2-dimensional, the right ideals of R are just $0, J$ and R. Thus, R is a right duo ring. However, Ku is a left ideal but not an ideal (since $J = Ku \oplus Kux$), so R is not a left duo ring.

Comment. The R above was constructed many years ago by K. Asano as an example of an artinian ring that is right uniserial but not left uniserial. ("Uniserial"= "uniqueness of composition series.") On the other hand, there is no lack of examples of (noncommutative) right/left duo rings. For instance, any valuation ring arising from a Krull valuation of a division ring is always (right and left) duo: see Exercise 19.9(2). The next exercise characterizes duo von Neumann regular rings.

Ex. 22.4B. Show that for any ring R, the following are equivalent:

(1) R is strongly regular;
(2) R is von Neumann regular and right duo;
(3) $I \cap J = IJ$ for any left ideal I and any right ideal J;
(4) $I \cap J = IJ$ for any right ideals I, J;
(5) $aR \cap bR = aRbR$ for all $a, b \in R$.

Solution. (1) $\Longrightarrow$ (2). It is enough to show that aR is an ideal for any $a \in R$. By Exercise 12.6A, $aR = eR$ where $e = e^2 \in Z(R)$. This is clearly an ideal.

(2) $\Longrightarrow$ (4). Let I, J be right ideals. Then J is also a left ideal, so by Exercise 4.14, $I \cap J = IJ$.

(4) $\Longrightarrow$ (5) is a tautology.

(5) $\Longrightarrow$ (1). Taking $a = 1$, we get $bR = RbR$. Taking $a = b$, we get

$$bR = bRbR = b(bR) = b^2 R,$$

so R is strongly regular.

(1) $\Longrightarrow$ (3). Under (1), we know R is both right and left duo. In particular, the I, J in (3) must be both ideals. The equation $I \cap J = IJ$ then follows from Exercise 4.14.

(3) $\Longrightarrow$ (2). Taking $I = R$ (resp. $J = R$), we see that R is right (resp. left) duo. Therefore $I \cap J = IJ$ also holds for *right* ideals I and *left* ideals J. By Exercise 4.14 again, R must be von Neumann regular.

Comment. The above characterizations of strongly regular rings came from the work of many authors, including Arens-Kaplansky, Andrunakievich, Lajos, Szász, Chiba-Tominaga, and others. Curiously enough, the notion of duo rings seemed to have appeared first in Hille's book on functional analysis and semigroups.

Ex. 22.5. Let S be the set of all central idempotents in a ring R. Define an addition $\oplus$ in S by

$$e \oplus e' = e + e' - 2ee' = (e - e')^2,$$

and define multiplication in S by the multiplication in R. Show that $(S, \oplus, \times)$ is a Boolean ring, i.e. a ring in which all elements are idempotents.

Solution. Since $e \oplus e' = e(1 - e') + e'(1 - e)$ and $e(1 - e) = e'(1 - e') = 0$, we see that $e \oplus e' \in S$. Also, an easy computation shows that $(e \oplus e') \oplus e''$ and $e \oplus (e' \oplus e'')$ are both equal to

$$e + e' + e'' - 2ee' - 2ee'' - 2e'e'' + 4ee'e'',$$

so $\oplus$ is associative. Clearly, 0 is the identity element for $(S, \oplus)$, and

$$e \oplus e = e + e - 2e = 0$$

shows that $(S, \oplus)$ is an elementary 2-group. It only remains to show that "$\times$" is distributive over "$\oplus$". This follows since, for $e, e', e'' \in S$:

$$e''(e \oplus e') = e''(e + e' - 2ee') = e''e + e''e' - 2e''ee'e' = e''e \oplus e''e'.$$

Comment. The ring S above is usually denoted by $B(R)$, and is called the *Boolean ring associated to R*. Here is a quick example: If $R = \mathbb{Z} \times \cdots \times \mathbb{Z}$, then

$$B(R) \cong \mathbb{Z}_2 \times \cdots \times \mathbb{Z}_2.$$

[The (central) idempotents are of the form $e = (e_1, \ldots, e_n)$, where each $e_i \in \{0, 1\}$. To see how these add under $\oplus$, take the simple case when $n = 3$. If $e = (1, 1, 0)$ and $e' = (0, 1, 1)$, then

$$e \oplus e' = (1, 1, 0) + (0, 1, 1) - 2(0, 1, 0) = (1, 0, 1),$$

as if we were adding components modulo 2.]

In a number of sources in the literature, the addition $\oplus$ in the Boolean ring $S = B(R)$ was wrongly given as $e \oplus e' = e + e' - ee'$. The confusion stems from the fact that $e + e' - ee'$ is indeed an idempont (if $ee' = e'e$). However, $e + e' - ee'$ is the "sup" of e and e' in a *Boolean algebra* structure on $B(R)$ (to be discussed in Ex. 7 below), and is *not* the sum of e and e' in the Boolean ring studied in this Exercise.

Ex. 22.6. For any ring R, let $S = B(R)$ be the Boolean ring of central idempotents in R, as defined in Exercise 5. For any central idempotent $e \in R$, show that e is centrally primitive in R *iff* e is (centrally) primitive in S.

Solution. Recall that e being centrally primitive in R means that $e \neq 0$ and e is not the sum of two nonzero orthogonal central idempotents in R. Assume this is the case. If $e = e_1 \oplus e_2$ with $e_i \neq 0$ in S and $e_1 e_2 = 0$, then

$$e = e_1 + e_2 - 2e_1 e_2 = e_1 + e_2$$

gives a contradiction. Conversely, assume e is (centrally) primitive in S. If $e = e_1 + e_2$ with $e_i \neq 0$ in R and $e_1 e_2 = 0$, then

$$e = e_1 + e_2 = e_1 + e_2 - 2e_1 e_2 = e_1 \oplus e_2$$

gives a contradiction.

Ex. 22.7. By definition, a Boolean algebra is a lattice $(S, \leq, \wedge, \vee)$ with 0 and 1 (such that $0 \leq s \leq 1$ for all $s \in S$) that is both distributive and complemented. For any ring R, show that $S = B(R)$ (defined in Ex. 5 above) is a Boolean algebra with respect to the partial ordering

$$e' \leq e \Longleftrightarrow e'e = e' \quad (e, e' \in S),$$

and the complement operation $e \mapsto e^* := 1 - e$.

Solution. To begin with, it is useful to note that, for e, $e' \in S$, we have $e = e'$ iff $eR = e'R$. (This is a special case of (1) $\Longleftrightarrow$ (3) in Ex. 21.4, and is clear in any case since e is the identity of the ring eR.) By "identifying" $e \in S$ with eR, we can think of S as the set of "ring direct factors" of R. Under this identification, $e' \le e$ corresponds to the inclusion relation $e'R \subseteq eR$. It is therefore clear that $(S, \le)$ is a poset. Next, we claim that, for e, $e' \in S$:

$$(1) \qquad eR \cap e'R = ee'R, \quad \text{and} \quad eR + e'R = (e + e' - ee')R.$$

For these, we only need to verify "$\subseteq$" as the reverse inclusions are clear. For $a = er \in e'R$, we have $a = e'a = e'er \in ee'R$, proving the first equation. For the second, it is enough to note that

$$(e + e' - ee')e = e + e'e - ee' = e$$

implies $e \in (e + e' - ee')R$, and similarly $e' \in (e + e' - ee')R$.

A simple calculation shows that, as long as e, e' are commuting idempotents, then ee' and $e + e' - ee'$ are idempotents. In particular,

$$e, e' \in S \Longrightarrow ee' \in S \quad \text{and} \quad e + e' - ee' \in S.$$

From (1), it follows immediately that $(S, \le)$ is a lattice, with

$$(2) \qquad e \wedge e' = ee' \quad \text{and} \quad e \vee e' = e + e' - ee',$$

and S has smallest element 0 and largest element 1. (In fact, under the identification $e \leftrightarrow eR$, S is a sublattice of the ideal lattice of the ring R.) Defining e^* to be $1 - e$, we have a unary operation $S \to S$ such that for every $e \in S$:

$$e \wedge e^* = e(1 - e) = 0 \quad \text{and} \quad e \vee e^* = e + (1 - e) - e(1 - e) = 1.$$

Therefore, $e \mapsto e^*$ is a complement operation. Finally, for all e, e', $e'' \in S$, we have

$$(3) \qquad e \wedge (e' \vee e'') = (e \wedge e') \vee (e \wedge e''),$$

since both sides are equal to $ee' + ee'' - ee'e''$. Thus, $(S, \le, \wedge, \vee, *)$ is a Boolean algebra. (The other distributive law can be checked similarly, but in general one distributive law suffices.)

Comment. Assuming Ex. 5, the conclusions of this exercise are entirely to be expected if the reader is familiar with the natural equivalence between the category of Boolean rings and the category of Boolean algebras. In general, given a Boolean ring $(B, \oplus, \cdot)$, we can make B into a Boolean algebra by defining $a \le b \Longleftrightarrow a = ab$, and taking

$$(4) \qquad a \wedge b = ab, \quad a \vee b = a \oplus b \oplus ab, \quad \text{and} \quad a^* = 1 \oplus a.$$

Conversely, given a Boolean algebra $(B, \leq, \wedge, \vee, *)$, we can turn B into a Boolean ring with addition and multiplication defined by

$$(5) \qquad ab = a \wedge b, \quad \text{and} \quad a \oplus b = (a \wedge b^*) \vee (a^* \wedge b).$$

These two structures share the same elements 0 and 1.

If we take the Boolean ring $(\mathrm{B}(R), \oplus, \cdot)$ defined in Ex. 5, then its associated Boolean algebra is the one defined in Ex. 7, since

$$e \oplus e' \oplus ee' = (e + e' - 2ee') \oplus ee'$$
$$= e + e' - 2ee' + ee' - 2(e + e' - 2ee')ee'$$
$$= e + e' - ee'.$$

Similarly, if we take the Boolean algebra in Ex. 7, then its associated Boolean ring is the one defined in Ex. 5.

Ex. 22.8. Given elements $e_1, \ldots, e_n$ in the Boolean algebra $S = \mathrm{B}(R)$ in Ex. 7 above, let $e = e_1 \vee \cdots \vee e_n \in S$. Show that there exist elements $x_1, \ldots, x_n \in S$ such that $e_i = x_i e$ for each i, and $x_1 R + \cdots + x_n R = R$.

Solution. We first handle the case $n = 2$. Here, we can take

$$x_1 = 1 - e_2 + e_1 e_2, \quad x_2 = 1 - e_1 + e_1 e_2,$$

both of which can be checked to be elements of S. Since $e_i e = e_i$, we have

$$x_1 e = (1 - e_2 + e_1 e_2)e = e - e_2 + e_1 e_2 = e_1,$$

and similarly $x_2 e = e_2$. Also

$$x_1 e_1 + x_2(1 - e_1) = e_1 + (1 - e_1)^2 = 1$$

implies that $x_1 R + x_2 R = R$.

For the general case, we induct on n. Suppose the elements $x_1, \ldots, x_n$ have been obtained as required for the set $\{e_1, \ldots, e_n\} \subseteq S$ with a supremum $e = e_1 \vee \cdots \vee e_n$. For another given element $e_{n+1} \in S$, we set

$$e' := e \vee e_{n+1} = e_1 \vee \cdots \vee e_n \vee e_{n+1},$$

and use the $n = 2$ case to find $x, y \in S$ such that

$$e = xe', \quad e_{n+1} = ye', \quad \text{and} \quad xR + yR = R.$$

Then, for $i \leq n$, we have $e_i = x_i e = x_i x e'$, with $x_i x \in S$ and

$$x_1 x R + \cdots + x_n x R + yR = x(x_1 R + \cdots + x_n R) + yR$$
$$= xR + yR = R.$$

Thus, $x_1 x, \ldots, x_n x$, and y in S are the elements we sought.

Ex. 22.9. Determine all integers $r \in \mathbb{Z}$ such that, for commuting idempotents e, e' in any ring R, $e + e' - ree'$ is an idempotent in R.

Solution. From the work on Ex. 5 and Ex. 7, we see that $r = 1$ and $r = 2$ are valid choices. (We don't need e, e' to be central; all we need is $ee' = e'e$.) Conversely, if $r \in \mathbb{Z}$ is a valid choice, then for $e = e' = 1$ in the ring $R = \mathbb{Z}$, we have $e + e' - ree' = 2 - r$, which is an idempotent in $\mathbb{Z}$ (if and) only if $r = 1$ or 2.

Comment. For a solution that is independent of Ex. 5 and Ex. 7, we can proceed as follows. Let $a = e + e' - ree'$, where e, e' are commuting idempotents in any ring R. Then

$$a^2 = (e + e' - ree')(e + e' - ree')$$
$$= (e + ee' - ree') + (ee' + e' - ree') - (ree' + ree' - r^2 ee')$$
$$= e + e' + (r^2 - 4r + 2) ee',$$

and so $a^2 - a = (r^2 - 3r + 2) ee' = (r - 1)(r - 2) ee'$. From this, we see that the "good" choices for r are precisely $r = 1$ and $r = 2$.

Ex. 22.10. If e, e' are noncommuting idempotents in a ring R, show that $e + e' - ree'$ may not be an idempotent of R for any $r \in R$.

Solution. Consider the noncommuting idempotents

$$e = \begin{pmatrix} 1 & 0 \\ 0 & 0 \end{pmatrix}, \quad e' = \begin{pmatrix} 0 & 0 \\ 1 & 1 \end{pmatrix} \quad \text{in} \quad R = \mathbb{M}_2(\mathbb{Z}).$$

Since $ee' = 0$, $e + e' - ree' = \begin{pmatrix} 1 & 0 \\ 1 & 1 \end{pmatrix}$ for any $r \in R$, and this is *not* an idempotent of R.

Chapter 8
Perfect and Semiperfect Rings

§23. Perfect and Semiperfect Rings

Even in the early days of ring theory, it was realized that semiprimary rings were worthwhile generalizations of artinian rings. In the 1950's, with the advent of homological algebra, a number of papers were written about the homological properties of semiprimary rings. In his seminal 1960 paper, H. Bass studied the classes of perfect and semiperfect rings as homological generalizations of semiprimary rings, and obtained striking characterizations of these rings, both in homological and non-homological terms.

By definition, a *semiperfect ring* is a semilocal ring R for which idempotents of $R/\mathrm{rad}\,R$ can be lifted to R. These rings are characterized by the property that the identity element $1 \in R$ can be decomposed into a sum of mutually orthogonal local idempotents (see *FC-(23.6)*).

A *left* (resp. *right*) *perfect ring* is a semilocal ring R whose Jacobson radical, rad R, is left (resp right) T-nilpotent. Here, a subset $A \subseteq R$ is called *left* (resp. *right*) *T-nilpotent* if, for any sequence of elements $\{a_1, a_2, \ldots\} \subseteq A$, there exists an integer $n \geq 1$ such that $a_1 a_2 \cdots a_n = 0$ (resp. $a_n \cdots a_2 a_1 = 0$). The use of such T-nilpotent properties can be traced back to the work of J. Levitzki. Since a (left or right) T-nilpotent ideal is nil, and idempotents modulo a nil ideal can always be lifted, a (left or right) perfect ring is always semiperfect. On the other hand, a nilpotent ideal is always (left and right) T-nilpotent, so a semiprimary ring is always

(2-sided) perfect. In more detail, we have, as in *FC*-p. 335:

$$\{\text{one-sided artinian rings}\}$$
$$\cap$$
$$\{\text{semiprimary rings}\}$$
$$\cap$$
$$\{\text{right perfect rings}\}$$
$$\cap$$
$$\{\text{local rings}\} \subset \{\text{semiperfect rings}\} \subset \{\text{semilocal rings}\}.$$

In the case of *commutative* rings, R is semiperfect iff it is a finite direct product of commutative local rings R_i (*FC*-(23.11)), and R is perfect iff, in this decomposition, the maximal ideal of each R_i is T-nilpotent (*FC*-(23.24)). In general, however, a right perfect ring need not be left perfect.

It may be said that the more interesting right perfect rings are the ones *without ACC* on right or left ideals. In fact, if a right perfect ring R is right (or left) noetherian, then, according to Exercise 20.6, R is already right (or left) artinian, and we are back to a more classical setting. For right perfect rings satisfying somewhat *weaker* ascending chain conditions, see Exercise 24.8.

While 1-sided artinian rings are always right perfect (by the above chart), the right perfect rings can in turn be characterized by certain kinds of descending chain conditions. A surprising theorem of Bass (*FC*-(23.20)) states that *a ring R is right perfect iff R satisfies DCC on principal left ideals, iff any left R-module satisfies DCC on its cyclic submodules.* (Note the switch from "right" to "left," which is not a misprint.) A further refinement of this characterization, due to J.E. Björk, is developed in Exercises 3 and 4.

A precursor to the study of *DCC* on 1-sided principal ideals was I. Kaplansky's work in the early 50's on *DCC* for chains of the special form

$$aR \supseteq a^2 R \supseteq a^3 R \supseteq \cdots \quad \text{and} \quad Ra \supseteq Ra^2 \supseteq Ra^3 \supseteq \cdots.$$

This work was continued later by G. Azumaya. However, it took almost 25 years before it was proved, by F. Dischinger, that the two resulting chain conditions (each imposed for all $a \in R$) are, in fact, equivalent! They characterize rings that are known as *strongly π-regular rings:* the relevant details are developed in Exercises 5 and 6 below. We have already had a brief encounter with these rings in Exercises 4.15 and 4.17; in the former, we saw that, among *commutative* rings, the strongly π-regular rings are just those with Krull dimension 0.

Exercises for §23

Ex. 23.1. Show that any left T-nilpotent set J is locally nilpotent.

Solution. To say J is locally nilpotent means that, for any finite subset $A = \{a_1, \ldots, a_n\} \subseteq J$, there exists a natural number N such that the product of any N elements from A is zero. Assume, on the contrary, that for A, no such integer N exists. Then, for any $i \geq 1$, there exists a product $a_{i1}a_{i2}\cdots a_{ii} \neq 0$ where $a_{ij} \in A$. Since $|A| < \infty$, there exists r_1 and an infinite set I_1 with $\min(I_1) \geq 1$ and such that $a_{i1} = a_{r_1}$ for all $i \in I_1$. Similarly, there exists r_2 and an infinite set $I_2 \subseteq I_1$ with $\min(I_2) \geq 2$ such that $a_{i2} = a_{r_2}$ for all $i \in I_2$. Inductively, we can define r_m and an infinite set $I_m \subseteq I_{m-1}$ with $\min(I_m) \geq m$ and $a_{im} = a_{r_m}$ for all $i \in I_m$. Consider now the sequence

$$a_{r_1}, a_{r_2}, \ldots \in A \subseteq J.$$

For any $m \geq 1$, fix an $i \in I_m$. Then $i \geq m$, and

$$a_{r_1} \cdots a_{r_m} = a_{i1} \cdots a_{im} \neq 0$$

since $a_{i1} \cdots a_{im} \cdots a_{ii} \neq 0$. This contradicts the fact that J is left T-nilpotent.

Comment. (1) The argument here amounts essentially to an application of the "König Tree Lemma" in graph theory. However, our *ad hoc* treatment above renders the proof self-contained.

(2) In the special case when J is a 1-sided ideal, a somewhat easier proof is possible: see the arguments in the proof of FC-(23.15).

(3) The converse for the Exercise is false in general, even if J is an ideal. In fact, let R be the commutative $\mathbb{F}_2$-algebra generated by $x_1, x_2, \ldots$ with the relations

$$x_1^2 = x_2^2 = \cdots = 0.$$

The ideal $J = \sum Rx_i$ is clearly locally nilpotent. However, since $x_1 \cdots x_n \neq 0$ for any $n \geq 1$, J is not T-nilpotent. The commutative ring R here is semiperfect (in fact local), but not perfect. (It was used once before in the *Comment* to Exercise 10.13.)

Ex. 23.2. Let $K \supseteq k$ be a field extension, and let A be a k-algebra such that rad A is right T-nilpotent. Show that $A \cap \text{rad}(A^K) = \text{rad } A$.

Solution. Without any conditions on A, we have always

$$A \cap \text{rad}(A^K) \subseteq \text{rad } A \quad (\text{see } FC\text{-}(5.14)).$$

Therefore, it suffices to show that rad $A \subseteq \text{rad}(A^K)$. Consider any element in $(\text{rad } A) \otimes_k K$, say

$$a_1 \otimes \alpha_1 + \cdots + a_n \otimes \alpha_n,$$

where $a_i \in \operatorname{rad} A$, $\alpha_i \in K$. By Exercise 1, there exists a natural number N such that the product of any N elements from $\{a_1, \ldots, a_n\}$ is zero. It follows that

$$(a_1 \otimes \alpha_1 + \cdots + a_n \otimes \alpha_n)^N = 0.$$

This means that $(\operatorname{rad} A) \otimes_k K$ is a nil ideal, so $(\operatorname{rad} A) \otimes_k K \subseteq \operatorname{rad}(A^K)$, as desired.

Ex. 23.3. (Björk) For any left module M over a ring R, show that if the cyclic submodules of M satisfy DCC, then the f.g. (=finitely generated) submodules of M also satisfy DCC.

Solution. Among submodules of M whose f.g. submodules satisfy DCC, there exists a maximal one, say N, by Zorn's Lemma. *Assume, for the moment, that $N \neq M$.* Then M/N contains a simple submodule. (For otherwise, there would exist an infinite chain

$$Rx_1 + N \supsetneq Rx_2 + N \supsetneq \ldots.$$

After re-choosing the x_i's, we may arrange that $x_{i+1} \in Rx_i$ for every i. But then $Rx_1 \supsetneq Rx_2 \supsetneq \cdots$ contradicts the hypothesis on M.) Fix an element $y \in M \backslash N$ such that, for $P := Ry + N$, P/N is a simple module. We shall derive a contradiction by showing that the f.g. submodules of P satisfy DCC.

Consider any chain $P_1 \supseteq P_2 \supseteq \cdots$ in P where each P_n is f.g. If some $P_n \subseteq N$, then of course the chain becomes stationary. Thus, we may assume that each $P_n \not\subseteq N$. Now choose $y_n \in P_n$ such that $y \equiv y_n \pmod{N}$ and that Ry_n is minimal. In the following, we shall construct f.g. submodules $Q_1 \supseteq Q_2 \supseteq \cdots$ in N such that

$(*)_n$ $\qquad\qquad P_n = Ry_n + Q_n = Ry_{n+1} + Q_n$ (for any n).

Once the Q_n's are constructed, we will have $Q_n = Q_{n+1} = \cdots$ for some n. From $(*)_n, (*)_{n+1}, \ldots$, we get therefore $P_n = P_{n+1} = \cdots$, as desired.

We finish now by constructing the Q_n's. To begin with, note that

$$P_1 = Ry_1 + P_1 \cap N.$$

By resolving a finite system of generators of P_1 with respect to this sum, we can write $P_1 = Ry_1 + Q_1$ for a suitable f.g. submodule $Q_1 \subseteq N$. Since $y_2 \equiv y_1 \pmod{N}$, we can write $y_2 = r_1 y_1 + q_1$ where $r_1 \in R$ and $q_1 \in Q_1$. Then

$$y \equiv y_2 \equiv r_1 y_1 \pmod{N},$$

and the minimal choice of Ry_1 implies that $Ry_1 = Rr_1 y_1$. Therefore, $P_1 = Ry_2 + Q_1$, as in $(*)_1$. From this, we have

$$P_2 = Ry_2 + P_2 \cap Q_1.$$

As before, we can write $P_2 = Ry_2 + Q_2$ for a suitable f.g. module $Q_2 \subseteq Q_1$, and we can repeat the above process to show that, for this Q_2, $(*)_2$ holds, etc.

The above proof shows that $M = N$, so f.g. submodules of M satisfy DCC!

Comment. This exercise is Theorem 2 in Björk's paper "Rings satisfying a minimum condition on principal ideals," J. reine angew. Math. *236* (1969), 112–119.

Ex. 23.4. Using Exercise 3, show that a ring R is right perfect iff the finitely generated submodules of any left R-module satisfy DCC.

Solution. First assume R is right perfect. By Bass' Theorem (FC-(23.20)), any left R-module M satisfies DCC on cyclic submodules. By Exercise 3, M also satisfies DCC on finitely generated submodules. Conversely, suppose any left R-module M satisfies DCC on finitely generated submodules. In particular, applying this to $M = {}_RR$, we see that principal left ideals of R satisfy DCC. By Bass' Theorem (FC-(23.20)) again, R must be a right perfect ring.

Ex. 23.5. (Dischinger) Let R be a ring in which any descending chain $aR \supseteq a^2R \supseteq \cdots$ ($\forall a \in R$) stabilizes. Show that any descending chain $Ra \supseteq Ra^2 \supseteq \cdots$ ($\forall a \in R$) also stabilizes. (Such a ring R is known as a *strongly π-regular* ring.)

Solution. The proof will be based on the following lemma.

Lemma. *Suppose $a = a^2b$ and $b = b^2c$ in the ring R given above. Then $a \in Ra^2$.*

Proof. Write $d = c - a \in R$, and note that

(1) $ac = a(ab)c = a(a^2b)bc = a^3b = a^2,$

(2) $abc = (a^2b)bc = a^2b = a,$

(3) $ad = a(c - a) = 0$ (by (1)),

(4) $d^2 = (c - a)d = cd$ (by (3)),

(5) $abd^2 = abcd = ad = 0$ (by (4), (2), (3)),

(6) $b^2d^2 = b^2cd = bd$ (by (4)).

Since $dR \supseteq d^2R \supseteq \cdots$ also stabilizes, we have $d^k = d^{k+1}r$ for some $r \in R$. Then, by repeated use of (6):

(7) $bd^2r = (b^2d^2)dr = b^2d^3r = \cdots = b^kd^{k+1}r = b^kd^k = bd.$

From this and (5), we have

(8) $0 = abd^2r = abd,$ and so

(9) $0 = abdr = ab^2d^2r = ab(bd) = ab^2d$ (by (8), (6), (7)).

Recalling that $d = c - a$, (8) now gives $aba = abc = a$ (by (2)), and (9) gives $ab^2a = ab^2c$. Therefore

$$a = aba = a(b^2c)a = (ab^2a)a \in Ra^2,$$

as desired.

To prove the assertion in the exercise, consider any descending chain $Ra_o \supseteq Ra_o^2 \supseteq \cdots$. Since $a_oR \supseteq a_o^2R \supseteq \cdots$ stabilizes, we have $a_o^n = a_o^{n+1}b_o$ for some $b_o \in R$, and similarly $b_o^m = b_o^{2m}c$ for some $c \in R$. Without loss of generality, we may assume that $m \geq n$. Since

$$a_o^n = a_o(a_o^n)b_o = a_o^{n+2}b_o^2 = \cdots = a_o^{n+m}b_o^m,$$

multiplication by a_o^{m-n} yields $a_o^m = a_o^{2m}b_o^m$. Letting $a = a_o^m$ and $b = b_o^m$, we have then $a = a^2b$ and $b = b^2c$. Applying the Lemma, we get $a \in Ra^2$. Hence, $a_o^m \in Ra_o^{2m}$, so the chain $Ra_o \supseteq Ra_o^2 \supseteq \cdots$ stabilizes.

Comment. The above highly intricate proof appeared in Dischinger's paper "Sur les anneaux fortement π-réguliers," C.R. Acad. Sc. Paris *283* (1976), Sér. A, 571–573. Another (allegedly somewhat more economical) proof can be found in Hirano's paper "Some studies on strongly π-regular rings", Math. J. Okayama Univ. *20* (1978), 141–149.

The somewhat clumsy name "strongly π-regular" is partly justified by the fact that *if R is strongly regular (in the sense of Exercise 12.5), then it is strongly π-regular* (since $aR = a^2R = a^3R = \ldots$ for all $a \in R$). More justification for the terminology is given in the *Comment* on the next exercise. A couple of elementary properties of strongly π-regular rings have been given earlier in Exercise 4.17.

A large class of strongly π-regular rings arise as endomorphism rings of modules. Let $R = \operatorname{End}(M_A)$, where M is a right module over some ring A. Armendariz, Fisher and Snider (Comm. Algebra *6* (1978), 659–672) have shown that *R is strongly π-regular iff M has the "Fitting Decomposition Property,"* i.e. for any $f \in R$, $M = \ker(f^n) \oplus \operatorname{im}(f^n)$ for some $n \geq 1$.[1] By the classical Fitting's Lemma (*FC*-(19.16)), any module M_A of finite length has this property. It follows therefore that *any such M has a strongly π-regular endomorphism ring.*

Applying the Armendariz-Fisher-Snider result to $R \cong \operatorname{End}(R_R)$ for any ring R, we see in particular that *R is strongly π-regular iff, for any $a \in R$, there exists $n \geq 1$ such that $R = \operatorname{ann}_r(a^n) \oplus a^nR$.*

Ex. 23.6. Recall that an element $b \in R$ is called (von Neumann) *regular* if $b \in bRb$.

(1) (Azumaya) Let R be any strongly π-regular ring, and let $a \in R$. Show that there exists an element $r \in R$ commuting with a such that $a^n = a^{n+1}r$ for some $n \geq 1$. From this, deduce that a^n is regular.

---

[1] This theorem is to be compared with similar results in Exercises 4.14A$_1$ and 4.14C on von Neumann regular and unit-regular endomorphism rings.

(2) (Kaplansky) Let R be an algebra over a field k, and $a \in R$ be algebraic over k. Show that $aR \supseteq a^2 R \supseteq \cdots$ stabilizes, and that a^n is regular for some $n \geq 1$. From this, deduce that any algebraic k-algebra is strongly π-regular.

Solution. As a motivation for (1), we begin by proving (2).

(2) Let $a^m + a^{m-1}c_{m-1} + \cdots + c_o = 0$ where $c_i \in k$, and m is chosen smallest. If $c_o \neq 0$, then $a \in U(R)$ (cf. Exercise 1.13), in which case the desired conclusions are obvious. We may now assume that, for some $n \geq 1$, $c_n \neq 0$, but $c_i = 0$ for all $i < n$. Then from

$$-a^n c_n = a^{n+1}c_{n+1} + \cdots + a^m,$$

we have $a^n = a^{n+1}r$ where $r \in R$ is a polynomial in a (and in particular commutes with a). This shows that $a^n R = a^{n+1}R = \cdots$. Since $ar = ra$, we have

$$(*) \qquad a^n = a^n ar = a^{n+1}rar = a^{n+2}r^2 = \cdots = a^{2n}r^n = a^n r^n a^n,$$

so $a^n \in R$ is regular. The last assertion in (2) is now obvious.

(1) The last assertion in (1) can be proved exactly as in $(*)$, once we find the element $r \in R$ such that $a^n = a^{n+1}r$ and $ar = ra$. Since (by Ex. 23.5)

$$aR \supseteq a^2 R \supseteq \cdots \quad \text{and} \quad Ra \supseteq Ra^2 \supseteq \cdots$$

both stabilize, there exist a large integer n such $a^n = a^{2n}x = ya^{2n}$ for suitable x, $y \in R$. Now let $b = a^n$. Then

(1) $\qquad b = b^2 x = yb^2$

(2) $\qquad bx = yb^2 x = yb$ $\qquad\qquad$ (by (1)),

(3) $\qquad bxb = yb^2 = b = b^2 x = byb$ $\qquad$ (by (2), (1)).

Setting $s = bx^2 \in R$, we have:

(4) $\qquad bs = bybx = bx = yb = ybxb = sb$ $\qquad$ (by (2), (3)),

(5) $\qquad b^2 s = bsb = bxb = b$ $\qquad\qquad$ (by (4), (3)),

(6) $\qquad bs^2 = ybs = ybx = bx^2 = s$ $\qquad\qquad$ (by (4)).

We have now from (5):

$$a^n = a^{2n}s = a^{n+1}r \quad \text{for} \quad r := a^{n-1}s.$$

Finally, *for any $c \in R$, we claim that*

(7) $\qquad\qquad\qquad ac = ca \Longrightarrow sc = cs.$

If so, then applying this to $c = a$, we see that a commutes with s, and hence with r, as desired. To prove (7), consider *any* $c \in R$ commuting with a (and hence with b). We have

(8) $sbc = scb = scb^2s = sb^2cs = bcs$ (by (5), (4)),

(9) $sc = bs^2c = s^2bc = sbcs$ (by (6), (4), (8))

(10) $cs = cs^2b = cbs^2 = sbcs$ (similarly).

Comparison of (9) and (10) shows $sc = cs$, as desired.

Comment. The proof for (1) is taken from Azumaya's paper "Strongly π-regular rings," J. Fac. Sci. Hokkaido Univ. *13* (1954), 34–39, while the argument for (2) is from Kaplansky's paper "Topological representations of algebras, II," Trans. Amer. Math. Soc. *68* (1950), 62–75.

McCoy defined a ring R to be *π-regular* if, for any $a \in R$, some a^n ($n \geq 1$) is regular. Using this definition, we have then

$$\begin{array}{ccc}
\text{strongly regular} & \Longrightarrow & \text{strongly } \pi\text{-regular} \\
\Downarrow & & \Downarrow \\
\text{regular} & \Longrightarrow & \pi\text{-regular}
\end{array}$$

In this chart, all implications are irreversible. However, for rings with a bounded index for their nilpotent elements, Azumaya (*loc. cit.*) has shown that the vertical implication on the right is an equivalence. In particular, this implies that *if a von Neumann regular ring R has a bounded index for its nilpotent elements, then R is strongly π-regular*. This result was rediscovered years later (via Pierce sheaves) by Burgess and Stephenson: see their paper in Comm. Algebra *4* (1976), 51–75, or Theorem 7.15 in Goodearl's book on von Neumann regular rings. For commutative rings R, *both* vertical implications above are equivalences. In this case, the conditions on the RHS just amount to K-dim $R = 0$: see Exercise 4.15.

In general, any von Neumann regular ring is *J*-semisimple. The analogous result for the π-case is that, *for any π-regular ring R, rad R is nil*. In fact, if $a \in$ rad R, then, choosing $n \geq 1$ and $x \in R$ such that $a^n = a^n x a^n$, we have $(1 - a^n x)a^n = 0$, which implies $a^n = 0$.

The relationship between right perfect rings and strongly π-regular rings is provided by Bass' Theorem *FC*-(23.20) characterizing the former. In view of this theorem, *any right perfect ring is strongly π-regular*. It follows (from Exercise 5) that, in such a ring R, any chain

$$aR \supseteq a^2R \supseteq a^3R \supseteq \cdots$$

stabilizes, although a chain

$$a_1R \supseteq a_2R \supseteq a_3R \supseteq \cdots$$

need not stabilize (since R may not be *left* perfect).

As we have pointed out in the *Comment* on Ex. 20.10E, P. Ara has proved in 1996 that any strongly π-regular ring has stable range 1.

Ex. 23.7. Show that a ring R is strongly regular iff it is reduced and strongly π-regular.

Solution. If R is strongly regular, we know that R is reduced by Ex. 12.6(A), and that R is strongly π-regular by the chart in the *Comment* on Ex. 23.6. Conversely, assume that R is reduced and strongly π-regular. For any $a \in R$, Ex. 23.6(1) yields an equation $a^n = a^{n+1}r$ for some $n \geq 1$ and some $r \in R$ commuting with a. We have then

$$(a - a^2 r)^n = [a(1 - ar)]^n = a^n(1 - ar)^n$$
$$= a^n(1 - ar)(1 - ar)^{n-1}$$
$$= (a^n - a^{n+1}r)(1 - ar)^{n-1} = 0.$$

Since R is reduced, this implies that $a = a^2 r$, so R is strongly regular.

Ex. 23.8. (Vasconcelos) Show that a commutative ring R has (Krull) dimension 0 iff all finitely generated left R-modules are cohopfian.

Solution. If $\dim R \neq 0$, there exist prime ideals $P \subsetneq Q$ in R. Fix an element $a \in Q \backslash P$. Then $f : R/P \to R/P$ given by multiplication by a is an injective R-homomorphism with image contained in $Q/P \subsetneq R/P$, so f is not surjective. This proves the "if" part.

For the "only if" part, assume that $\dim R = 0$, and let f be an injective endomorphism of a finitely generated left R-module M. We apply here the same technique used in the hopfian case (Ex. 20.9). View M as a left $R[t]$-module by letting t act via f, and let $J \subseteq R[t]$ be the annihilator of M. Pick $m_1, \ldots, m_n \in M$ such that $M = \sum_{i=1}^{n} R m_i$. We can write

$$t m_i = f(m_i) = \sum_j r_{ij} m_j, \quad \text{where} \quad r_{ij} \in R.$$

By the "determinant trick" again, this gives $\det(t I_n - (r_{ij})) \cdot M = 0$. Thus J contains a *monic* polynomial in t, and hence $R' := R[t]/J$ is an integral extension of the 0-dimensional ring $R/R \cap J$. By the Cohen-Seidenberg Theorem, it follows that $\dim R' = 0$. Now $\bar{t}$ is not a 0-divisor in R', since for any polynomial $\alpha(t) \in R[t]$:

$$\bar{t} \cdot \overline{\alpha(t)} = 0 \implies f(\alpha(f)M) = 0 \implies \alpha(f)M = 0 \implies \alpha(t) \in J.$$

Since $\dim R' = 0$, Ex. 4.15 and Ex. 4.17 imply that $\bar{t}$ is a unit in R'. From this, it is clear that f is an automorphism.

Comment. The result in this exercise comes from Vasconcelos's paper in Proc. Amer. Math. Soc. *25* (1970), 900–901. This result was extended to P.I. rings (rings satisfying a polynomial identity) by Armendariz, Fisher, and Snider, who proved that, *for any P.I. ring R, all finitely generated left*

R-modules are cohopfian iff all prime ideals of R are maximal; see their paper in Comm. Algebra *6* (1978), 659–672.

Ex. 23.9. (Dischinger) For any ring R, show that all finitely generated left R-modules are cohopfian iff all matrix rings $\mathbb{M}_n(R)$ are strongly π-regular.

Solution. First assume that all finitely generated left R-modules are cohopfian. We view $S = \mathbb{M}_n(R)$ as $\mathrm{End}_R(R^n)$, acting on the right of R^n. Given any $f \in S$, let $T = \bigcup_{i=1}^{\infty} \ker(f^i)$. Then $Tf \subseteq T$, so f induces an $\overline{f} \in \mathrm{End}_R(R^n/T)$. This endomorphism is injective, since

$$(x + T)\overline{f} = 0 \Longrightarrow xf \in T \Longrightarrow x \in T.$$

Thus, $\overline{f}$ is also surjective. Write $e_i + T = (x_i + T)\overline{f}$, where $e_1, \ldots, e_n$ are the unit vectors in R^n. Then for a sufficiently large integer k, we have $e_i - x_i f \in \ker(f^k)$ for all i. Let $g \in S$ be defined by $e_i g = x_i$ $(1 \leq i \leq n)$. Then

$$e_i(f^k - gf^{k+1}) = e_i(1 - gf)f^k = (e_i - x_i f)f^k = 0$$

for all i, so we have $f^k = gf^{k+1}$. This means that the chain $Sf \supseteq Sf^2 \supseteq \cdots$ stabilizes, so S is strongly π-regular.

Conversely, assume that all matrix rings $\mathbb{M}_n(R)$ are strongly π-regular. Given an injective $f \in \mathrm{End}(_R M)$ where $_R M$ is finitely generated, we must prove f is surjective. We shall first treat the case where $_R M$ is a *cyclic* module, say $M = R/A$, where $A \subseteq R$ is a left ideal. Say $(1 + A)f = r + A$. Since R is strongly π-regular, there exists an equation $r^k = sr^{k+1}$ for some $k \geq 1$ and $s \in R$. Thus,

$$(1 + A)f^k = r^k + A = sr^{k+1} + A = (s + A)f^{k+1},$$

and so $Mf^k \subseteq Mf^{k+1}$. Since f is injective, this implies that $M \subseteq Mf$, so f is surjective.

For the general case, let $M = \sum_{i=1}^{n} Rm_i$. Writing the elements of $M^{(n)}$ as column vectors $(x_1, \ldots, x_n)^t$, we can view $M^{(n)}$ as a left module over $S = \mathbb{M}_n(R)$ via matrix multiplication. This S-module is *cyclic*, with generator $(m_1, \ldots, m_n)^t$. Given an injective $f \in \mathrm{End}(_R M)$, we can define an injective $f^* \in \mathrm{End}(_S M^{(n)})$ by the equation

$$(x_1, \ldots, x_n)^t f^* = (x_1 f, \ldots, x_n f)^t \quad (x_i \in M).$$

Since the matrix rings $\mathbb{M}_\ell(S) \cong \mathbb{M}_{\ell n}(R)$ are all strongly π-regular, the cyclic case covered before implies that f^* is surjective. It follows that f is also surjective.

Comment. The result in this exercise is a theorem proved in Dischinger's 1977 doctoral dissertation at the Universität München; see also his paper in C. R. Acad. Sci., Sér. A *283* (1976), 571–576. The proof above follows that given by Armendariz, Fisher, and Snider in their paper cited in the *Comment* on Ex. 8. In view of Dischinger's other theorem (Ex. 5) that strong π-regularity is a left-right symmetric notion, it follows from this exercise that, for any ring R, all finitely generated left R-modules are cohopfian iff all finitely generated right R-modules are cohofian. (This statement is *not* true if "cohopfian" is replaced by "hopfian": see the *Comment* on Ex. 20.9.)

To some readers, an "annoying" feature of the statement of this exercise might be that, for the sufficiency part, you have to assume strong π-regularity *not only for R, but for all matrix rings* $\mathbb{M}_n(R)$. This raised the question whether the two conditions in this exercise are equivalent to R itself being strongly π-regular. Dischinger observed in his paper (*loc. cit*) that, if the property of strong π-regularity always passed to matrix rings, then the famous Köthe Conjecture (see Ex. 10.25) would have an affirmative answer. However, Cedó and Rowen have shown that strong π-regularity *does not* pass to 2×2 matrix rings; see their paper in Israel J. Math. *107* (1998), 343–348. Thus, in general, Ex. 23.9 cannot be further improved.

For certain classes of rings, strong π-regularity does pass to matrix rings. We shall only cite one result in this direction. Recall that a ring R is *2-primal* if its prime radical $\mathrm{Nil}_*(R)$ consists of all nilpotent elements of R (see *FC*-§12, or Ex. 12.18, Ex. 12.19(0)); examples of 2-primal rings include commutative rings and reduced rings. Building on the work of Tominaga (Math. J. Okayama Univ. *4* (1955), 135–144), Hirano has proved that, if R is 2-primal, then R is π-regular iff it is strongly π-regular, iff all matrix rings $\mathbb{M}_n(R)$ are strongly π-regular; see his paper in Math. J. Okayama Univ. *20* (1978), 141–149. The commutative case of this result is covered by Ex. 4.15 and the exercise below.

Ex. 23.10. Let S be a commutative ring of dimension 0, and R be an S-algebra that is finitely generated as an S-module. Show that any matrix ring $\mathbb{M}_n(R)$ is strongly π-regular.

Solution. By Ex. 8, any finitely generated left S-module is cohopfian. In view of Ex. 4.16*, this implies that any finitely generated left R-module is also cohopfian. Applying Ex. 9, we see that any matrix ring $\mathbb{M}_n(R)$ is strongly π-regular.

Ex. 23.11. Let R be a right perfect ring. Show that (1) any matrix ring $\mathbb{M}_n(R)$ is right perfect, and (2) any finitely generated left or right R-module is cohopfian.

Solution. (1) By *FC*-(20.4), $S = \mathbb{M}_n(R)$ is also semilocal, so it suffices to show that $J = \mathrm{rad}\, S = \mathbb{M}_n(\mathrm{rad}\, R)$ is right T-nilpotent. Accordingly to *FC*-(23.16), this will follow if we can show that, for any right S-module M, $MJ = M \implies M = 0$. Let $\{e_{ij}\}$ be the matrix units in S. From $MJ = M$, we have

$$Me_{11} = MJe_{11} = M\left(\sum_i e_{i1} \cdot \text{rad } R\right)$$
$$= M \cdot \left(\sum_i e_{i1}e_{11} \cdot \text{rad } R\right)$$
$$\subseteq Me_{11} \cdot \text{rad } R.$$

Since Me_{11} is a right R-module, and rad R is right T-nilpotent, this implies that $Me_{11} = 0$. Right multiplying this by e_{1i}, we see that $Me_{1i} = 0$, and hence

$$Me_{ii} = M(e_{i1}e_{1i}) \subseteq Me_{1i} = 0.$$

Summing over i, we get $M = 0$, as desired.

(2) Since $\mathbb{M}_n(R)$ is right perfect, its principal left ideals satisfy DCC (by FC-(23.20)), so in particular $\mathbb{M}_n(R)$ is strongly π-regular. By Ex. 10 (and the left-right symmetry of strong π-regularity), it follows that finitely generated left, right R-modules are all cohopfian.

Comment. Part (1) is folklore in Bass's theory of (right) perfect rings, and is usually proved by using the Morita equivalence of module categories over R and $\mathbb{M}_n(R)$, and categorical characterizations of right perfect rings (such as "flat right modules are projective"). Since we do not have Morita's theory at our disposal, we gave a direct (and more elementary) proof. Part (2) of the exercise was an observation of Armendariz, Fisher, and Snider; see Cor. (1.3) in their paper cited in the *Comment* on Ex. 8.

§24. Homological Characterizations of Perfect and Semiperfect Rings

As we have mentioned in the Introduction to §23, Bass' study of perfect and semiperfect rings was partly inspired by homological algebra. In order to retrace his path of discovery, we have to first review some of the module-theoretic notions introduced in FC-§24.

A submodule S of an R-module M is said to be *small* (or *superfluous*) (written $S \subseteq_s M$) if, for any submodule $N \subseteq M$,

$$S + N = M \Longrightarrow N = M.$$

The sum of all small submodules of M is called the *radical* of M, and is denoted by rad M. This radical is also the intersection of all maximal submodules of M (FC-(24.4)). (If there are no maximal submodules, this intersection is understood to be M.) Thus, the radical of a module is just a straightforward generalization of the Jacobson radical of a ring.

For any M_R, a *projective cover* of M means an epimorphism $\theta : P \to M$ where P_R is a projective module and $\ker \theta \subseteq_s P$. (Sometimes we shall loosely refer to P as a projective cover of M, suppressing the role of θ.) In general, M_R may not have a projective cover, but whenever it does, as above, then the projective cover is essentially unique (FC-(24.10)). The question is: when does *every* M_R admit a projective cover?

One of Bass' main theorems is that *every M_R has a projective cover iff R is right perfect* (FC-(24.18)), *and every finitely generated M_R has a projective cover iff R is semiperfect* (FC-(24.16)). A pleasant consequence of this is that if R is right perfect (resp. semiperfect), then so is R/I for any ideal $I \subseteq R$ (FC-(24.17), FC-(24.19)).

The homological connection does not stop here, as there is also another surprising link to the notion of flat modules. A module M_R is called *flat* if the functor $M \otimes_R -$ is exact on the category of left R-modules. (The notion of flat modules generalizes the notion of torsion-free abelian groups.) Another theorem of Bass states that *a ring R is right perfect iff every flat module M_R is projective*[2] (FC-(24.25)). In fact, it is this homological result which first led Bass to the characterization of right perfect rings as rings satisfying DCC on principal *left* ideals. Later, a nonhomological proof of this characterization was found by R. Rentschler, but the proof was by no means easy.

The exercises in this section deal mainly with maximal, minimal, and superfluous submodules, and their connections to projective covers and perfect and semiperfect rings. Projective covers of simple modules are characterized in Exercise 4. Commutative rings for which every nonzero module has a maximal submodule are characterized in Exercise 9. Since we do not have the full homological machinery at our disposal, no exercises on flat modules are included in this section.

Exercises for §24

Ex. 24.1. For any finitely generated right R-module M, show that rad $M \subseteq_s M$.

Solution. Assume rad M is not superfluous in M. Then there exists a submodule $N \subsetneq M$ such that $N + \operatorname{rad} M = M$. Since M is finitely generated, Zorn's Lemma implies that $N \subseteq N'$ for some maximal submodule N' of M. But then $N' \supseteq \operatorname{rad} M$, and hence

$$N' \supseteq N + \operatorname{rad} M = M,$$

a contradiction.

Ex. 24.2. Let S, M be right R-modules such that $S \subseteq_s M$. Show that M has a projective cover iff M/S does.

---

[2] Of course, every projective module is always flat.

Solution. For the "only if" part, let $f : P \to M$ be a projective cover of M. Let $T = f^{-1}(S)$. *We claim that* $T \subseteq_s P$. To see this, suppose N is a submodule of P such that $N + T = P$. Then

$$M = f(P) = f(N) + f(T) = f(N) + S$$

implies that $f(N) = M$. This in turn implies that $N + \ker(f) = P$. Since $\ker(f) \subseteq_s P$, we conclude that $N = P$. Having now proved that $T \subseteq_s P$, it follows that $\overline{f} : P \to M/S$ is a projective cover of M/S.

Conversely, let $g : P \to M/S$ be a projective cover of M/S. Since P is projective, there exists an R-homomorphism $f : P \to M$ such that $\pi \circ f = g$ where π is the projection map $M \to M/S$. From

$$\ker(f) \subseteq \ker(g) \subseteq_s P,$$

it follows that $\ker(f) \subseteq_s P$. We are done if we can show that $f(P) = M$. But

$$M/S = g(P) = \pi f(P)$$

implies that $f(P) + S = M$. Since $S \subseteq_s M$, we have indeed $f(P) = M$.

Ex. 24.3. Show that a ring R is semiperfect iff every simple right R-module has a projective cover.

Solution. Recall that R is semiperfect iff every finitely generated right R-module has a projective cover (FC-(24.16)). Our job is therefore to show that, *if every simple right R-module has a projective cover, then so does every finitely generated right R-module M.* Let $\{M_i : i \in I\}$ be the family of maximal submodules of M. For each $i \in I$, fix a projective cover $g_i : P_i \to M/M_i$. Since P_i is projective, there exists $f_i \in \operatorname{Hom}_R(P_i, M)$ "lifting" g_i. It is easy to see that $\sum_i f_i(P_i) = M$. (For otherwise $\sum_i f_i(P_i) \subseteq M_j$ for some j, which contradicts $M_j + f_j(P_j) = M$.) Since M is finitely generated, it follows that

(1) $$f_{i_1} \oplus \cdots \oplus f_{i_n} : \quad P_{i_1} \oplus \cdots \oplus P_{i_n} \longrightarrow M$$

is onto for suitable indices $i_1, \ldots, i_n \in I$. To simplify the notation, let us write $1, \ldots, n$ for $i_1, \ldots, i_n$ in the following. From (1), we have an epimorphism

(2) $$\bigoplus_{i=1}^{n} P_i/\mathrm{rad}\, P_i \cong \left(\bigoplus_{i=1}^{n} P_i \right) \Big/ \mathrm{rad} \left(\bigoplus_{i=1}^{n} P_i \right) \longrightarrow M/\mathrm{rad}\, M.$$

Since $\ker(g_i)$ is maximal in P_i, we have $\mathrm{rad}\, P_i \subseteq \ker(g_i)$. On the other hand, $\ker(g_i) \subseteq_s P_i$ implies that $\ker(g_i) \subseteq \mathrm{rad}\, P_i$ (see FC-(24.4)). Therefore, equality holds, and we have $P_i/\mathrm{rad}\, P_i \cong M/M_i$ for every i. It follows from (2) that $M/\mathrm{rad}\, M$ is a finitely generated semisimple module. Since each simple right R-module has a projective cover, so does $M/\mathrm{rad}\, M$ (cf. FC-(24.11)(3)). By Exercise 1, $\mathrm{rad}\, M \subseteq_s M$, so by Exercise 2, M also has a projective cover.

Ex. 24.4. For any projective right R-module $P \neq 0$, show that the following are equivalent:

(1) P is a projective cover of a simple R-module;
(2) P has a superfluous and maximal submodule;
(3) $\mathrm{rad}\, P$ is a superfluous and maximal submodule;
(4) every maximal submodule of P is superfluous;
(5) $E := \mathrm{End}_R(P)$ is a local ring;
(6) $P \cong eR$ for some local idempotent $e \in R$.

Solution. $(1) \Leftrightarrow (2)$ is clear.

$(4) \Longrightarrow (2)$ is clear since (by FC-(24.7)) P always has a maximal submodule.

$(2) \Longrightarrow (3)$. Let $K \subset P$ be both superfluous and maximal. Then $K \subseteq \mathrm{rad}\, P$ by FC-(24.4), and hence $K = \mathrm{rad}\, P$, proving (3).

$(3) \Longrightarrow (5)$. First note that P must be indecomposable. (For, if $P = P_1 \oplus P_2$ with $P_i \neq 0$, then $\mathrm{rad}\, P_i \neq P_i$ by FC-(24.7), and

$$\mathrm{rad}\, P = \mathrm{rad}\, P_1 \oplus \mathrm{rad}\, P_2$$

would not be maximal.) Since P is projective, it follows that an endomorphism of P is an automorphism iff it is an epimorphism. To check that E is a local ring, it suffices to show that, whenever $f + g = 1 \in E$ and $f \notin \mathrm{U}(E)$, then $g \in \mathrm{U}(E)$. Now, from $f \notin \mathrm{U}(E)$, we have $f(P) \neq P$. The assumptions on $\mathrm{rad}\, P$ then imply that $f(P) \subseteq \mathrm{rad}\, P$. On the other hand, $f + g = 1$ implies that $f(P) + g(P) = P$. Therefore, $g(P) = P$ and we have $g \in \mathrm{U}(E)$.

$(5) \Longrightarrow (4)$. Let N be any maximal submodule of P. To prove that $N \subseteq_s P$, suppose $N + M = P$, where M is another submodule of P. Let f be the composition of

$$P \to P/N \cong M/M \cap N,$$

and choose a homomorphism $g : P \to M$ which lifts f. We shall view g as an element of E. For any $p = m + n$ where $m \in M$ and $n \in N$, we have $f(p) = \overline{m}$, so $m - g(p) \in N$. This implies that $(1 - g)(P) \subseteq N$, so $1 - g \notin \mathrm{U}(E)$. Since E is a local ring, it follows that $g \in \mathrm{U}(E)$, and in particular $M = P$, as desired.

We have now proved the equivalence of (1)–(5).

$(6) \Longrightarrow (5)$ is clear, since $\mathrm{End}_R(eR) \cong eRe$ by FC-(21.7).

(1) $\Longrightarrow$ (6). Let $\alpha : P \to S$ be a projective cover, where S is a simple R-module. Fix a surjection $\pi : R \to S$. Since P is projective, there exists $\beta \in \mathrm{Hom}_R(R, P)$ such that $\alpha\beta = \pi$. Then $\beta(R) + \ker(\alpha) = P$, so we must $\beta(R) = P$. This implies that $P \cong eR$ for some $e = e^2 \in R$. Since we already know (1) $\Rightarrow$ (5), $eRe \cong \mathrm{End}_R(P)$ is a local ring, so e is a local idempotent.

Ex. 24.5. Show that a semiprime right perfect ring R is semisimple.

Solution. This follows without any work from FC-(10.24) if we use the characterization of right perfect rings as rings satisfying DCC on principal left ideals ((1) $\Longleftrightarrow$ (2) in FC-(24.25)). An alternative proof is given below. Since rad R is right T-nilpotent, FC-(23.15) implies that

$$\mathrm{rad}\, R \subseteq \mathrm{Nil}_* R \quad \text{(the lower nilradical of } R\text{)}.$$

The fact that R is semiprime means that $\mathrm{Nil}_* R = 0$. Therefore, rad $R = 0$. It follows from the definition of a right perfect ring that $R(= R/\mathrm{rad}\, R)$ is semisimple.

Ex. 24.6. Let R be a ring satisfying one of the following conditions:

(a) Every nonzero right R-module has a maximal submodule;
(b) Every nonzero left R-module has a simple submodule.

Show that $J = \mathrm{rad}\, R$ is right T-nilpotent.

Solution. First assume (a) holds. For any right R-module $M \neq 0$, (a) implies that rad $M \subsetneq M$. Since $MJ \subseteq \mathrm{rad}\, M$ by FC-(24.4), it follows that $MJ \subsetneq M$. The fact that this holds for any $M_R \neq 0$ implies that J is right T-nilpotent, according to FC-(23.16).
    Next assume (b) holds. Then for any left R-module $N \neq 0$,

$$\mathrm{ann}_N(J) : = \{x \in N : Jx = 0\}$$

is necessarily nonzero. This fact implies that J is right T-nilpotent, again by FC-(23.16).

Ex. 24.7. For any ring R, show that the following are equivalent:

(1) R is right perfect;
(2) R is semilocal and every right R-module $M \neq 0$ has a maximal submodule;
(3) R is semilocal and every left module $N \neq 0$ has a simple submodule.

Solution. (2) or (3) $\Longrightarrow$ (1). Assume (2) or (3). By Exercise 6, rad R is right T-nilpotent. Since R is semilocal, it follows by definition that R is right perfect.

(1) $\Longrightarrow$ (2) + (3). Assume R is right perfect. By definition, R is semilocal and $J : = \mathrm{rad}\, R$ is right T-nilpotent. Let $M \neq 0$ be any right R-module.

By the last part of FC-(24.4), $MJ = \operatorname{rad} M$, and by FC-(23.16), $MJ \neq M$. Therefore, $\operatorname{rad} M \neq M$, which means that M has a maximal submodule. Next, consider any left R-module $N \neq 0$. By Exercise 4.18,

$$\operatorname{soc}(N) = \operatorname{ann}_N(J),$$

and by FC-(23.16) again, $\operatorname{ann}_N(J) \neq 0$. Therefore, $\operatorname{soc}(N) \neq 0$, which means that N has a simple submodule.

Comment. In (2) and (3), one may try to replace "R is semilocal" by the weaker condition that "there is no infinite set of nonzero orthogonal idempotents in R." Let (2)$'$ and (3)$'$ denote the resulting conditions. In FC-(23.20) and (24.25), it was shown that (1) $\Longleftrightarrow$ (3)$'$: this was part of Bass' Theorem in his original 1960 paper (Trans. AMS *95*). Of course, we also have (1) $\Rightarrow$ (2)$'$, so it is tempting to ask if (2)$'$ $\Rightarrow$ (1). This was raised as an open question in p. 471 of Bass' paper (although it was misquoted as a "conjecture" in a few other papers). In the case when R is a *commutative* ring, (2)$'$ $\Rightarrow$ (1) was proved later by Hamsher, Renault, and Koifman. (This implication is an easy consequence of Exercise 9 below.) However, a *noncommutative* example was constructed by Koifman to show that (2)$'$ $\Rightarrow$ (1) *does not* hold in general: see his paper (in Russian) in Mat. Zametki *7* (1970), 359–367 (translated into English in Math. Notes *7*(1970), 215–219).

Ex. 24.8. Let R be a right perfect ring which satisfies ACC on right annihilators of ideals. Show that $J := \operatorname{rad} R$ is nilpotent (so R is a semiprimary ring).

Solution. Fix an integer $n \geq 1$ such that

$$\operatorname{ann}_r(J^n) = \operatorname{ann}_r(J^{n+1}).$$

(Here, "ann_r" denotes "right annihilator.") If $\operatorname{ann}_r(J^n) = R$, then $J^n = 0$ and we are done. Assume, for the moment, that $\operatorname{ann}_r(J^n) \neq R$. Since $\operatorname{ann}_r(J^n)$ is an ideal, we can view $R/\operatorname{ann}_r(J^n)$ as a *left* R-module. By Exercise 7, there exists a left ideal $N \supsetneq \operatorname{ann}_r(J^n)$ such that $N/\operatorname{ann}_r(J^n)$ is a simple left R-module. But then $JN \subseteq \operatorname{ann}_r(J^n)$, so

$$0 = J^n(JN) = J^{n+1}N.$$

Since $\operatorname{ann}_r(J^n) = \operatorname{ann}_r(J^{n+1})$, this implies $N \subseteq \operatorname{ann}_r(J^n)$, a contradiction.

Comment. There do exist semiprimary rings satisfying ACC on right (and left) annihilators that are not right noetherian. The most immediate examples are nonnoetherian local rings $(R, \mathfrak{m})$ for which $\mathfrak{m}^2 = 0$. (The only annihilators are (0), R, and $\mathfrak{m}$.) This is an observation of Carl Faith; see his paper in Proc. Amer. Math. Soc. *112* (1991), 657–659.

Ex. 24.9. (Hamsher, Renault) Let R be a commutative ring. Show that every nonzero R-module has a maximal submodule iff $\operatorname{rad} R$ is T-nilpotent and $R/\operatorname{rad} R$ is von Neumann regular.

Solution. First assume every nonzero R-module has a maximal submodule. By Exercise 6, rad R must be T-nilpotent. After factoring out rad R, we may assume that rad $R = 0$. In particular, R is now a reduced ring; we must show that R is von Neumann regular. By Exercise 4.15, it suffices to show that any prime ideal $\mathfrak{p} \subset R$ is maximal. After factoring out $\mathfrak{p}$, we may further assume that R is an integral domain, and must show that R is a field. Let K be the quotient field of R, and assume $R \neq K$. Let M be a maximal submodule of K_R, and fix an R-isomorphism

$$\varphi : R/\mathfrak{m} \to K/M,$$

where $\mathfrak{m}$ is a maximal ideal of R. Since $R \neq K$, there exists a nonzero element $r \in \mathfrak{m}$. Let $\varphi(1 + \mathfrak{m}) = k + M$ and pick $s \in R$ such that $\varphi(s + \mathfrak{m}) = k/r + M$. Then

$$\varphi(sr + \mathfrak{m}) = (k/r + M)r = k + M = \varphi(1 + \mathfrak{m})$$

implies that $1 \in sr + \mathfrak{m} \subseteq \mathfrak{m}$, a contradiction.

For the converse, assume now that $J = \text{rad } R$ is T-nilpotent and R/J is von Neumann regular. Let $M_R \neq 0$ be any R-module. By FC-(23.16), we have $MJ \neq M$. It suffices to show that M/MJ has a maximal submodule. Since M/MJ is a right module over R/J, we may replace R by R/J to assume that R is von Neumann regular. To find a maximal submodule of M, we follow here an argument suggested by K. Goodearl. Fix a maximal ideal $\mathfrak{m} \subset R$ at which the localization $M_\mathfrak{m} \neq 0$. By Exercise 4.15, the localization $R_\mathfrak{m}$ is a field. Therefore, as a nonzero vector space over $R_\mathfrak{m}$, $M_\mathfrak{m}$ admits an $R_\mathfrak{m}$-homomorphism f onto $R_\mathfrak{m}$. Composing f with the natural map $M \to M_\mathfrak{m}$, we get an R-homomorphism $g : M \to R_\mathfrak{m}$, which is easily seen to be nonzero. As an R-module, $R_\mathfrak{m}$ is just the simple module $R/\mathfrak{m}$. Since $g \neq 0$, it must be onto, so $\ker(g)$ is a maximal submodule of M.

Comment. For the relevant literature, see R. Hamsher's paper in Proc. Amer. Math. Soc. *17* (1966), 1471–1472, and G. Renault's paper in C. R. Acad. Sc. Paris *264* (1967), Sér. A, 623–624.

Ex. 24.10. (Hamsher) Let R be a commutative noetherian ring. Show that every nonzero R-module has a maximal submodule iff R is artinian.

Solution. If R is artinian, then R is perfect, so by Exercise 7, every nonzero R-module has a maximal submodule. Conversely, assume R has this property. By Exercise 9, $J = \text{rad } R$ is T-nilpotent and R/J is von Neumann regular. Since R/J is also noetherian, FC-(4.25) implies that it is semisimple. It follows now from Exercise 20.6 that R is an artinian ring.

Comment. The above result can be generalized to the noncommutative setting. In fact, Renault has shown that *a right noetherian ring R is right artinian iff every nonzero right R-module has a maximal submodule.* The proof requires considerably more work: see Renault's paper in C. R. Acad. Sc. Paris *267* (1968), Sér. A, 792–794.

§25. Principal Indecomposables and Basic Rings

The theory of principal indecomposable modules and basic rings was classically developed for right artinian rings, but can be easily adapted to the case of semiperfect rings. If R is semiperfect, any primitive idempotent $e \in R$ is local (FC-(23.5)), and eR is called a *principal indecomposable module*. The isomorphism classes of principal indecomposables are in one-one correspondence with the isomorphism classes of simple right R-modules, by the map

$$[eR] \longmapsto [eR/eJ]$$

where $J = \text{rad } R$. Moreover, any finitely generated projective right R-module is uniquely a direct sum of principal indecomposables (FC-(25.3)).

A *basic idempotent* in a semiperfect ring R is an idempotent of the form

$$e = e_1 + \cdots + e_r$$

where the e_i's are orthogonal primitive idempotents in R such that $e_1 R, \ldots, e_r R$ represent a complete set of isomorphism classes of the principal indecomposables. For any such basic idempotent e, eRe is called a *basic ring* of R. This ring is also semiperfect, and its isomorphism type is uniquely determined (FC-(25.6)). The ultimate justification for the formation of the basic ring is that it provides a "canonical minimal representative" for the class of rings that are "Morita equivalent" to R. However, since we do not have the details of the Morita Theory at our disposal, we shall not pursue this connection here.

Exercise 4 below shows that a basic idempotent of a semiperfect ring R is determined up to a conjugation in R. The other exercises deal with the classical case of finite-dimensional algebras (over fields), and are largely motivated by the representation theory of such algebras.

Exercises for §25

Ex. 25.1. Let R be a finite-dimensional algebra over a field k, and let M be any finite-dimensional right R-module. For any idempotent $e \in R$, show that $\dim_k \text{Hom}_R(eR, M) = \dim_k Me$.

Solution. By FC-(21.6), we have an additive group isomorphism

$$\lambda : \text{Hom}_R(eR, M) \longrightarrow Me$$

defined by $\lambda(f) = f(e)$ for any $\lambda \in \text{Hom}_R(eR, M)$. It is easy to check that $\text{Hom}_R(eR, M)$ and Me are k-vector spaces, and that λ is a k-homomorphism. Therefore, λ is a k-isomorphism, and we have

$$\dim_k \text{Hom}_R(eR, M) = \dim_k Me.$$

Ex. 25.2. In Exercise 1, assume k is a splitting field for R. Let $e \in R$ be a primitive idempotent and let $J = \operatorname{rad} R$. Show that the number of composition factors of M_R isomorphic to eR/eJ is given by $\dim_k Me$.

Solution. Since eRe is finite-dimensional over k and has no nontrivial idempotents, it is a local k-algebra by *FC*-(19.19). Therefore, e is in fact a *local* idempotent, so by *FC*-(21.18), eR/eJ is a *simple* R-module. Fix a composition series

$$M = M_1 \supsetneq M_2 \supsetneq \cdots \supsetneq M_n \supsetneq M_{n+1} = 0.$$

Suppose

$$M_{i_1}/M_{i_1+1}, \ldots, M_{i_r}/M_{i_r+1} \quad (1 \le i_1 < \cdots < i_r \le n)$$

are the composition factors isomorphic to eR/eJ. Recall that

$$M_i/M_{i+1} \cong eR/eJ \iff M_i e \nsubseteq M_{i+1}$$

(see *FC*-(21.19); of course this is also a special case of Exercise 1). Therefore, there exist $m_{i_j} \in M_{i_j}$ such that $m_{i_j} e \notin M_{i_j+1}$ for $1 \le j \le r$. After replacing m_{i_j} by $m_{i_j} e$, we may assume that $m_{i_j} \in Me$. From $M_{i_j} = m_{i_j} R + M_{i_j+1}$, we have

(A) $\qquad M_{i_j} e \subseteq m_{i_j} Re + M_{i_j+1} = m_{i_j}(eRe) + M_{i_j+1}.$

Let us write "bar" for the residue map from R to $\overline{R} = R/J$. Since k is a splitting field for R, $\overline{e}\overline{R}\overline{e} \cong \operatorname{End}_{\overline{R}}(\overline{e}\overline{R})$ is 1-dimensional over k. Therefore, $eRe = ek + eJe$, so from (A) we have

(B) $\qquad M_{i_j} e \subseteq m_{i_j} ek + m_{i_j} eJe + M_{i_j+1} \subseteq m_{i_j} k + M_{i_j+1}.$

We shall now complete the proof by showing that $m_{i_1}, \ldots, m_{i_r}$ *form a k-basis for Me.* Since they are clearly k-linearly independent, it suffices to show that they span Me. Consider any $m \in Me$. Since $M_i e \subseteq M_{i+1}$ whenever $i \notin \{i_1, \ldots, i_r\}$, the fact that $m = me$ implies that $m \in M_{i_1}$. In view of (B), there exists $\alpha_1 \in k$ such that

$$m' := m - m_{i_1} \alpha_1 \in M_{i_1+1},$$

and therefore $m' = m'e \in M_{i_1+1}e$. This implies as before that $m' \in M_{i_2}$, and so $m' \in M_{i_2} e$. We then have

$$m' - m_{i_2} \alpha_2 \in M_{i_2+1}$$

for a suitable $\alpha_2 \in k$. Continuing like this, we obtain in r steps an equation

$$m = m_{i_1} \alpha_1 + \cdots + m_{i_r} \alpha_r + \varepsilon,$$

with an "error term" $\varepsilon \in M_{i_r+1}$. But now $\varepsilon = \varepsilon e \in M_{i_r+1} e = 0$, so we have

$$m = m_{i_1} \alpha_1 + \cdots + m_{i_r} \alpha_r,$$

as desired.

Ex. 25.3. Construct a basic idempotent e for the group algebra $R = \mathbb{Q} S_3$, and determine the corresponding basic ring eRe.

Solution. Referring to Exercise 8.3, we have

$$R \cong \mathbb{Q} \times \mathbb{Q} \times M_2(\mathbb{Q}),$$

with the three centrally primitive idempotents e_1, e_2, e_3 as worked out in that exercise. The first two of these are already primitive idempotents in R, but e_3 is not. To find a primitive idempotent in the block $e_3 R$, we fix an irreducible $\mathbb{Q}$-representation belonging to this block, say the one afforded by the module

$$V = (\mathbb{Q} v_1 \oplus \mathbb{Q} v_2 \oplus \mathbb{Q} v_3)/\mathbb{Q} \cdot (v_1 + v_2 + v_3),$$

on which S_3 acts by permuting the v_i's. Using $\bar{v}_1, \bar{v}_2$ as basis for V, we have

$$(1) \mapsto \begin{pmatrix} 1 & 0 \\ 0 & 1 \end{pmatrix}, \ (12) \mapsto \begin{pmatrix} 0 & 1 \\ 1 & 0 \end{pmatrix}, \ (13) \mapsto \begin{pmatrix} -1 & 0 \\ -1 & 1 \end{pmatrix}, \ (123) \mapsto \begin{pmatrix} 0 & -1 \\ 1 & -1 \end{pmatrix}.$$

These four matrices are linearly independent, so we have a unique linear combination, namely

$$\alpha = \tfrac{1}{3}[(1) - (12) - 2(13) - (123)],$$

which maps to the primitive idempotent $\begin{pmatrix} 1 & 0 \\ 0 & 0 \end{pmatrix} \in M_2(\mathbb{Q})$. Of course, α may not lie in the block $e_3 R$. But we can simply replace it by [3]

$$\beta = \alpha e_3 = \tfrac{1}{3}[(1) - (13) + (23) - (123)] \in e_3 R.$$

Since our representation gives an isomorphism φ from $e_3 R$ to $M_2(\mathbb{Q})$, β is a primitive idempotent in $e_3 R$. A basic idempotent for R is therefore

$$\begin{aligned} e &= e_1 + e_2 + \beta = 1 - e_3 + \beta \\ &= 1 - \tfrac{1}{3}[2(1) - (123) - (132)] + \tfrac{1}{3}[(1) - (13) + (23) - (123)] \\ &= \tfrac{1}{3}[2(1) - (13) + (23) + (132)]. \end{aligned}$$

To compute eRe, note that e_1, e_2, β are mutually orthogonal, and that $e_3 \beta = \beta e_3 = \beta$. Thus, $e(e_i R)e = \mathbb{Q} e_i$ for $i = 1, 2$, and

$$e(e_3 R)e = ee_3 Re_3 e = \beta(e_3 Re_3)\beta = \mathbb{Q}\beta$$

---

[3] If we had chosen to work with the matrices representing (1), (13), (23) and (123), we would have gotten this β in one shot, but we didn't know better.

in view of the isomorphism φ. A basic ring for R is therefore

$$eRe = (\mathbb{Q}e_1) \times (\mathbb{Q}e_2) \times (\mathbb{Q}\beta) \cong \mathbb{Q} \times \mathbb{Q} \times \mathbb{Q},$$

which is, of course, consistent with the predictions of the general theory (cf. (25.7)(3)).

Comment. For the general symmetric group $G = S_n$, there is a well-known method, due to Alfred Young, by which one can construct explicitly a primitive idempotent associated with each irreducible $\mathbb{Q}G$-representation, from the so-called Young diagram of the representation. This method is described in detail in §28 of "Representation Theory of Finite Groups and Associative Algebras" by Curtis and Reiner. In concluding this discussion, Curtis and Reiner stated: "An interesting unsolved problem in this connection is whether the sort of algorithm used in the case of S_n, which gives a direct construction of the minimal (right) ideals in the group algebra from the conjugacy classes in the group, is available for other classes of (finite) groups."

In the case of $G = S_3$, the representation used in the solution of our exercise arises from the Young tableau:

$$\begin{array}{|c|c|}
\hline
1 & 2 \\
\hline
3 \\
\cline{1-1}
\end{array}$$

Following through Young's algorithm (as given in §28 of Curtis-Reiner), we obtain the primitive idempotent

$$\gamma = \tfrac{1}{3}[(1)(1) - (1)(13) + (12)(1) - (12)(13)]$$
$$= \tfrac{1}{3}[(1) - (13) + (12) - (132)].$$

This leads to a new basic idempotent

$$e' = e_1 + e_2 + \gamma = 1 - e_3 + \gamma = \tfrac{1}{3}[2(1) - (13) + (12) + (123)].$$

Note that the basic idempotent e we constructed earlier is just the conjugation of e' by (13). This is not surprising in view of our next (and last) exercise!

Ex. 25.4. Let e be a basic idempotent of a semiperfect ring R, and e' be another element of R. Show that e' is a basic idempotent for R iff $e' = u^{-1}eu$ for some $u \in U(R)$.

Solution. Let $e = e_1 + \cdots + e_r$, where the e_i's are orthogonal primitive idempotents such that $e_1 R, \ldots, e_r R$ represent a complete set of isomorphism classes of principal indecomposables. Let $e' = u^{-1}eu$ where $u \in$

$U(R)$, and write $e_i' = u^{-1}e_i u$. In view of Exercise 21.14, the e_i''s have the same properties as the e_i's. Therefore, $e' = e_1' + \cdots + e_r'$ is also a basic idempotent of R.

Conversely, let e' be any basic idempotent of R. Then $eR \cong e'R$, since both modules are isomorphic to the direct sum of a complete set of principal indecomposables. Also, both rings eRe, $e'Re'$ are semiperfect $(FC\text{-}(25.6))$ and hence semilocal. Therefore, by Exercise 21.16, e, e' must be conjugate in R.

Ex. 25.5. Let R be the ring of upper triangular $n \times n$ matrices over a division ring k, and let $P = k^n$ be the space of row n-tuples, viewed as a right R-module by the natural R-action. Show that P has a unique composition series (of length n).

Solution. We first construct a composition series for P. Let $e_1, \ldots, e_n \in P$ be the unit vectors, and let V_i be the span of $\{e_i, \ldots, e_n\}$ $(1 \le i \le n+1)$ (so that $V_{n+1} = 0$). These are clearly R-submodules of P, with

$$(*) \qquad P = V_1 \supsetneq V_2 \supsetneq \cdots \supsetneq V_n \supsetneq V_{n+1} = 0,$$

and it is easy to see that the n filtration factors are precisely the n different simple right R-modules. Therefore, $(*)$ is a composition series for P. (This is essentially the same as the composition series constructed for P_1 in FC-(25.11), upon identifying P_1 there with our P.) Our job is to show that any R-submodule $V \subseteq P$ is one of the modules in $(*)$.

Let i be the largest integer such that $V_i \supseteq V$. We may assume $i \le n$ (for otherwise $V = 0$). Pick a vector $v \in V \backslash V_{i+1}$. Then

$$v = (0, \ldots, 0, a_i, \ldots, a_n) \quad \text{with} \quad a_i \ne 0.$$

If we right multiply v by the matrix (in R) with ith row

$$(0, \ldots, 0, a_i^{-1}x_i, a_i^{-1}x_{i+1}, \ldots, a_i^{-1}x_n)$$

and 0's elsewhere, we get $(0, \ldots, x_i, x_{i+1}, \ldots, x_n) \in V_i$. Thus, $V_i \subseteq vR \subseteq V$, so $V = V_i$, as desired.

Name Index

Subject Index

Problem Books in Mathematics

Algebraic Logic
by *S.G. Gindikin*

Unsolved Problems in Number Theory (2nd ed.)
by *Richard K. Guy*

An Outline of Set Theory
by *James M. Henle*

Demography Through Problems
by *Nathan Keyfitz and John A. Beekman*

Theorems and Problems in Functional Analysis
by *A.A. Kirillov and A.D. Gvishiani*

Exercises in Classical Ring Theory (2nd ed.)
by *T.Y. Lam*

Problem-Solving Through Problems
by *Loren C. Larson*

Winning Solutions
by *Edward Lozansky and Cecil Rosseau*

A Problem Seminar
by *Donald J. Newman*

Exercises in Number Theory
by *D.P. Parent*

Contests in Higher Mathematics:
Miklós Schweitzer Competitions 1962–1991
by *Gábor J. Székely (editor)*